INDUSTRY 4.0 AND
CIRCULAR ECONOMY

INDUSTRY 4.0 AND
CIRCULAR ECONOMY

Towards a Wasteless Future or a Wasteful Planet?

ANTONIS MAVROPOULOS
D-WASTE™
Athens
Greece

ANDERS WAAGE NILSEN
Bergen
Norway

Registered Offices
John Wiley & Sons, Inc., 111 River Street, Hoboken, NJ 07030, USA
John Wiley & Sons Ltd., The Atrium, Southern Gate, Chichester, West Sussex, PO19 8SQ, UK

Editorial Office
The Atrium, Southern Gate, Chichester, West Sussex, PO19 8SQ, UK

For details of our global editorial offices, customer services, and more information about Wiley products visit us at www.wiley.com.

Wiley also publishes its books in a variety of electronic formats and by print-on-demand. Some content that appears in standard print versions of this book may not be available in other formats.

Library of Congress Cataloging-in-Publication Data

Names: Mavropoulos, Antonis, author. | Nilsen, Anders Waage, 1975– author.
Title: Industry 4.0 and circular economy : towards a wasteless future or a
    wasteful planet? / Antonis Mavropoulos, Anders Waage Nilsen.
Other titles: Industy four point zero and circular economy
Description: Hoboken, NJ : Wiley, 2020. | Includes bibliographical
    references and index.
Identifiers: LCCN 2020018139 (print) | LCCN 2020018140 (ebook) | ISBN
    9781119699279 (cloth) | ISBN 9781119699286 (adobe pdf) | ISBN
    9781119699330 (epub)
Subjects: LCSH: Sustainable development. | Technological
    innovations–Environmental aspects. | Salvage (Waste, etc.) | Human
    ecology.
Classification: LCC HC79.E5 M565 2020 (print) | LCC HC79.E5 (ebook) | DDC
    338.9/27–dc23
LC record available at https://lccn.loc.gov/2020018139
LC ebook record available at https://lccn.loc.gov/2020018140

Cover Design: Nikolaos Rigas, D-Waste
Cover Image: © Nikolaos Rigas, D-Waste

Set in 9/13pt Ubuntu Regular by SPi Global, Pondicherry, India

Printed and bound by CPI Group (UK) Ltd, Croydon, CR0 4YY

10 9 8 7 6 5 4 3 2 1

Contents

Contents

Foreword by Ad Lansink

TOWARDS SOCIAL CIRCULARITY

Is it really a hard choice between a wasteless future and a wasteful planet? The question mark behind the subtitle of *Industry 4.0 and Circular Economy*, the comprehensive and fascinating book of Antonis Mavropoulos and Anders Waage Nilsen, shows that a straightforward answer is difficult. The same is valid for defining circular economy. The question mark gives room for further consideration, looking for a sustainable road to circularity, at the same time addressing the issues of global warming, resources scarcity, and biodiversity losses. The large economic differences between rich and poor countries also require a consistent policy. Until 2020, globalization was a main topic for economy and, often as a counterpart, ecology. Trade and transport became issues for politicians and policy makers, developing treaties to defuse tensions and barriers. Since the international community is going through turbulent times by the outbreak of Covid-19 and the grim fight against this pandemic, a social and even cultural component must be added to the paradigm of globalization. 2020 will be referred as a historical year, because of the long-term impact of the Covid-19 pandemic. The consequences for the international community will outweigh the effects of previous crises. Therefore, combined efforts on climate policy, circularity, and disease control require a worldwide basis of public support, governance, leadership, and cooperation. Spreading knowledge is as important as sharing responsibilities, both in the field of fundamental research, applied sciences, and practical innovation. International cooperation will be stimulated by new technologies, especially in the implementation of digitalization, networking, and tools such as blockchain technology, artificial intelligence, and worldwide communication channels, presenting society a sustainable future. But social aspects such as solidarity, social justice, and human rights are even essential. Inclusive system thinking seems the right way to defeat all challenges.

Talking about challenges: about four decades ago, concerns about scarcity of raw materials and fossil energy besides increasing waste streams brought me to the idea of the waste hierarchy. The original proposal contained five steps: prevention, source separation, post-separation, incineration with energy recovery, and functional landfilling. During legislation, source and post-separation were replaced by product and material reuse. Recycling does not occur in the preference order but is merely a means of facilitating material reuse. So waste hierarchy and circular economy are greater than recycling. Nowadays, the waste hierarchy is a universal model, an essential route map to circularity, otherwise respecting thermodynamic laws. Progress in effective waste management, especially waste prevention and product reuse requires "rethinking" society as a whole: a tough task due to the habituation to prosperity and freedom of choice. Therefore, action is needed. Abandoning unnecessary items and services should be a key priority. However, that's a difficult task in a market-driven society. Extending the lifespan of products and (weight) savings of materials are easier to achieve, as is eco-design with a focus on quantitative prevention, optimal reuse, and high-quality recycling. All transitions require a solid basis of support in society. Communication remains an indispensable instrument for increasing environmental awareness, against vested interests and opportunistic media violence. The return of a strong environmental awareness is necessary to maintain and strengthen public and political support for the transition to a circular economy.

After all, chain management implies a revolution of thinking and doing in services and industry. History learned us that water and steam power founded the first industrial revolution. Thereafter, electrical power facilitated by fossil energy allowed mass production of goods during the second one. Meanwhile, the transition to sustainable energy, forced by active climate policy, requires attention for eco-design of materials and products, with large lifetime and reuse possibilities. Meanwhile, automation and digitalization founded the third industrial revolution. The fourth version is no simple continuation of the third one. Characteristic of Industry 4.0 is the demolition of boundaries between physical, chemical, and biological domains. Also, scale and velocity are higher than in previous times, sometimes leading to disruptive innovations: a new term in the social-economic vocabulary. But a real disruption is not necessary and even impossible due to the large international differences. However, population growth and depletion of raw materials require a clear circular approach, changing the chain of production, distribution, and consumption of the well-known waste hierarchy to a new resource hierarchy. Of course, governments need to set clear goals and ensure transparent and enforceable legislation, if needed with applicable financial instruments. Meanwhile, Industry 4.0 must facilitate major changes in commodity policy, waste management, resource policy, and value creation. These changes are influenced by three principal context factors: time, place, and function. In other words, the speed of the transition is partly defined by the geographical location and function of the product and material flows.

At the end of their interesting and important book, the authors of Industry 4.0 and Circular Economy ask another black-white question: will Industry 4.0 be "a stairway to heaven or a highway to hell"? Hopefully by then, the reader will have already found a series of answers: useful guides on the successful road to circular economy, paved with indispensable building blocks of social justice, employment, selective growth, environmental awareness, and real governance: international, national, and local.

Dr. Ad Lansink

Founder of the waste hierarchy

Author of the book *Challenging Changes – Connecting Waste Hierarchy and Circular Economy*

Former Member of Parliament, the Netherlands

Foreword by Dimitris Kaliampakos

SEARCHING FOR REAL "OUT OF THE BOX" ANSWERS TO PRESSING QUESTIONS ABOUT WASTE AND CIRCULAR ECONOMY

Currently, I am lucky (or maybe unlucky) to wear two hats. I am the dean of the School of Mining and Metallurgical Engineering at the National Technical University of Athens. I am also the President of the Associated research Centers for the Urban Underground Space (ACUUS), a pioneer international organization that promotes the systematic utilization of the underground space of cities. While reading the fascinating book of Antonis Mavropoulos and Anders Waage Nilsen, I witnessed the continuous confrontation of two schools of thought, a fact that, among other things, comprises an interesting personal experiment.

What is, therefore, the response of a mining engineer to the intriguing questions posed throughout the book? It is widely known that mining is one of the most unwanted activities at a local level (local unacceptable land uses). Perhaps, not unjustly. For many years, the mining sector, being the visible backbone of at least the first and the second industrial revolution, entrenched itself, with a pinch of arrogance, behind the proverb "you can't make an omelet without breaking eggs." The problem is that during this time, when we falsely believed that the "eggs" were unlimited, we were breaking more eggs than necessary for the "omelet," making a mess of our "home" (our planet) at the same time. These times have irrevocably passed. Today, the prevailing thinking of the public is to reduce mining activity. In addition, one point of view considers mining no longer necessary, since circular economy is able to essentially marginalize it. Even more so if it is combined, as we should do, with a reasonable constrain to the rampant consumerism, a disease that developed world suffers from. But, let me say, this is a false view. Not only it is not based on actual proof (Industry 4.0 is thirsty for more quantities and increasingly more rare raw materials), but it is also socially unfair, if not half blind, seen from the developed world's perspective. Have we considered how huge quantities of raw materials are required to satisfy the basic needs of the 2.2 billion people of this planet that do not have access to safe-drinking water and the 4.2 billion of those who lack safe sanitation services?

What are the limits of circular economy? How feasible is a goal, e.g. for a zero waste future? Does it reflect modern and real technological capabilities or is it another arrogant statement of intent, without a real meaning? If we consider the second thermodynamic law, any energy transformation for a pursued purpose is accompanied by a loss of energy. In my opinion, the most general term of waste is exactly this loss of energy. Even if we exploit this loss of energy, the "waste," in another process, there will be a new loss of energy and so on. This does not mean that the core idea of circular economy is wrong. We are living in a world of finite resources, which in many ways is reaching its limits. However, the clarity of goals is the first factor in achieving them.

Wearing my other hat this time, I had the opportunity to reflect on some of the critical observations made in the book. One of them is the rebound effect. An iconic example of our emerging era is e-commerce. Indeed, the modern consumer can buy almost everything from home and also benefit from the reduced prices that cost savings entails. In practice, however, this has led to a rapid increase in consumption and thus a linear increase in the relating environmental impacts. It is very interesting to analyze the impact of e-commerce on the everyday life of cities and especially the megacities. While one would expect an improvement in the environmental conditions due to reductions in unnecessary commuting, the situation is very different. Take China, for example. China's online retail market is set to hit US $1.99 trillion by the end of 2019. How does this translate in terms of waste and the environment? Delivery packaging is accounting for 93% of solid waste growth in China's megacities. Moreover, delivery trucks (~10% of all automotive in China) are contributing to more than 75% of harmful particulate matter released into the atmosphere. Last mile deliveries have a significant impact, unconceivable just a few years ago. The scientific community that I work with is systematically working towards underground, automatic transportation of goods to cities using many of the technologies of Industry 4.0 (e.g. big data analysis). Then again, only a "paradigm shift," as is often referred to in the book, can protect us from a second rebound effect.

Our planet will be more urbanized than ever. The megacities, 34 today, are in danger of collapsing under their own weight, like the high mountains. It seems that the Biking–Metro–Walking model, the BMW model, as urban planners playfully have named it, is perhaps the only solution for a livable city, supporting a more natural way of life. However, that requires again enormous quantities of raw materials and huge public interest investments in order to build an efficient subway system. The problem of social inequalities is also evident here. Africa, for example, is a glaring proof. The International Association of Public Transport (Union Internationale des Transports Publics [UITP]) reports that only two of the 54 African countries (Algeria and Egypt) have a metro system.

The book of Antonis Mavropoulos and Anders Waage Nilsen has the advantages of, first of all, asking the right questions, of the multidimensional analysis, and of searching for real, "out of the box" answers. If you are looking for a "handbook" on waste management in the new era, you may feel uncomfortable with the number of definitive answers in relation to the questions raised. But if you, like me, are tired of the same answers served with a different wrapper each time, you will enjoy the fresh oxygen of original thinking.

Dimitris Kaliampakos
Dean of the School of Mining and Metallurgical Engineering at the
National Technical University of Athens
President of the Associated research Centers for the
Urban Underground Space (ACUUS)

Foreword by Erik Solheim

GETTING OUT OF THE COVID-19 CRISIS

A New, Better, and More Circular World Is Possible and Necessary

Moving from a linear to a circular economy is a no brainer. It is not a matter of *if*, but of *how*.

By the midst of this century we will be 10 billion people on our planet, Earth. We strive to bring everyone into the global middle class, to enjoy life, free of extreme poverty, in good health, educated, and with a decent living standard. Our aim is to leave no one behind. The planet simply doesn't provide the resources for the realities of this ambition at the present level of inefficiency in our economies.

We need to treat waste as resources for new products, rather than, yes, waste.

Whales, sea birds, and turtles are dying from plastic waste. Slum dwellers perish from sewage systems choked with plastics. Nations declare waste emergencies and see plastic pollution as a major economic threat because tourists are not flocking to beaches overflown with plastics. The miracle material of plastics, which once was heralded as the eighth wonder of the world, is turned into a menace.

In this important book (*Industry 4.0 and Circular Economy*), Antonis Mavropoulos and Anders Waage Nilsen present us the choice between a wasteless future and a wasteful planet.

The authors set us on the track to the wasteless future of a true circular economy. The book answers practical questions as regards waste management. At the same time, their book guides us through the wider societal issues of social justice, creation of much needed jobs, and an ecological civilization. The benefits of the circular economy are there to be reaped.

Let's just do it.

The twentieth century was marked by a rather unfruitful debate between economists and ecologists. The economists would point to the fantastic global progress since the first industrial revolution occurred in the United Kingdom in the 1780s. At that time life expectancy on the planet was around, or may be even below, 30. Now it's 72. Also, extreme poverty has gone from being the global rule to the fate of a suffering but small minority. Experts will argue the price to be paid was environment degradation, pollution, the destruction of ecosystems, and climate change.

The very good news is that this is a debate of the past. There is no longer a choice to be made between environment and prosperity.

When the price of renewables like solar and wind creeps below the price of coal, we are on to a stellar shift. A green, sustainable economy is for the first time in human history open to us. If we want to reap the benefits the technologies bring us and at the same time protect Mother Earth, the circular economy is an important part of this puzzle.

Why do we waste plastics everywhere rather than see used plastics as a raw material for new plastics, or for fuel or for energy? How long will we buy a computer or cellphone built from virgin material when the old one goes broke or we desire a new fashion? Why don't we recover the components of the old equipment and use them in the new, establishing at least in theory an endless loop of innovation and improvements?

Today a third of all food produced globally is wasted, at a time where still too many humans go to bed hungry. Of course, we can avoid food loss, and we just need to see it as a priority.

Time has come to establish a truly circular economy. It's also time we avoid and abandon products we simply don't need.

I am writing this in the midst of the corona crisis. The virus has brought some unintended environment benefits. Climate emissions haven't been so low in decades. Pollution is way down in cities around the globe; Indian skies are so clear that many Indians can even see the Himalayas for the first time. Wild animals are roaming our cities. Penguins are seen walking in empty Cape Town streets and wild bears in Barcelona; wild Turkeys are coming out in California, deer in Japanese towns, and wild goats in the United Kingdom. American conservationists speak about wildlife having a party in empty national parks like the Yosemite.

Emerging out of the crisis, our challenge is not to restore the world as it was when we embarked upon the critical year of 2020 but like the authors of this book to impress upon ourselves that a new and better world is possible.

The economic stimulus packages and global politics should focus on the green exit from the crisis. Moving toward a circular economy is at the core of that.

Mavropoulos and Waage Nilsen are right in showing the many opportunities arising from artificial intelligence, big data, block chain, and every other corner of the digitized economy of the future.

The circular economy is critically linked to the fourth industrial revolution. The main technological and economical quantum leaps happening in front of our eyes give major tools for a circular economy.

The current crisis enhances the current trends. We get used to get clothes, electronics, and pizza delivered by the Alibaba and the Amazon of this world. We get used to consult medical expertise over the net rather than by face-to-face encounter. We now know from experience that meetings conducted at Zoom, Teams, or WeChat will not replace all physical meetings, but surprisingly many of them. While staying at home, we did not buy new vehicles or a new dress for every party. There were none. It is likely that the sharing economy will take off. We have experienced the pains of social distancing, but the crisis has also opened new and exciting opportunities for the future.

Of course, there is no real choice between a circular economy and wasteful planet. A circular economy is feasible, desirable, and necessary. The wasteful planet does not work for us. In the end it makes our beautiful planet inhabitable.

The wealth of ideas, numbers, and arguments in this book will help us get there.

Erik Solheim
Former Minister of Environment and Development, Norway
Former Executive Director, UN Environment Programme
CEO, Plastic REVolution Foundation

Series Preface

As the world's population, and with it the amount of resources we consume, continues to grow, it becomes ever more important to properly manage the "waste" that is generated by this growth. With the expanding volume and complexity of discarded domestic and industrial waste, and changing household consumption patterns, waste management is fast becoming one of the key challenges for the modern society.

According to some estimates, up to 2 billion people lack access to sound waste management. Uncollected municipal waste that ends up at illegal or improper dumpsites close to waterways and coasts generates marine litter, affecting marine ecosystems as well as the fishing and tourism sectors. Marine litter is primarily caused by the improper management of waste on land, which finds its way into the marine environment. Dumpsites, again caused by the improper collection and management of waste, can be sources of greenhouse gases and other short-lived climate pollutants. It is therefore recognized that the inadequate removal and treatment of waste poses multiple threats to human health and the environment. Particularly in low- and middle-income countries, open dumping and burning close to urban centers represent a substantial threat to human health and climate. However, advanced economies have shown that through effective waste management it is possible to significantly reduce these impacts and even, through recirculation of the materials in waste back into the production of new goods, minimize the broader impacts associated with consumption. Hence, the proper management of waste and resources is critical to successfully realizing several of the Sustainable Development Goals set out by the UN in 2015.

A sound and sustainable waste management system is a fine balance of a number of elements – technical, legislative, financial - carefully planned to unlock the economic potential of waste, including the creating of new jobs and development of new enterprises.

This series of books will address a range of topics that are integral to sound waste management systems. This will include technical solutions for waste collection and treatment, as well as addressing financing opportunities and the value of waste management, organizational and management challenges, and policy development and enforcement. The series will also explore global trends like the 4th Industrial Revolution and how it can transcend boundaries and can have an impact on both high- and low-income settings when it comes to waste management. Moreover, the books will emphasize the need to adopt a holistic view of waste management by considering the total waste system, and then developing the most appropriate mix of infrastructure and services to manage the specific waste streams.

This series will stress the importance of well-planned and well-managed waste management systems and will highlight the centrality of waste management within all spheres of human activity. It will attempt to bring into focus how waste management links to the climate, human health, resources, the environment and various other aspects.

Björn Appelqvist
Chair of the ISWA Scientific and Technical Committee (STC)

Any opinions expressed in this publication are solely those of the authors. They do not necessarily reflect the opinions or standpoints of ISWA, or its members, on any specific issue – unless explicitly stated.

Preface

The idea for this book was firstly conceived during a wet but not so cold afternoon in Bergen, Norway by the end of November 2016. I (Antonis) had been invited to make a lecture on the fourth industrial revolution and the future of waste management by Avfall Norge, the national member of ISWA in Norway, and after my lecture I got involved in a very interesting discussion with Toralf Igesund, head of planning at BIR, Bergen's waste management company. The discussion was so interesting that we decided to continue it in a nice but rather dark coffee shop, few blocks behind the Radisson hotel in Bergen. Since then, and for the next four years, the idea for this book was getting more mature but also more difficult, because the fourth industrial revolution was unfolding so fast and, at the same time, the circular economy narrative was actually reshaping the waste management policies. But by September 2019, the final decision was made to proceed with a book, although not with the same content we were imagining in 2016. The decision was to combine the fourth industrial revolution with circular economy and try to portray the new landscape that is formulated not only for the waste sector but also for the whole world. In this crucial moment, Toralf introduced me to Anders Waage Nilsen, a business developer, design strategist, investor, technology columnist, and public speaker, who likes the combination of big ideas and hands-on practical work. After some minutes of discussion, we easily concluded that the book will be written by me and Anders, although Toralf has also contributed a lot with ideas and information.

Writing this book was a great adventure for both of us, a trip to the unknown, which seems to be known but is not, a rush through the Symplegades rocks. On the one side the anxiety and the responsibility to keep the deadlines – on the other side the anxiety and the responsibility to write something good enough, well documented, and valuable.

The completion of this book in time was an achievement. As all human achievements, it is by definition imperfect. But we also went further than we expected, discovering things about science, and ourselves, that we did not know. In this effort, we had the pleasure to work with Nick Rigas, a very creative designer who takes deadlines very seriously. Nick was not only patient enough to manage several cycles of changes and corrections, but he also contributed substantially with his visual creativity to the final outcome and the graphics of this book. Without his contribution, the book would be less interesting and certainly, less visually attractive.

Writing this book, we had in mind that we have to express our opinion, as Gustave Flaubert suggested: "The art of writing is the art of discovering what you believe." The result is now in your hands or in your screens and you will make your own evaluation about the book – from our side, we hope that you will enjoy it and find it useful.

Athens and Bergen, March 2020

Antonis Mavropoulos
Anders Waage Nilsen

How to Cite This Book:

To cite this book, please use the following format:
Industry 4.0 and Circular Economy: Towards a Wasteless Future or a Wasteful Planet?, Antonis Mavropoulos, Anders Waage Nilsen, 9781119699279, © 2020 John Wiley & Sons Ltd.

About the Authors

Antonis Mavropoulos is a chemical engineer that was prepared for a career in quantum chemistry and optoelectronics but, in a mysterious way that is still subject of scientific research in psychology, he found himself in the field of waste management. He is obsessed with social change and innovation, and he believes that the fourth industrial revolution provides an historical opportunity to reshape human societies and their environmental footprint. Professionally, he is a waste management consultant with working experiences in 30 countries, and he has founded D-Waste, a company that aims to make waste management services accessible to everyone that needs them. He has invented the Waste Atlas, designed several mobile apps and information systems, and written many papers and reports. He was privileged to serve the International Solid Waste Association (ISWA) as the chair of the Scientific and Technical Committee (2008–2016) and as the president (2016–2020). He is also a member of the advisory board of UNEP's International Environmental Technology Centre. All his writings can be found at his blog wastelessfuture.com.

Anders Waage Nilsen is a business developer, design strategist, investor, technology columnist, and public speaker. He likes the combination of big ideas and hands-on practical work, because the big picture is made from the important details. During his career, he founded several companies, events, and not-for-profit initiatives. He started out as a newspaper journalist and then founded Fri Flyt skiing magazine, which turned into the leading publisher of sports/outdoor magazine in Norway. He also founded Fjellfilmfestivalen (mountain film festival), Stormkastkonferansen (business conference), and Netlife Bergen (design agency). For some years, he worked with the business cluster projects NCE Tourism and MediArena (currently known as NCE media). In 2018, he co-founded NEW, a business building new start-ups with industrial partners. He focuses on high-impact business ideas. Currently, he is developing WasteIQ – a digital platform for public waste management, based on open standards and IoT integrations. He believes that a data-driven resource management is necessary to create circular business models and new behavioral incentives.

About the Graphic Designer

Nick Rigas is a graphic designer, with a master of arts and a bachelor's degree in animation and interactive media. Besides being a creative designer, he used to play the guitar in rock music groups, and when he was younger, he was playing basketball too. He has designed almost everything, from websites, logos, books, reports, and infographics till mobile applications and T-shirts. For the last eight years he is working as the master designer in D-Waste.

Endorsements

This book is most timely given the urgent need for societies worldwide to deliver net zero carbon. Mavropoulos has for many years been at the forefront of circular economy thinking and is recognized as a futurist in the resources sector, someone that is looking at the bigger social and industrial trends – Nielsen is a digital strategist and deals with the ways digitalization is challenging our current systems and ways of thinking in terms of our ability to cope with the changes that are coming. In this book the authors draw on years of both research, consultancy, and digital experiments to get right under the skin of some of these big trends, in particular what is the 4th industrial revolution and what opportunities and risks might it pose for a world trying to design and deliver more circular solutions and systems (the circular economy).

Mavropoulos and Nielsen rightly identify that for the circular economy to be the reality that we all hope for, it must be digitized and work in tandem with the 4th industrial revolution, but for that to happen we must join the agendas of industry, governments, and consumers, which is easier said than done at a global scale. The book has plenty of examples from the past, case studies of innovation, and presents many areas of optimism, but always with the overlying concern of what if things don't go to plan. A valuable addition to any university library or a good read for those of us interested in the next revolution in resources and waste management.

Dr. Adam Read, PhD, FRSA, FCIWM, FRGS
External Affairs Director
SUEZ Recycling & Recovery UK

This book is a must-read for everyone in the waste and resources sector because of three reasons. First, because this book creates a sense of addiction, the more you read, the more you need to read the rest of the book. Second because, with clarity, the book presents the key questions that must be answered as soon as possible to reverse the fatal tendency to surpass the planet boundaries. Fortunately, but also coincidentally, the so much needed paradigm shift to circular economy would be determined by Industry 4.0, hopefully stimulating a sustainable way of living. And last but not least, because the book presents in details the central and indispensable role of sustainable waste management in the transition towards circular economy. In this transition path to a more fair and sustainable society, more and not less waste management is needed.

Atilio Savino
ISWA Board member, Vice-president of Asociación para
el Estudio de los Residuos Sólidos
Former Secretary of Environment and Sustainable Development of Argentina

Our world is changing at an unprecedented rate. Global consumption of finite resources continues to escalate, resulting in increased waste generation and environmental degradation. Mavropoulos and Nielsen provide a very timely perspective on what a transition to a circular economy means and how it will transform the waste management sector, from the novel perspective of Industry 4.0. Providing both a theoretical and very practical approach, the authors point out how "Industry 4.0 provides the technological, economic and social framework in which a circular economy will flourish or fail." They challenge us to acknowledge and address this interface between a circular economy and Industry 4.0 and consider whether "In the era of Industry 4.0, the circular economy will either be digitized, automated and augmented or it will not prevail at all."

Prof Linda Godfrey
Manager Waste RDI Roadmap Implementation Unit
Principal Scientist
CSIR, South Africa

The agricultural revolution, and three subsequent industrial ones, each dramatically increased the rate at which humans were able to extract resources from our environment, do great things with them, and generate waste as a byproduct. In this book, Mavropoulos and Nielsen speculate, with a wealth of interesting examples and perspectives, as to whether the ongoing "fourth" industrial revolution will continue this trend or perhaps enable transition to a "circular" economy – in which instead of unwanted waste, the byproducts of human activities become our primary resource supply, obviating the need to unsustainably extract them from our environment. It is a very good question, and the eventual answer will define the planetary impact of the Anthropocene era.

Keith Alverson
Director, UNEP International Environmental Technology Centre
Osaka, Japan

Waste is like a river. It flows downhill…to the cheapest price. Landfills have historically been very cheap and often underpriced. They are like big vacuum cleaners – the cheaper the price, the larger the suction they exert on materials in the economy, whether that is truly residual waste or recyclable materials by mistake. Landfills don't discriminate. Nor do energy from waste facilities without the right policy prescriptions. Resource conservation and waste management are all about policy and proactive intervention. They are about choices.

   In a resource constrained world, the sooner we make the choice to stop wasting resources, the less likelihood there is that scarcity and rationing will force us to do it.

Antonis and Anders understand these dynamics. This book plots a course that we can choose to take or a course that we will inevitably be forced to take, by demand and scarcity pricing. The choices about which policies and when are ours to make.

Essential reading.

Mike Ritchie
Managing Director MRA Consulting Group
New South Wales, Australia

My major takeaway from Mavropoulos and Nielson is that "the implementation of a circular economy is a matter of political struggle." I am left with critical reflections on where informal waste pickers fit in the circular economy. Would a transition to the CE render these workers invisible or even allocate them a deeper subaltern role? Can we strike a balance with our pursuits for alternative economic models and our ethical dilemmas regarding social inclusion? It's time for waste pickers organizations throughout the world to unite with their allies in a struggle for a truly inclusive circular economy for all!

Sonia Dias, WIEGO's waste specialist

Glossary

Actuators   An actuator is a component of a machine that is responsible for moving and controlling a mechanism or system.

Business model   A business model describes the rationale of how an organization creates, delivers, and captures value, in economic, social, cultural, or other contexts. The process of business model construction and modification is also called business model innovation and forms a part of business strategy.

Cascading   Cascading is a term that is used to demonstrate that material cycles are not perfect. Each time a material has to be recycled, its quality is downgraded. Energy recovery is only the last step to terminate the cascade for biogenic materials. The challenge is to define the optimal cascade in order to minimize resource and energy consumption as well as environmental impact. Source: From ISWA [1]

Circular economy   There is no single commonly accepted definition of the term "circular economy," but different definitions share the basic concept of decoupling of natural resource extraction and use from economic output, having increased resource efficiency as a major outcome. The authors prefer this definition. "Circular economy is an economic system that replaces the 'end-of-life' concept with reducing, alternatively reusing, recycling and recovering materials in production/distribution and consumption processes. It operates at the micro level (products, companies, consumers), meso level (eco-industrial parks) and macro level (city, region, nation and beyond), with the aim to accomplish sustainable development, thus simultaneously creating environmental quality, economic prosperity and social equity, to the benefit of current and future generations. It is enabled by novel business models and responsible consumers." Source: From Kirchherr et al. [2].

Closed loops   Closed-loop recycling is focused on supply chain sustainability. It is a process where waste is collected, recycled, and used to make new products that are the same as before. It is common in specialized industries such as computing that use complex parts that cannot easily be broken down post-consumption into individual materials. Closed-loop recycling focuses on bringing the products back to the manufacturer so that they can be reused or refurbished without a loss of material.

Cloud   Cloud computing is the on-demand availability of computer system resources, especially data storage and computing power, without direct active management by the user. The term is generally used to describe data centers available to many users over the Internet. Large clouds, predominant today, often have functions distributed over multiple locations from central servers.

Digital platforms   Digital platforms are online businesses that facilitate commercial interactions between at least two different groups – with one typically being suppliers and the other consumers. Airbnb, Amazon, BlaBlaCar, Deliveroo, Facebook, Google, TaskRabbit, Uber, and Xing are all platforms, but they have different business models, and they interact with end users and other businesses in different ways. Consequently, each platform has created different rules to optimize these interactions. Some important distinctions are the degree to which a platform relies on advertising revenue versus fees, its rules for managing suppliers and content, and its relationship with consumers. Source: From ITIF Technology Explainer [3].

Disposal   Disposal means any operation that is not recovery even where the operation has as a secondary consequence the reclamation of substances or energy. Source: EU Directive 2008/98 [4].

Dumpsite   The term "open dump" (or dumpsite) is used to characterize a land disposal site where the indiscriminate deposit of solid waste takes place with either no, or at best very limited measures to control the operation and to protect the surrounding environment. Source: From ISWA [5].

Entropy   Entropy is a measure of the randomness or disorder of a system. The value of entropy depends on the mass of a system. A highly ordered system has low entropy.

Final sink   The term final sink applies to the following: sinks are defined as water, air, and soil receiving anthropogenic material flows. They are necessary for accommodating emissions and residues from anthropogenic activities such as primary production (e.g. tailings), processing and manufacturing, and consumption. The permissible capacity of a sink for holding a particular substance is determined by the geogenic concentration of the material in the sink: For example, the atmosphere is an appropriate sink for molecular nitrogen, the oceans for water and chlorides, and the soil for carbonates. For trace metals such as cadmium or lead, the capacity of air, water, and soil as a sink is very limited. Source: From Baccini and Brunner [6].

Industry 4.0 and fourth industrial revolution   A new generation of smart and partly autonomous production systems are evolving around us. The fourth industrial revolution is a complex transformation with multiple outcomes for every sector and industry. An important part of this revolution is the concept called Industry 4.0. Industry 4.0 is not a historical fact. It is a possible outcome, an invention in the making, something that is still ahead of us. The term itself was coined in a 2011 research project initiated by the German Federal Ministry of Education and Research. They were asked to identify how new technologies and trends could improve the world and boost productivity. The result was a conceptual invention of a process that connects mechanical and virtual into holistic systems.

Informal recyclers of informal recycling sector The informal solid waste sector refers to individuals or enterprises who are involved in recycling and waste management activities but are not sponsored, financed, recognized, or allowed by the formal solid waste authorities, or who operate in violation of or in competition with formal author.

Internet of Things (IoT) The Internet of Things (IoT) is a system of interrelated computing devices, mechanical and digital machines provided with unique identifiers (UIDs) and the ability to transfer data over a network without requiring human-to-human or human-to-computer interaction.

Material flow analysis Material flow analysis is a quantitative procedure for determining the flow of materials and energy through the economy. It uses input/output methodologies, including both material and economic information. It captures the mass balances in an economy where inputs (extractions + imports) equal outputs (consumption + exports + accumulation + wastes). It is based on the laws of thermodynamics. MFA asks whether the flow of materials is sustainable in terms of the environmental burden it creates. Urban metabolism simply quantifies the flows; it does not engage in analysis of environmental burdens. Source: From Pincetti [7].

Open innovation Open innovation is a term used to promote an information age mindset toward innovation that runs counter to the secrecy and silo mentality of traditional corporate research labs. The benefits and driving forces behind increased openness have been noted and discussed as far back as the 1960s.

Open loops Open-loop recycling is where the recycled materials are converted into both new raw material and waste product typically in the manufacturing sector. This usually involves processing various types of product of similar material make-up and changes the properties of the material itself often through heat, chemical reactions, or crushing. Open-loop recycling is often referred to as "downcycling" as each time a product is recycled it degrades the material.

Recovery Recovery means any operation the principal result of which is waste serving a useful purpose by replacing other materials that would otherwise have been used to fulfil a particular function or waste being prepared to fulfil that function, in the plant or in the wider economy. Source: From EU Directive 2008/98 [4].

Recycling Recycling means any recovery operation by which waste materials are reprocessed into products, materials, or substances whether for the original or other purposes. It includes the reprocessing of organic material but does not include energy recovery and the reprocessing into materials that are to be used as fuels or for backfilling operations. Source: From EU Directive 2008/98 [4].

Reuse   Reuse means any operation by which products or components that are not waste are used again for the same purpose for which they were conceived. Source: From EU Directive 2008/98 [4].

Sanitary landfill   Sanitary landfill is an acceptable waste management method, with controlled emissions and limited health and environmental impacts, while open dumps or dumpsites are exactly the opposite. In between an open dump and a sanitary landfill there is a grey area usually named as "controlled dump" with varying levels of engineering and environmental controls. Source: From ISWA [1]

Sharing economy   The sharing economy refers to the sharing of goods or other resources by multiple people. Sharing allows existing goods and resources to be used more fully, rather than letting them lay dormant, and depends greatly on either access to goods via a membership (car sharing, resource libraries), or peer-to-peer interaction (Airbnb [sic!], ride sharing, clothing swaps). The circular economy is more about goods as they are manufactured and as they are taken apart for reuse and reconstruction as new goods. It deals with the raw materials, ensuring that they never become waste or pollution.

Smart containers   Containers equipped with different sensors that are parts of the Internet of Things.

Smart sensors   A digital transducer or actuator combined with a processing unit and a communication interface.

Stocks   Stocks are accumulations of things (not necessarily physical) that change over time through the actions of inflows and outflows.

Thermodynamics first law   When energy passes, as work, as heat, or with matter, into or out of a system, the system's internal energy changes in accord with the law of conservation of energy. Equivalently, perpetual motion machines of the first kind (machines that produce work with no energy input) are impossible.

Thermodynamics second law   The second law of thermodynamics states that the total entropy of an isolated system can never decrease over time and is constant if and only if all processes are reversible. Isolated systems spontaneously evolve towards thermodynamic equilibrium, the state with maximum entropy.

Urban metabolism   Urban metabolism is a model to facilitate the description and analysis of the flows of the materials and energy within cities, such as undertaken in a material flow analysis of a city. It provides researchers with a metaphorical framework to study the interactions of natural and human systems in specific regions. Source: From Pincetl et al. [8].

Waste   Waste means any substance or object which the holder discards or intends or is required to discard. Source: From EU Directive 2008/98 [4].

Waste hierarchy   The waste hierarchy shall apply as a priority order in waste prevention and management legislation and policy: (i) prevention (ii) preparing for reuse (iii) recycling (iv) other recovery (v) disposal. Source: From EU Directive 2008/98 [4].

Waste management   Waste management means the collection, transport, recovery, and disposal of waste, including the supervision of such operations and the aftercare of disposal sites and including actions taken in the role of principal to purchase and sell waste or actions arranging the recovery or disposal of waste. Source: EU Directive 2008/98 [4].

Waste prevention   Prevention means measures taken before a substance, material, or product has become waste, which reduce the quantity of waste, including trough the reuse of products or the extension of the lifespan of products, the adverse impacts of the generated waste on the environment and human health, or the content of harmful substances in materials and products. Source: From EU Directive 2008/98 [4].

Waste to energy   Facilities where waste is thermally converted with energy recovery, generating electricity, heat, bottom, and fly ash.

Waste treatment   Treatment means recovery or disposal operations, including preparation prior to recovery or disposal. Source: From EU Directive 2008/98 [4].

REFERENCES

1. ISWA (2015). *Circular Economy: Cycles, Loops and Cascades*, ISWA. https://www.iswa.org/fileadmin/galleries/Task_Forces/Task_Force_Report_2.pdf (accessed 9 May 2020).
2. Kirchherr, J., Reike, D., and Hekkert, M. (2017). Conceptualizing the circular economy: an analysis of 114 definitions. *Resources, Conservation and Recycling* 127: 221–232.
3. ITIF (2018). ITIF technology explainer: what are digital platforms. https://itif.org/publications/2018/10/12/itif-technology-explainer-what-are-digital-platforms (accessed 9 May 2020).
4. EU Directive 2008/98/EC (2008). Directive 2008/98/EC of the European Parliament and of the Council of 19 November 2008 on waste and repealing certain directives. https://eur-lex.europa.eu/legal-content/EN/TXT/?uri=celex%3A32008L0098 (Accessed 3 May 2020)
5. ISWA (2016). *A Roadmap for Closing Waste Dumpsites The World's Most Polluted Places*. ISWA. https://www.iswa.org/fileadmin/galleries/About%20ISWA/ISWA_Roadmap_Report.pdf (accessed 9 May 2020).
6. Baccini, P. and Brunner, P. (2012). *Metabolism of the Anthroposphere: Analysis, Evaluation, Design*. Cambridge, MA: The MIT Press.

7. Pincetti, S. (2012). A living city, using urban metabolism analysis to view cities as life forms. In: *Metropolitan Sustainability Understanding and Improving the Urban Environment*, Series in Energy (ed. F. Zeman), 3–25. Woodhead Publishing https://doi.org/10.1533/9780857096463.1.3.
8. Pincetl, S., Bunje, P., and Holmes, T. (2012). An expanded urban metabolism method: toward a systems approach for assessing urban energy processes and causes. *Landscape and Urban Planning* 107: 193–202.

Acronyms

3Rs	Reduce, reuse, recycle
AEB	Afval Energie Bedrijf Amsterdam
AI	Artificial intelligence
APC	Air pollution control
API	Application programming interface
AVAC	Automated vacuum collection
AVL	Anstalt für Verbrennungskraftmaschinen List
BFA	Bisphenol-A
BIM	Building information model
BIR	Bergensområdets Interkommunale Renovasjonsselskap
CD	Compact disc
CE	Circular economy
Cefic	European Chemical Industry Council
CEO	Chief Executive Officer
CEPS	Centre for European Policy Studies
CERN	European Organization for Nuclear Research
CNN	Cable News Network
COP	Conference of parties
CRM	Customer relationship management
DDT	Dichlorodiphenyltrichloroethane
DEFRA	Department for Environment, Food, and Rural Affairs
DR	Deposit return
ECC	Essentially contested concepts
EMF	Ellen MacArthur Foundation
EPR	Extended producer responsibility
ERP	Enterprise resource planning
ESRI	Environmental Systems Research Institute
EUCIA	European Composites Industry Association
EURELCO	European Enhanced Landfill Mining Consortium
EUROSTAT	European Statistical Office
EVs	Electric vehicles
FAO	Food and Agricultural Organization
GDL	Grand challenge for development
GDP	Gross domestic product
GHGs	Greenhouse gases
GPS	Global positioning system

GW	Gigawatt
HREEs	Heavy rare earth elements
IBM	International Business Machines Corporation
ICT	Information and Communication Technologies
IDC	International Data Corporation
IDs	Digital identities
IND4.0	Fourth industrial revolution
IPCC	Intergovernmental Panel on Climate Change
ISWA	International Solid Waste Association
ITU	International Telecommunication Union
LCA	Life cycle analysis
LCE	Lithium carbonate equivalent
LREEs	Light rare earth elements
MBT	Mechanical biological treatment
MIT	Massachusetts Institute of Technology
MP3	MPEG-1 audio layer III or MPEG-2 audio layer III
MRW	Materials recycling week
MSW	Municipal solid waste
MTOE	Million tonnes of oil equivalent
MVP	Minimum viable product
NASA	National Aeronautics and Space Administration
NGO	Nongovernmental organization
NIR	Near infrared
NOx	Nitrogen oxides
OECD	Organisation for Economic Co-operation and Development
OS	Operating system
PAHs	Polycyclic aromatic hydrocarbons
PAYT	Pay as you throw
PE	Polyethylene
PET	Polyethylene terephthalate
PGM	Platinum group metals
PMS	Project management software
PP	Polypropylene
PVC	Polyvinyl chloride
RDF	Residual derived fuel
REDA	Rizhao Eco-Industrial Park
RFID	Radio-frequency identification
RIR	Recycled input rate
RoAF	Romerike Avfallsforedling
SDGs	Sustainability development goals
SWM	Solid waste management

TSCA	Toxic Substances Control Act
UNCTAD	United Nations Conference on Trade and Development
UNEP	United Nations Environmental Program
UNU	United Nations University
USD	United States dollars
USSR	Union of Soviet Socialist Republics
VDI	Verein Deutscher Ingenieure
WEEE	Waste electrical and electronic equipment
WGR	Waste generation rate
WtE	Waste to energy

Chapter 1
The End of Business as Usual

Prediction is very difficult, especially if it's about the future!

— Niels Bohr

Recommended Listening
Kind of Blue, **Miles Davis**

Due to the fact that it is melancholic, romantic, cool, but still remarkably inspiring for forward thinking.

Recommended Viewing
Wall-E **by Disney**

Because it shows the future of our world if IND4.0 and the circular economy are not combined.

Industry 4.0 and Circular Economy: Towards a Wasteless Future or a Wasteful Planet?, First Edition.
Antonis Mavropoulos and Anders Waage Nilsen.
© 2020 John Wiley & Sons Ltd. Published 2020 by John Wiley & Sons Ltd.

1.1 THE TRILLION-DOLLAR QUESTION

This is a book about circular economy (CE) and the fourth industrial revolution (IND 4.0) and about their interlinkage and the way their interaction will determine the future of our planet, as the authors of this book assert. Despite the many ongoing discussions on both the circular economy and IND 4.0, rarely is it recognized that they are deeply interconnected, and in real life they cannot be discussed separately, and should not be considered to be mutually exclusive.

In spite of the widespread discussion and countless policy initiatives regarding circular economy, it is usually ignored the fact that IND 4.0 provides the technological, economic, and social framework in which a circular economy will flourish or fail. As an example, a well-referenced report [1] estimated that the implementation of the circular economy in the European Union (EU) will create 1.2–3 million additional jobs by 2030. However, the report seems to ignore that the essential works required in preparation for reuse, repair, and disassembly either will be automated and robotized or will not become economically viable. In the era of IND4.0, the circular economy will either be digitized, automated, and augmented, or it will not prevail at all.

On the other hand, discourse around IND 4.0 usually focuses on the advances relevant to resource and labor productivity, the radical changes to business models, and the social challenges involved. It is rarely discussed that IND4.0, unless it diverts from the "business as usual" linear approach, will also stimulate and accelerate resource depletion and pollution in an era in which Earth is fast approaching, or has already surpassed, some of its planetary limits. IND4.0 will either meet the circular economy or accelerate environmental deterioration and potential collapse of ecosystems and human societies. This will be covered in more detail in Chapter 2.

IND4.0 and the circular economy represent challenges that are valued, quite literally, in the many trillions of dollars. One of the key aspects that should be considered is that a shift to a circular economy might be the only possible means of sustaining economic growth in the long term through a serious rearrangement of the economic inputs and outputs in all the industrial supply chains. An online analysis [2] of the economic benefits of circular economy reveals that in the EU the annual material cost savings are estimated up to $630 billion, and that is only for the sectors of complex medium-lived products such as mobile phones and washing machines. For fast consumption products such as those for household cleaning, the expected material cost savings are estimated at $700 billion/year. Accenture [3] has assessed that the shift to a circular economy has the potential to create an additional $4.5 trillion economic output by 2030, if circular business models are rapidly adopted. In addition, such a shift can contribute to closing the eight billion-tonne material gap between supply and demand that is expected in 2030. In the long term, an additional $25 trillion economic output is forecast by 2050. These economic benefits do not include the monetization of the expected environmental benefits such as the reduction of carbon dioxide emissions; the gradual decline in consumption of primary materials, which can reach to 32% within the next 10 years; and the benefits from the reduction of land degradation that already costs more than $40 billion annually. McKinsey has calculated [4] that for the EU an additional $1.3 trillion of benefits annually is expected in non-resource and externality costs.

Moving to IND4.0, for the period 2015–2025, the World Economic Forum assessed [5] the economic and social benefits of the digital transformation to almost $100 trillion through digital consumption, digital enterprise, societal implications, and platform governance. By 2025, the digital economy will have a share of 24.3% of the global economy and a value of $23 trillion, compared with 15.5% and $11.5 trillion in 2016, according to a report prepared by Oxford Economics [6], providing extra income of $500/year for the average worker. The same report assessed that the digital economy is growing 2.5 times faster than the global economy and that, on average over the past three decades, a US $1 investment in digital technologies has led to a US $20 rise in gross domestic product (GDP).

Since both the circular economy and IND4.0 represent trillions of dollars in opportunities, this book poses the trillion-dollar question: will IND 4.0 and the circular economy converge, hence delivering not only better resource efficiency but a more sustainable future for everyone on the planet? Or will IND4.0 evolve according to the business as usual linear model, leaving the circular economy a mere flight of fancy, resulting in faster resource depletion, acceleration of environmental degradation, and deeper inequality? The authors of this book believe that any answer to the trillion-dollar question involves the transformation of the waste management sector. However, before delving deeper into this discussion, it is beneficial to outline the planetary framework in which the circular economy and IND4.0 are discussed. A warmer, continuously urbanized, and more resource-demanding planet sets the scene from which our future will evolve.

1.2 THE FUTURE IS WARMER, URBANIZED, POLLUTED, AND RESOURCE-HUNGRY

It has always been difficult to predict the future, but in a world that is becoming con-tinuously more interconnected and complex, discontinuity and abrupt change become the rule and not the exception, as the 2008–2013 financial crisis proved. Still, there are some basic trends that will continue to shape our planet for the next 20–30 years.

The first trend is global warming. Although there are some differing opinions, there is a very strong consensus about the roots of the problem and the way to deal with it. About 97% of the studies published by climate scientists agree that global warming is extremely likely to be linked to human activities [7]. Almost all of the major scientific organizations worldwide [8] endorse such a view. Regarding extreme weather phenomena, between 2015 and 2018, more than 100 papers have been published on this subject, and 75% of them concluded [9] that extreme weather is related to global warming. In October 2018, the Intergovernmental Panel on Climate Change (IPCC) warned [10] that only a dozen years are available to keep global warming to a maximum of 1.5°C, beyond which even half a degree will significantly worsen the risks of drought, floods, extreme heat, and poverty for hundreds of millions of people. In November 2019, more than 11 000 scientists from all over the world declared that "planet Earth is facing a climate emergency" [11].

Global warming is directly linked with both IND4.0 and circular economy. The development of innovative low-carbon technologies, especially in the energy sector, which will allow the immediate reduction of carbon dioxide emissions and the hasty decarbonization of our economies, is an urgent necessity. IND4.0 not only brings the promise of unimaginable innovation, but it also provides new tools to coordinate the global responses on global warming, increase energy efficiency and reduce losses, as well as actively forecast and link energy supply and demand [12]. However, according to the International Resource Panel [13], by 2050 almost 1.5 billion tonnes of metals will be required to develop low-carbon infrastructure and wiring. The World Bank has predicted that the development of green low-carbon technologies will definitely drive a substantial increase in the demand for several minerals and metals such as aluminum, copper, lead, lithium, manganese, nickel, silver, steel, zinc, and other rare earth minerals [14]. Thus, the progress in circular economy becomes a condition for a decarbonized future, especially as some of the metals required (indium, tellurium) are already characterized by severe risks of medium and long term supply deficits [15]. More about the role of circular economy and climate change will be discussed in Chapter 3.

Another pressing issue that must be considered in the context of the planet's future is the serious damage and loss of biodiversity, a clear signal that the world is approaching its natural physical limits. The human footprint has become dominant on Earth: 75% of the land-based environment and about two third of the marine environment have been significantly altered by human actions, and almost one million animal and plant species are now threatened with extinction, many within decades, more than ever before in human history [16]. At least 680 vertebrate species have been driven to extinction since the sixteenth century, and more than 9% of all domesticated breeds of mammals used for food and agriculture became extinct by 2016, with at least 1000 more breeds still under threat. Land degradation has reduced the productivity of 23% of global land surface, up to $577 billion in annual global crops are at risk from pollinator loss, and 100–300 million people are at increased risk of flooding and hurricanes due to loss of coastal habitats and protection. IND4.0 provides a lot of tools to better monitor, prevent, and perhaps reverse biodiversity losses, and if the circular economy is applied on a global and regional scale, it will reverse some of the causes that stimulate biodiversity losses.

Urbanization is reshaping our planet. Between 1900 and 2015, the share of the world's urban population rose from 14 to 54%, according to the estimations of the UN Department of Economic and Social Affairs, Population Division. By around 2050, the share of the urban population is expected to reach at 66%. For example, from 2014 to 2019, the urban population increased by 9.62% (almost 70% faster than the mean population growth), from 3.9 to 4.27 billion. This means that between 2014 and 2019, the population in urban areas was increasing by 75 million people per year or by almost 200 000 people per day. Over the next 30 years, it is expected that practically all of the world's population growth is going to be concentrated in urban areas in the developing world. Cities in Asia and Africa are expected to absorb 90% of the world's 2.5 billion new urban residents by 2050 [17]. The rapid urban growth in the developing world creates huge health and environmental pressures. It is directly linked with air, water, and soil pollution, and it is one of the main factors in the rise of urban waste generation. At the same time, a health emergency is rising in poor urban areas. According to the World Health Organization (WHO) [18], 7 million premature deaths annually are linked to air pollution, and 90% of the urban residents breathe air containing high levels of pollutants. Another report [19] found that deaths related to improper solid waste management (SWM) are between 0.4 and 1 million annually. In 2010, it was found that in India, Indonesia, and Philippines alone, almost 8.7 million people were at high risk of exposure to industrial and hazardous waste pollutants, mainly lead and hexavalent chromium, from 373 dumpsites. In addition, according to the latest progress report on water and sanitation [20], almost 2.2 billion people lack access to safe drinking water, 4.2 billion people lack safe sanitation services, and 3 billion lack basic handwashing facilities.

Cities are, undeniably, the hubs of both IND4.0 and circular economy. IND4.0 is already transforming cities, especially in the developed countries, and the concept of "agile cities" has been suggested as a key concept that allows the merging of "the biological, physical and digital worlds" through innovations such as artificial intelligence (AI), the Internet of Things (IoT), and 5G Internet [21]. It is hoped that IND4.0 will offer the technological means to allow cities to proceed with smart buildings and construction, reduce the carbon footprint of their transportation and logistics, deliver green energy, improve their resiliency to global warming, and reduce pollution in an affordable way. However, all these hopes are in direct contrast with the reality in the developing world. Cities that do not have the administrative, institutional, and financial capacity to protect the health of their residents by proper management of their waste and wastewater will not be able to explore and utilize the opportunities of IND4.0. Despite this, even in rather poor or emerging cities, the proper use of mobile phones and apps provides serious opportunities to improve urban governance and deliver improvements in environmental protection, resource recovery, and waste management [22].

Circular economy approaches and policies are becoming a key element for the sustainability of cities. Today, cities generate roughly 85% of the world's GDP, and their material consumption is expected to skyrocket by 125% between 2010 and 2050, from 40 to 90 billion tonnes/year [23]. Urban municipal solid waste is expected to rise from 1.3 to 2.2 billion tonnes of waste annually between 2012 and 2025, and total municipal solid waste is expected to be around 3.4 billion tonnes in 2050 [24]. Figure 1.1 shows the municipal waste management practices on a global scale, for the year 2016, according to the data sets of the latest Word Bank Report [24] and the findings about open burning. It is obvious that with such an increase, the traditional SWM practices such as sanitary landfills and waste to energy plants, though very important as starting points, are not capable of absorbing the rising wave of urban waste. Hence the need for a circular economy and waste prevention measures is becoming central to the livelihood of cities. Urbanization, supply and price risks, ecosystem degradation, environmental accountability, consumer behavior, and digital advances are considered the key drivers for the circular economy application in cities [25].

Our world is also becoming relatively richer, at least in terms of the world average GDP/cap. According to the International Monetary Fund [26], between 2000 and 2019,

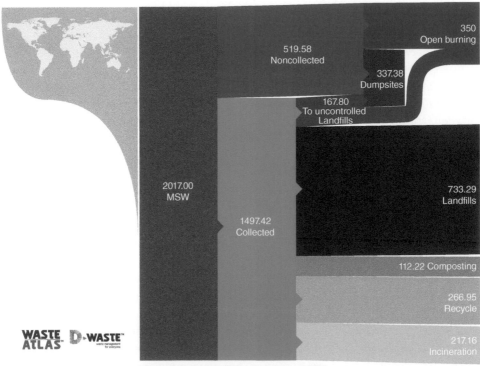

All numbers are in million tonnes/year or 2016

Figure 1.1 Sankey diagram that presents the current municipal solid waste management practices worldwide.
Source: With kind permission from Nick Rigas, D-Waste.

the GDP/cap (in current prices) in emerging markets and developing economies rose from $1.41 to 5.38 thousand (380% increase), while the world's GDP/cap rose from $5.67 to 11.46 thousand (202% increase). According to some calculations [27], from September 2018, just over 50% of the world's population, or some 3.8 billion people, live in households with enough discretionary expenditure to be considered "middle class," marking a global tipping point. The middle class is already the largest and most rapidly growing segment of demand in the global economy, projected to reach some 4 billion people by 2020 and 5.3 billion people by 2030. Although the rising middle class in the developing world seems to be more vulnerable [28] than the one in OECD countries, one can expect that sales in motors and automobiles, white goods such as refrigerators and kitchen sets, televisions, personal computers, mobile phones, and televisions will definitely continue to rise. As an engine of economic growth in the developing world, the rising middle class will certainly deliver increasing waste quantities, making the shift to circular economy more essential and more urgent. As one of the basic beneficiaries of the daily advances of IND4.0, mainly through the day-to-day use and the growing dependence of our daily lives on mobile phones and Internet services, the emerging middle class is also expected to utilize billions of household sensors, invest massively in interconnectivity applications, and produce a new rising wave of e-waste.

The rise of the population, the increase of the share of urban residents, and the emerging middle class are also strong drivers that stimulate more resource extraction. According to the data presented at the report "Global Resources Outlook 2019" [29] since 2000, the growth in extraction rates has accelerated to 3.2% per year, driven largely by major investments in infrastructure and higher material living standards in developing and transitioning countries, especially in Asia. Between 2014 and 2019 alone, the global resource extraction rose from 83.70 to 97.98 billion tons/year, a dramatic 17.06% increase. The extraction and processing of materials, fuels, and food make up about half of total global greenhouse gas (GHG) emissions and more than 90% of biodiversity loss and water stress. While extraction and consumption are growing in upper-middle-income countries, high-income countries continue to outsource resource-intensive production. An average person living in developed countries consumes 60% more and over 13 times the level of the upper-middle- and low-income groups, respectively. A study [30] on the metabolism of the global economy between 1900 and 2015 found that the global material extraction rose from 12 to 89 billion tonnes/year and that in 2015 the accumulated anthropogenic material stocks were roughly 961 billion tonnes. On an annual basis, in 2015, each human on Earth contributes to stocks by 7.1 tonnes. The average solid waste generation per capita, including all the types of solid waste, was found to be at 2.6 tonnes per capita per year for 2015. In practice, all these figures mean that the rising middle class brings an increase in all the waste streams but also in the anthropogenic stocks. These stocks, at the end of their useful life, will become the future waste streams that we need to manage for a truly forward-thinking circular economy approach. In addition, this serious acceleration of the resource extraction rates makes the resource efficiency and the circular economy concepts much more important, raising the bar for the expectations of IND4.0 advances.

Similar patterns are shown in energy consumption. The world's total energy consumption increased by 43% between 2000 and 2018, reaching 14 391 MTOE [31]. In the same period, it rose by 2.4% in OECD countries, and it was almost doubled in Asia, meaning that today Asia's share in total energy consumption is 40.7%, while the OECD countries' share is 37.6%. Regarding renewables, in 2017, renewable energy accounted for an estimated 18.1% of total final energy consumption, with modern renewables supplying 10.6% and traditional use of biomass for cooking and heating in developing countries accounting for the remaining share [32]. According to the World Bank database [33], the energy use in kilograms of oil equivalent per capita rose by 43.7% between 1960 and 2014 and 17.4% compared with 2000. Regarding the future, the International Energy Agency forecasts [34] a steady growth in energy consumption between 1 and 1.3% each year until 2040 and an increasing importance of electricity, since it is the vital condition for the continuous digitalization of our world. Africa is expected to play a special role in the future global energy trends as more than half a billion people are expected to be added in Africa's urban population by 2040, causing a dramatic increase in the need for cooling services and air conditioning.

The rise of the middle class creates a serious impact on food consumption and supply chains, more explicitly when considering meat consumption. Globally speaking, meat consumption per capita has increased by approximately 20 kg since 1961, from 23 to 43 kg of meat in 2014 [35]. Thus, global meat production has been growing much faster than the growth of the population. A further increase is expected by FAO [36], resulting in an annual meat consumption of 45 kg/cap/year in 2019. In addition, the share of global meat types has changed significantly over the last 50 years, with poultry meat rising from 12 to 35% between 1961 and 2013 and beef–buffalo meat declining to almost half share (22% today). This shift to meat creates serious environmental impacts [37]. It is hoped that IND4.0 can provide more viable alternatives by advancing cultured meat techniques (meat produced in vitro, using tissue engineering techniques) as this requires [38] almost 745% lower energy use, 7896% lower GHG emissions, 99% lower land use, and 8296% lower water use, depending on the product compared. The shift to meat is also accompanied by a rise in food waste, a real challenge for any circular economy approach. It seems that food waste, considering all the losses in different phases of the food supply chain, is between a third and a half of the food that was intended for human consumption [39], with a total annual direct cost in the vicinity of a trillion dollars, or $2 trillion if environmental and social costs are taken into account [40]. The importance of a circular economy for food and biowaste has been highlighted by a study [41], which showed that closing the loop for food waste in Amsterdam can save 0.6 million tonnes of carbon dioxide equivalents per year and 75 000 tonnes of raw materials.

To conclude, it seems that the continuous growth of urban population, resource extraction, energy consumption, and food demands are pushing our planet toward the end of *business as usual* and closer to some of its natural limits. Our economic model delivers a rather catastrophic impact on the livelihood of Earth. What happened in all previous industrial revolutions simply cannot happen again.

1.3 IT CAN'T HAPPEN AGAIN

All industrial revolutions have rearranged economies by creating and wiping out economic sectors and millions of jobs. They have stimulated urbanization and the creation of cities and created the material basis and structures of modern political and governance systems. They have also transformed the way humans interact with their environment. By changing the key technologies, industrial revolutions have used different primary resources and raw materials, produced new products, and created different environmental and health impacts. All of the industrial revolutions, besides delivering magnificent innovations that radically changed the way we live, have resulted in increased pollution and environmental problems. According to researcher Sjur Kasa [42], the first industrial revolution was characterized by the smog and the creation of slums, the second one by the rise of water pollution and the gradual emergence of acid rain, soil pollution, and the problem of global warming that remains with us today. However, Kasa suggests that this is not necessarily the fate of every industrial revolution and that in recent history there were alternatives that did not prevail in the end. He refers to the example of Japan, which, after the oil crisis in 1970s, was actually able to decouple energy demand and pollution from economic growth. Growth was mostly fueled by information and not by fossil fuel-intensive industries, while very high investments in energy efficiency and massive transportation were made. As Kasa notes, "The re-emergence of the US as a leader in the ICT economy also, combined with the demise of Japan and the rise of China as a chief supplier of consumer durables to the US market, appears to have blocked the generalization of the opportunities emerging in the Japanese model."

Another common characteristic of all the industrial revolutions is the well-known **rebound effect and Jevons paradox** [43]. In economics, Jevons paradox occurs when technological progress increases the efficiency with which a resource is used, but the rate of consumption of that resource rises due to increasing demand. This paradox is perhaps the most widely known in environmental economics. By delivering innovative technologies that made it possible to deliver more products with fewer resources, labor, and energy per unit [44], industrial revolutions resolved many problems and substantially improved resource efficiency while at the same time stimulating massive consumption of the new products. In most cases, the potential environmental benefits of the resource efficiency gains were outweighed by the exponential rise of consumption that drove further resource depletion and pollution. Terry Barker and Athanasios Dagoumas have studied the macroeconomic rebound effects [45] in energy efficiency policies, and they found out that the proportion of avoided energy savings due to changes in the consumers' and industries' perceptions of energy efficiency could be up to 52% for decades of stable and consistent energy efficiency policies. The rebound effect was also studied in pricing by bag policies in SWM in Japan [46], and it was found out that the initial benefits were gradually canceled after almost 20 years.

The potential rebound effect on circular economy and sustainable consumption [47] has also been studied. Circular economy rebound occurs when such an economy's activities, with lower per-unit-production impacts, finally cause increased levels of production, reducing their resource and environmental benefits [48]. According to Trevor Zink and Roland Geyer, "Circular economy activities can lead to rebound by either failing to compete effectively with primary production or by lowering prices, thereby increasing and shifting consumption. Pricing re-used products and re-cycled materials lower to make up for real or perceived technical deficiencies is very likely to produce rebound. Even if secondary products are not discounted, their increased production can depress their own price and that of all substitutes, leading to rebound. Secondary products that compete in either low-end or high-end niches simply grow the 'pie' rather than taking slices from primary production and also result in rebound."

The problem is that in our era, additional pollution due to IND4.0 and a massive rebound effect due to circular economy are hardly affordable because there are a lot of signs that our planet is fast approaching some of its natural physical limits. Global warming and the loss of biodiversity have already been mentioned, but here we will present two more concrete and systematic approaches. The concept of planetary boundaries was proposed in 2009 by Johan Rockström, former director of the Stockholm Resilience Center, with a team of 28 scientists. In brief, they considered that there are nine quantitative planetary boundaries that humanity should not cross, as a condition for the world's future development. The nine boundaries [49] are carbon dioxide concentration in the atmosphere, ocean acidification, stratospheric ozone, biogeochemical nitrogen cycle, phosphorus cycle, global freshwater use, land system change, the rate at which biological diversity is lost, chemical pollution, and atmospheric aerosol loading. This concept has been widely adopted by UN and EU officers, the European Environment Agency, and the World Business Council for Sustainable Development and recently by the Ellen MacArthur Foundation. According to the latest published assessment [50], anthropogenic perturbation levels of four of the boundaries (climate change, biosphere integrity, biogeochemical flows, and land system change) exceed the proposed safe values. Figure 1.2 presents the nine planetary boundaries as they were visualized in 2015 by J. Lokrantz/Azote based on the article "Planetary boundaries: Guiding human

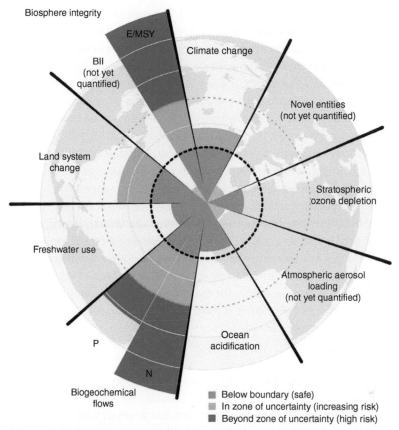

Biosphere integrity

E/MSY

Climate change

BII
(not yet
quantified)

Novel entities
(not yet quantified)

Land system
change

Stratospheric
ozone depletion

Freshwater use

Atmospheric aerosol
loading
(not yet quantified)

P

Ocean
acidification

N

Biogeochemical
flows

■ Below boundary (safe)
■ In zone of uncertainty (increasing risk)
■ Beyond zone of uncertainty (high risk)

Figure 1.2 The nine planetary boundaries visualized by the Stockholm Resilience Center.

Source: With kind permission from J. Lokrantz/Azote [50].

development on a changing planet" [50] written by Steffen et al. and published in the journal *Science*.

A similar though less quantified approach has been adopted by the famous report "Limits to Growth," [51] which was published for the first time by the Club of Rome in 1972. The main message of the book was that the global system of nature probably cannot support present rates of economic and population growth much beyond the year 2100. The 30-year update of the book [52] warned that there are "...symptoms of a world in overshoot, where we are drawing on the world's resources faster than they can be restored, and we are releasing wastes and pollutants faster than the Earth can absorb them or render them harmless. They are leading us toward global environmental and economic collapse – but there may still be time to address these problems and soften their impact." A comparison of the predictions of the first report with the historical data for the period 1970–2000 showed [53] that "The data compare favorably with key features of a business-as-usual scenario called the 'standard run' scenario, which results in collapse of the global system midway through the 21st century."

Considering all of our experiences and the scientific knowledge available, the authors of this book believe that there are specific natural barriers that limit the planet's economic expansion. In other words, infinite economic growth is not feasible in a planet with finite resources. The logical consequence is that a circular economy cannot resolve – or it may only partially resolve – the conflict between environmental impacts and economic activity, even if it is 100% adopted worldwide. As the planet gets close to its natural limits, this industrial revolution should evolve in a different way. A systemic adoption of circular economy combined with the advances of IND4.0 can be substantially advantageous and useful, as well as helping to redress the balance in the relationship between human activities and the Earth's ecosystems. However, this can only happen if we are able to avoid a new exponential rise in pollution and massive rebound effects that will cancel out the resource efficiency gains. A circular economy and IND4.0 are limited and at the same time stimulated by the existing limits to growth. Waste prevention, reuse, recycling, and recovery activities are becoming more and more crucial for certain materials, just because the limits to growth exist! With this in mind, it is time to consider the role of waste management and its increasing importance.

1.4 IT'S ABOUT PEOPLE, NOT WASTE

In this chapter we will discuss in detail the relationship between circular economy and waste management, while later in Chapter 4 we will examine how the term "waste management" is redefined in the context of IND4.0. However, in this intro-ductory part, it is necessary to highlight three main points regarding waste man-agement that are usually ignored or misunderstood, and they will have a crucial role in the future.

The first point is about pollution, which according to the National Geographic "is the introduction of harmful materials into the environment. These harmful materials are called pollutants." Although in developed countries this is something frequently forgotten, it is important to remember that the first priority of any waste management system is to protect human health and the environment from the health and environmental impacts of pollution related to uncontrolled waste disposal. Of course, in many cases, waste management creates by-products like materials or energy recovery, but the primary driver for waste management is health and environmental protection. It is useful to recall a small part of the famous Chadwick report [54], which investigated sanitation and ways to improve it in 1842, in a period where the United Kingdom was moving from the first to the second industrial revolution, and described the pollution and the living conditions of the working class: "That the various forms of epidemic, endemic, and other disease caused, or aggravated, or propagated chiefly amongst the laboring classes by atmospheric impurities produced by decomposing animal and vegetable substances, by damp and filth, and close and overcrowded dwellings prevail amongst the population in every part of the kingdom, whether dwelling in separate houses, in rural villages, in small towns, in the larger towns – as they have been found to prevail in the lowest districts of the metropolis." It is important to note that what we know as modern SWM is the response of human societies to the pollution of the first and second industrial revolution and the vast health impacts it created, as Chadwick's report describes. David Wilson has written that [55] "Waste management is one of the essential utility services underpinning society in the 21st century, particularly in urban areas. Waste management is a basic human need and can also be regarded as a 'basic human right.' Ensuring proper sanitation and solid waste management sits alongside the provision of potable water, shelter, food, energy, transport and communications as essential to society and to the economy as a whole. Despite this, the public and political profile of waste management is often lower than other utility services."

Here is the important issue that is sometimes underestimated or ignored: both circular economy and IND4.0 will bring new forms of pollution, and they will generate new types of waste. This will be dealt in detail in Chapters 3 and 4, but for the time being it is important to highlight that pollution and waste are essentially the by-products of the way humanity transforms and utilizes the natural world for its own purposes. IND4.0 technologies will use new raw materials and produce new products that will finally become new types of waste, besides the ones we already know of. The rising stream of e-waste and the difficulties we have to manage them are a snapshot of the future. Extracting the raw materials and manufacturing them will create both old and new forms of pollution. We do expect that, generally, the resource efficiency (the resource used per unit of product) will be increased, so we will be able to produce more (products) with less (labor, resource, energy, and waste); however this expected improved efficiency will not eliminate waste, but it will only reduce it per unit of production. As already discussed, it might also exponentially stimulate more consumption that will outweigh the benefits gained and/or create rebound effects.

As for circular economy, it is sure that some of the readers will easily say that a circular economy aims to eliminate waste and pollution, so at least we should expect a serious reduction of both. More extreme approaches believe that the waste management sector will gradually disappear because it will not be required anymore, since everything will be recycled in closed loops. These thoughts are certainly misleading. As Costas Velis and Paul Brunner have written [56], "magine a world where everything we use is eventually recycled. Sooner or later, this means also a world where everything we use contains recycled materials...In a world full of products of human innovation, such as highly engineered materials, synthetic chemical compounds and complex products such as electrical and electronic equipment (EEE), and run at a fast forward pace, being inspired by natural environments remains paramount, but can also prove misleading." When we talk about recycling activities, we should also consider that all the potentially useful materials that can be recycled bring a certain amount of pollution both due to their use and due to the way they are constructed (paintings, additives, hazardous waste substances) to serve their function. In addition, all the potential recyclables can be recycled for a certain, usually small, number of cycles, which means that in the end, they will themselves become a type of waste. Paul Brunner has described this straightforwardly [57]: "If products are recycled, the entrained hazardous materials are included in the cycles, too. Hence, over time the recycling of waste that contains hazardous materials will lead to a new material stock that poses not only risks for its users but also problems for future waste management."

The International Solid Waste Association (ISWA) states that [58] "Dumpsites spread pollutants across our atmosphere and our oceans, they damage the health and violate the human rights of the hundreds of millions of people who are living on or around them," and it is campaigning globally to consider dumpsites as a serious health emergency and develop a proper international response. We have to consider that we are not living in a world where waste is properly managed. In contrast, 40% of the world lacks access to waste collection services, while at least 35% of municipal waste generated is driven to dumpsites, rivers, and oceans or is openly burned. This means that the pollution created by IND4.0 and circular economy efforts will be an additional load globally, and not the only one we have to manage, so it is necessary to advance our efforts to manage it.

It is also worthy remembering that the waste management sector already makes an important contribution by delivering closed loops for specific materials, recovering energy potential, and ensuring safe disposal for nonrecyclables. Many of the ideas for circular loops are based on decades of waste management experiences, successes, and failures, with different materials, chemicals, and practices that provide the "soil" required for planting innovations. It is therefore time to revisit and rethink the importance of waste management in the framework of IND4.0 and the efforts toward a circular economy. In fact, the circular economy and zero waste approaches require much more and not less waste management [59]. Their implementation requires advanced management of multiple streams of materials, before they become waste but after they have been discarded from the main production. Those streams must be as clean as possible in order to have high added value, so advanced treatment for removal of residuals will in one way or another be required. Moreover, there will always be residuals looking for appropriate final sinks, including energy recovery where this is possible. For that reason, the road to a circular economy passes through substantial improvements of the waste management sector, and IND4.0 advances can deliver the required shifts.

With this said and having introduced our viewpoint of the key issues that have guided us, it is time to present the whole book and its structure.

1.5 ABOUT THIS BOOK

This book is written on the premise that a circular economy and IND4.0 will determine the future of our planet, the livability of our cities, and the way humans will further develop their societies and environmental footprint. This seems a very bold statement, as the two terms – circular economy and IND4.0 – lack a consistent and widely accepted definition, something that the authors of the book will try to clarify in Chapters 2 and 3. It is hoped that the readers that have read Chapter 1 are already persuaded that this statement is very close to reality.

This book aims to achieve three objectives. Firstly, it aims to introduce the readers to two key trends that are reshaping our world, namely, the circular economy and IND4.0. Neither are well defined, so we prefer to introduce them as flowing ideas that will create tangible results, depending on their final content and social interpretation.

The second aim is to shed light on the interlinkage and interaction between IND4.0 and circular economy and explain why these two factors combined can make or break the Earth's ecosystems and human societies. Indeed, the authors of this book believe that if they are combined, then a more sustainable and wasteless future is possible, and a planet full of waste seems to be inevitable.

The third objective is to highlight that the role of waste management is the key to unlocking the benefits of IND4.0 and circular economy and to demonstrate the ongoing transformation of this crucial industry and its potential to stimulate a wasteless future.

Initially our plan was to write a book that should be read as a complete document, chapter by chapter. However, it became clear that it was necessary to leave it up to the reader to focus on their preferences and the issues they prioritize. We have therefore developed the chapters in a way that they can be read as stand-alone documents.

Thus, in case you have not read it yet, Chapter 1 aims to introduce the reader to the dialectics between the concepts of IND4.0 and circular economy and to put both concepts in the broader context of the changes that are reshaping our world. Explaining the vital role of waste management and its interdependence with circular economy and IND4.0 was also considered necessary. In addition, our efforts were geared toward presenting the main views that guided us in the development of this book. Finally, the scope and the content of the book are presented in brief, as a kind of navigation for the readers.

Chapter 2 aims to introduce the reader to the basic content and the ongoing evolution of IND4.0 from the view of waste management and circular economy. It also aims to portray the broader context of IND4.0 and its economic and social implications, especially the ones related with environmental challenges.

Chapter 3 aims to present the reader the concept of circular economy and to outline the basic ongoing debate surrounding it. The chapter continues by detailing the connections between waste management and circular economy. Chapter 3 is ending by explaining the social context of circular economy and resource efficiency.

Chapter 4 describes in detail how IND4.0 and circular economy are interconnected and redefine the terms waste and resources. It goes on with discussing several concepts like the waste hierarchy and the new role of recycling in a circular economy and ends with the importance of final sinks.

Chapter 5 guides the reader through the advances of IND4.0 that are already transforming the waste sector. It also provides a conceptual framework to understand their impact and poses several questions about the next steps and the future of the waste sector. The chapter concludes by presenting and discussing selected case studies that demonstrate the importance of new business models.

Chapter 6 builds on the experiences gained from the pioneers of digitalization in waste management. It presents the role of leadership and management as a key tool for successful digitalization and provides key tips for transforming the business models and historical examples about the transition period. Lastly, it deals with the new labor needs and skills.

Chapter 7 concerns the rise of a new science and its opportunities and challenges. This chapter describes how the use of mobile phones, sensors, and the Internet of Things can stimulate the rise of a new science, the science of urban sustainability in which online data can be used to redefine urban environmental footprints.

Finally, Chapter 8 provides basic conclusions from the whole book through selected cases that are well connected to all the chapters, concluding with a new vision for the role of the waste industry.

We sincerely hope that the readers will enjoy and appreciate the content of this book. It was written in the knowledge that our readers will benefit by understanding the huge opportunities and challenges involved in IND4.0 and circular economy and that they will be able to develop their own views regarding their importance and their interdependence, which is usually underestimated. We also hope that our readers will appreciate our efforts in explaining IND4.0 and the circular economy not only from a technocratic perspective but also as social trends that will create tangible social impacts, besides their environmental footprint. Last but not least, our book aims not only to inform but also to request the readers to become involved and support our call to combine IND4.0 with a circular economy as the only reasonable way to reduce the human environmental footprint and advance a better, more fair, and sustainable global society.

Key Take-Outs of Chapter 1

Key take-outs	Why is it important?
The trillion-dollar question: will IND 4.0 meet with circular economy delivering not only better resource efficiency but a more sustainable future for everyone in the planet? Or will IND4.0 evolve according to the business as usual linear model, restricting the circular economy to a mere flight of fancy, resulting in faster resource depletion, acceleration of environmental degradation, and deeper inequality?	Understanding the interlinkages and interdependence of circular economy and IND4.0 is the only way to forecast their impact in real life and advance proper policy responses
Global warming and the continuous growth of urban population, resource extraction, energy consumption and food demands are pushing our planet toward the end of business as usual	IND4.0 and circular economy should substantially reduce the human footprint on earth and redress the balance of our relationship with the natural world toward a more sustainable future
There are specific planetary limits that restrict economic expansion. Infinite economic growth in a planet with finite resources is not possible	Circular economy and IND4.0 are limited and at the same time stimulated by the existing limits to growth. Waste prevention, reuse, recycling, and recovery activities are crucial for certain materials just because the limits to growth exist
IND4.0 should not follow the path of the previous industrial revolutions that, besides the benefits they delivered, stimulated exponential increase of air, soil, and water pollution worldwide. Human societies and ecosystems cannot afford such a scenario	The only way to avoid the path of the previous industrial revolutions is to combine IND4.0 and a circular economy
Circular economy and IND4.0 will bring new forms of pollution and generate new types of waste. All the potentially useful materials that can be recycled bring a certain amount of pollution and can be recycled for a certain, usually small, number of cycles	Waste management has the vital role (i) to protect human health and environment as IND4.0 is advancing and it will produce new types of waste besides the existing ones and (ii) to provide final sinks for the waste streams that are not recyclable. Circular economy requires much more and not less waste management

REFERENCES

1. Hollins, O., Lee, P., Sims, E. et al. (2017). *Towards a Circular Economy – Waste Management in the EU*; Study IP/G/STOA/FWC/2013-001/LOT 3/C3. EPRS | European Parliamentary Research Service.

2. Ellen MacArthur Foundation. The circular economy in detailm. https://www.ellenmacarthurfoundation.org/explore/the-circular-economy-in-detail (accessed 23 December 2019).

3. ACCENTURE (2016). The circular economy could unlock $4.5 trillion of economic growth, finds new book by Accenture. *ACCENTURE Newsroom*. https://newsroom.accenture.com/news/the-circular-economy-could-unlock-4-5-trillion-of-economic-growth-finds-new-book-by-accenture.htm (accessed 23 December 2019).

4. McKinsey (2015). *Growth Within: A Circular Economy Vision for a Competitive Europe*, 98. Ellen MacArthur Foundation.

5. Oliver Cann (2016). $100 trillion by 2025: the digital dividend for society and business. *World Economic Forum*, 2016. https://www.weforum.org/press/2016/01/100-trillion-by-2025-the-digital-dividend-for-society-and-business (accessed 22 January 2016).

6. Oxford Economics (2017). *Digital Spillover - Measuring the True Impact of the Digital Economy*. Huawei.

7. Cook, J., Oreskes, N., Doran, P.T. et al. (2016). Consensus on consensus: a synthesis of consensus estimates on human-caused global warming. *Environmental Research Letters 11* (4): 048002.

8. Mavropoulos, A. (2019). *ISWA: In Defense of Science – Global Warming Is Not an Opinion*. International Solid Waste Association. https://www.iswa.org/home/news-detail/browse/7/article/in-defense-of-science-global-warming-is-not-an-opinion/109 (accessed 17 January 2019).

9. Union of Concerned Scientists (2018). *The Science Connecting Extreme Weather to Climate Change*; Fact sheet. Union of Concerned Scientists.

10. Masson-Delmotte, V., Zhai, P., Pörtner, H.-O. et al. (2018). Summary for policymakers. In: *Global Warming of 1.5°C. An IPCC Special Report on the Impacts of Global Warming of 1.5°C Above Pre-Industrial Levels and Related Global Greenhouse Gas Emission Pathways, in the Context of Strengthening the Global Response to the Threat of Climate Change, Sustainable Development, and Efforts to Eradicate Poverty*. IPCC.

11. Ripple, W.J., Wolf, C., Newsome, T.M. et al. (2019). World scientists' warning of a climate emergency. *BioScience*: biz088. https://doi.org/10.1093/biosci/biz088.

12. Riasat, N. (2019). *Tackling Climate Change Through Fourth Industrial Revolution*. MIT Climate.

13. Suh, S., Bergesen, J., Gibon, T.J. et al. (2017). *Green Technology Choices: The Environmental and Resource Implications of Low-Carbon Technologies*, 76. Nairobi, Kenya: International Resource Panel, UN.

14. World Bank Group (2017). *The Growing Role of Minerals and Metals for a Low Carbon Future*. World Bank.

15. Dawkins, E., Chadwick, M., Roelich, K., and Falk, R. (2012). *Metals in a Low-Carbon Economy: Resource Scarcity, Climate Change and Business in a Finite World*; Project Report, 56. Stockholm Environment Institute.

16. Díaz, S., Settele, J., Brondízio, E. et al. (2019). *Summary for Policymakers of the Global Assessment Report on Biodiversity and Ecosystem Services – Unedited Advance Version*, 39. IPBES.

17. United Nations; Department of Economic and Social Affairs; Population Division (2019). *World Population Prospects Highlights, 2019 Revision Highlights, 2019 Revision*. United Nations.

18. WHO. Air pollution. https://www.who.int/westernpacific/health-topics/air-pollution (accessed 22 December 2019).

19. Tearfund (2019). *No Time to Waste Tackling the Plastic Pollution Crisis before It's Too Late*. Tearfund.

20. UNICEF WHO (2019). *Progress on Household Drinking Water, Sanitation and Hygiene 2000–2017, Special Focus on Inequalities*. New York: UNICEF WHO.

21. Joshi-Ghani, A. and Ratti, C. (2018). Agile Cities: Preparing for the Fourth Industrial Revolution. Global Future Council of Cities and Urbanization; WORLD ECONOMIC FORUM, 2018; p. 36.

22. Mavropoulos, A., Tsakona, M., and Anthouli, A. (2015). Urban waste management and the mobile challenge. *Waste Management and Research 33* (4): 381–387.

23. Swilling, M., Robinson, B., Marvin, S. et al. (2013). *City-Level Decoupling: Urban Resource Flows and the Governance of Infrastructure Transitions*. United Nations Environment Programme.

24. Kaza, S., LisaYao, P.B.-T., and Van Woerden, F. (2018). *What a Waste 2.0 A Global Snapshot of Solid Waste Management to 2050*. World Bank Group.

25. WEF, PwC (2018). Circular Economy in Cities Evolving the Model for a Sustainable Urban Future. *White Paper; World Economic Forum*, 2018; p. 29.

26. IMF (2019). GDP per capita, current prices U.S. dollars per capita. https://www.imf.org/external/datamapper/NGDPDPC@WEO/OEMDC/ADVEC/WEOWORLD (accessed 10 April 2020).

27. Kharas, H. and Hamel, K. (2018). A Global Tipping Point: Half the World Is Now Middle Class or Wealthier. *Brookings*, 2018. https://www.brookings.edu/blog/future-development/2018/09/27/a-global-tipping-point-half-the-world-is-now-middle-class-or-wealthier (accessed 10 April 2020).

28. Mario, P. (2012). An Emerging Middle Class – OECD Observer. *OECD Observer*, 2012.

29. Oberle, B., Bringezu, S., Hatfield-Dodds, S. et al. (2019). *Global Resources Outlook 2019: Natural Resources for the Future We Want*. Nairobi, Kenya: International Resource Panel UN.

30. Krausmann, F., Lauk, C., Haas, W., and Wiedenhofer, D. (2018). From resource extraction to outflows of wastes and emissions: the socioeconomic metabolism of the global economy, 1900–2015. *Global Environmental Change 52*: 131–140.
31. Enerdata (2018). Global energy statistical book 2019. https://yearbook.enerdata.net (accessed 23 December 2019).
32. REN21 (2019). *Renewables 2019 Global Status Report*. Paris: REN21.
33. World Bank. Energy use (kg of oil equivalent per capita)|Data. https://data.worldbank.org/indicator/EG.USE.PCAP.KG.OE (accessed 23 December 2019).
34. IEA (2019). *World Energy Outlook 2019 Executive Summary*; WORLD ENERGY OUTLOOK, 11. International Energy Agency.
35. Ritchie, H. and Roser, M. (2017). *Meat and Dairy Production*. Our World in Data.
36. OECD, FAO (2016). *Meat, in OECD-FAO Agricultural Outlook 2016–2025*. Paris: OECD.
37. de Vries, M. and de Boer, I.J.M. (2010). Comparing environmental impacts for livestock products: a review of life cycle assessments. *Livestock Science 128* (1–3): 1–11.
38. Tuomisto, H.L. and Teixeira de Mattos, M.J. (2011). Environmental impacts of cultured meat production. *Environmental Science and Technology 45* (14): 6117–6123.
39. Stuart, T. (2009). *Waste: Uncovering the Global Food Scandal*. W.W. Norton Company.
40. Gustavsson, J., Cederberg, C., and Sonesson, U. (2011). *Global Food Losses and Food Waste: Extent, Causes and Prevention; Study Conducted for the International Congress Save Food! At Interpack 2011, [16–17 May], Düsseldorf, Germany*. Rome: Food and Agriculture Organization of the United Nations.
41. Circle Economy (2016). *Circular Amsterdam a Vision and Action Agenda for the City and Metropolitan Area*. TNO.
42. Kasa, S. (2008). Industrial revolutions and environmental problems. In: *Confluence Interdisciplinary Communications 2007/2008* (ed. W. Østreng), 140. Oslo, Norway: Centre for Advanced Study at the Norwegian Academy of Science and Letters.
43. Freeman, R. (2018). A theory on the future of the rebound effect in a resource-constrained world. *Frontiers in Energy Research 6*: 81.
44. Schettkat, R. (2011). Analyzing rebound effects. In: *International Economics of Resource Efficiency* (eds. R. Bleischwitz, P.J.J. Welfens and Z. Zhang), 253–278. Heidelberg: Physica-Verlag HD.
45. Barker, T., Dagoumas, A., and Rubin, J. (2009). The macroeconomic rebound effect and the world economy. *Energy Efficiency 2* (4): 411–427.
46. Usui, T. (2008). Does a rebound effect exist in solid waste management–panel data analysis of unit-based pricing.
47. Font Vivanco, D., Kemp, R., and van der Voet, E. (2016). How to deal with the rebound effect? A policy-oriented approach. *Energy Policy 94*: 114–125.
48. Zink, T. and Geyer, R. (2017). Circular economy rebound: circular economy rebound. *Journal of Industrial Ecology 21* (3): 593–602.

49. Rockström, J., Steffen, W., Noone, K. et al. (2009). Planetary boundaries: exploring the safe operating space for humanity. *Ecology and Society 14* (2) https://doi.org/10.5751/ES-03180-140232.

50. Steffen, W., Richardson, K., Rockstrom, J. et al. (2015). Planetary boundaries: guiding human development on a changing planet. *Science 347* (6223): 1259855. The infographic is reprinted from https://www.stockholmresilience.org/research/planetary-boundaries.html.

51. Meadows, D.H., Meadows, D.L., Randers, J., and Behrens, W. (1972). *The Limits to Growth: A Report for the Club of Rome's Project on the Predicament of Mankind*. New York: Universe Books.

52. Meadows, D. (2002). *Limits to Growth*. Chelsea Green Publishing.

53. Turner, G.M. (2008). A comparison of the limits to growth with 30 years of reality. *Global Environmental Change 18* (3): 397–411.

54. Chadwick Edwin. Chadwick's report on sanitary conditions http://www.victorianweb.org/history/chadwick2.html (accessed 23 December 2019).

55. Wilson, D., Rodic, L., Modak, P. et al. (2015). *Global Waste Management Outlook*. UNEP – ISWA.

56. Velis, C.A. and Brunner, P.H. (2013). Recycling and resource efficiency: it is time for a change from quantity to quality. *Waste Management and Research 31*: 539–540.

57. Brunner, P.H. (2010). Clean cycles and safe final sinks. *Waste Management and Research 28* (7): 575–576.

58. ISWA (2019). Let's close the world's biggest dumpsites! closingdumpsites.iswa.org (accessed 24 December 2019).

59. Mavropoulos, A. (2015). Circular Economy Needs More Waste Management than Linear One! *Wasteless Future*, 2015.

Chapter 2
Understanding Industry 4.0

You cannot wait until a house burns down to buy fire insurance on it. We cannot wait until there are massive dislocations in our society to prepare for the Fourth Industrial Revolution.

— Robert J. Shiller, Professor of Economics, Yale University

Recommended Listening
The Four Seasons, **Vivaldi**

Because it shows how different segments within a musical structure can change and yet be interrelated to form a whole.

Recommended Viewing
2001: A Space Odyssey, **Stanley Kubrick**

Because it can be interpreted as both darkly apocalyptic and optimistic of the hopes of mankind and humanity.

Industry 4.0 and Circular Economy: Towards a Wasteless Future or a Wasteful Planet?, First Edition.
Antonis Mavropoulos and Anders Waage Nilsen.
© 2020 John Wiley & Sons Ltd. Published 2020 by John Wiley & Sons Ltd.

2.1 THE FOUR INDUSTRIAL REVOLUTIONS

Before we start exploring the future, we need to understand the past. Human history is not merely a linear curve of progress. It is a story of rise and fall, of thriving civilizations, of devastating wars, of climate adaptation, of depleted ecosystems, of new narratives about the world, and of ever-changing power dynamics. It is, however, also a story of practical creativity, of sudden leaps in technology, and of the close relationship between societal structures and access to material resources and energy.

And then, at times, the emergence of an innovative tool changes everything. Some scientists consider the first technological revolution to be the Upper Paleolithic Revolution – around 50 000 years ago [1]. The invention of groundbreaking tools such as knife blades as well as the ability to control and use fire expanded the capacity of mankind and with it the introduction of behavioral modernity. These events laid the foundation for the first waves of migrations – when humans started moving from Africa and settling on other continents.

There are numerous examples of similar leaps: the agricultural techniques that paved the way for the establishment of the first major civilizations on the river plains of the Middle East and Asia 13 000 years ago; the invention of modern weapons, the printing press, and the mechanical inventions of Leonardo da Vinci and his peers in the Renaissance; the commercial revolution in the sixteenth century when Europeans expanded through colonialism; and the scientific revolution in the sixteenth century, with a systematic collection and distribution of knowledge. These are all profound transformations that changed the course of civilization, at least partly driven by technological breakthroughs. With this in mind it must also be considered that these transformations have not always necessarily been advantageous.

A technological revolution will often be caused by some sort of trigger innovation, causing a series of subsequent changes. Different innovations reinforce each other and gradually cause a transition from one form of society to another. It is not as much the specific technologies as the cascading effects they cause that lead to structural changes. To use a geological metaphor, the tectonic plates of the economy break loose and start drifting, forming new continents. These changes can sometimes have dramatic effects. Over the span of only a few decades, production systems, educational systems, political regimes, and dominant business models can be transformed beyond recognition.

2.1.1 The First Industrial Revolution

The early 1760s saw the introduction of a revolution in Great Britain that would change the trajectory of mankind. The first wave of industrialization is widely considered an event that would affect the history of humanity as much as the domestication of animals and plants [2].

The historical backdrop of this was the Enlightenment – an intellectual movement in the seventeenth and eighteenth century. Philosophers and scientists of this age challenged religious ideas and power structures of previous ages, seeking reason and logic and paving the way for a more rational and scientific exploration of the world. Great Britain was a global superpower at the time with access to both markets and raw materials. Business owners and philosophers started exploring how new scientific insight could improve the material foundation of society.

Times were indeed exciting. Natural laws and elements were mapped out. Old truths were abandoned. Concrete had recently been rediscovered after lingering for 1300 years in relative obscurity. Simple machines run by donkeys created a proof of concept for more efficient production of textiles. The initial proto-experiments in utilizing energy from coal and running water were promising. The enlightened British bourgeoisie started dreaming of a future where nothing seemed impossible anymore. Large-scale factories? Machinery run by steam? Wagons without horses? The sky was the limit.

Machine tools meant the more efficient, precise, and accelerated manufacturing of metal, which in turn enabled the construction of bigger and better mechanical production systems. Horse railways were introduced toward the end of the eighteenth century, and within a few decades the horses were replaced by steam locomotives. Improved sail technologies boosted average sailing speed by 50% between 1750 and 1830 [3].

The term "industrial revolution" seems to have been used for the first time in 1799 in a letter written by a French diplomat, announcing that France now had entered the race to industrialize [4]. The ideas and concepts spread quickly, both in Europe and in the United States. In a process that lasted over 80 years, the Western societies went from primarily hand production methods to mechanized and streamlined manufacturing systems.

The first industrial revolution also entailed chemistry, with large-scale production of chemicals such as sulfuric acid and sodium carbonate. Access to chemicals in large quantities led to a range of new and improved techniques for production of metals, paper, glass, textile, and soap. The invention of gas lights (gas made from coal) illuminated streets, factories, and stores, increased opening hours, and even enabled an urban nightlife.

The new mechanized production tools lead to a rethink of the processes and physical facilities for production. In the new factories the workers were organized collectively, and the work was streamlined to speed up production.

Societal structures changed, with fewer people working with agriculture and more in factories. Cities were growing rapidly. People moved in huge numbers from villages to urban areas, a societal transformation that created huge problems with housing, food, and sanitation for poor families. Due to overpopulated areas and poor living conditions, there were severe outbreaks of diseases such as cholera and tuberculosis. There were also subsequently several periods of social unrest, as workers started organizing themselves in order to demand better conditions.

The most important impetus for the first industrial revolution was the use and availability of fossil fuels. As big coal reserves were discovered, there was a massive increase in available energy. But the coal-driven steam engines came at a cost. Cities were often enveloped with smoke, and the air pollution created severe health problems. Exposure to hazardous materials could be devastating for the working families near factories, though the effects remained mainly localized. There was no technology available to handle the waste products – apart from the solution that is sometimes still in use today – to disperse pollutants more widely by increasing the height of the chimneys.

Partly inspired by the hygiene movement that followed new insights in medicine, a circular philosophy developed in the industrial cities: "In industry, there must not be any actual scrap, and everything must be used either for industry itself or for agriculture," the hygienist Henri Napias summarized in 1882.

There was generally a massive increase in the material throughput in society, and recycling was seen as a sign of modernity. "Growth in the recycling sector can be considered an indicator of the spread of the first wave of industrialization," writes Sabine Barles in her historical review of waste management [5]. In new thriving cities during the first industrial revolution, there was great demand for resources that could be used for industrial purposes. Many by-products had a great agricultural and industrial value, and the informal recycling market was booming. Profitable circular value chains were especially related to the manufacture of paper, candles, dyes, and other products. The absence of synthetic fertilizers and a lack of knowledge of fossil fertilizers created a natural market for biological waste. The new industrialized processes opened up for organic waste being systematically collected or raided and sold at attractive prices. Vegetable rags had for centuries been used for traditional papermaking, but due to the efficient machines for industrial paper production, higher volumes were demanded. In the nineteenth century, rags became a strategic industrial issue in several countries. Animal bones were used in manufacturing products such as grease and glue. Gelatin was used for both food preparation and later for photographic negatives. Phosphorus was used to make matches ignited by friction. Coal ash was important for both soil improvement and brickmaking.

In this period, the first consumer goods were introduced, and waste volumes in urban areas increased as population in cities grew. Lack of sanitation and refuse collection in the new and densely populated urban areas soon became a problem. As early as 1750, Corby Morris of London was concerned with the health of the commoners and arranged that the garbage and scraps should be diverted into the Thames river [6].

The discovery of microbes and the rise of the "hygiene moevement" lead to a rapid development in living conditions and better treatments for many medical conditions.

In 1842, a report entitled "The Sanitary Condition of the Labouring Population" was published by Edwin Chadwick, a social activist. A devastating cholera outbreak brought the issue of public health to the public's attention. Systems started to emerge. The Disease Prevention Act of 1846 inaugurated the regulation of waste management, and in 1855 physician John Snow discovered that a cholera outbreak was related to a water pump. His study is considered the founding event of the science of epidemiology. In 1874, "the destructor," the first incineration technology for refuse, was invented in Nottingham, UK. With the Public Health Act of 1875, households were obliged to store their waste in a "movable receptacle." The modern dustbin was born. The same year, New York introduced the first public waste management system [7].

A new industrial chapter was about to open.

2.1.2 The Second Industrial Revolution

As the first mechanized revolution continued to spread, a new invention was introduced around 1870 – electricity. Within a few years several innovators contributed to the different parts of a modern electricity system, such as generators and lightbulbs. Thomas Edison is arguably the most celebrated inventor, but the first electrical street lighting in the world was established in Britain, in Mosley Street in Newcastle upon Tyne. This used the incandescent lightbulbs of Joseph Swan, who was also involved in developing the first large-scale power station, located in London.

With the commercialization of electricity came a subsequent swift transformation. The high-voltage alternating current enabled the assembly line and mass production principles. A mere six years after the first lightbulb illuminated the streets of Newcastle, the telephone was invented, and in 1901 the first radio waves were transmitted over the Atlantic. What is now known as the second industrial revolution was a revolution of mobility, electricity, and communication. Railroads enabled faster routes for overland travel and expanded geographical markets. The ability to communicate in real time over huge distances enabled new processes and global interaction. Cheap paper made from wood fiber in combination with the rotary press opened up for wide distribution of both books and newspapers. With the internal combustion engine came the first cars, and with the availability of liquid fossil fuels, the new mechanized vehicles were increasingly available to the public.

During the early stages of this revolution, the productivity gains were in fact not substantial. However, an acceleration in labor productivity growth allowed the economy to withstand rapid population growth without a discernible decline in living standards [8]. In his much-cited paper from 1989, Paul David explores the reasons for "productivity paradox" – comparing what he calls general-purpose machines (such as steam engines, dynamo, and computers). These innovations were highly revolutionary, but in the second industrial revolution, a considerable impact on productivity from electrification did not materialize in statistical form until the 1920s. The initial effects are not always captured by conventional productivity measures [9].

Ideas, products, and people started flowing faster and further. There was a significant growth in both economy and productivity in the first decades of this revolution. Improved transportation infrastructure increased the size of the farmland serving the booming cities, reducing vulnerability to crop failures. The new inventions helped reduce prices of goods dramatically. Living conditions improved.

It was in this period that public health systems in their contemporary form were developed. The cities acquired more modern sewage systems, and in industrialized Western cities, infrastructure was developed to secure minimum standards of water quality.

It was also at this time that waste really started to emerge as a societal issue. Thanks to the access to petroleum, new mining sources, and plentiful materials, the second industrial revolution rapidly undermined the scarcity-driven circular philosophy and recycling value chains of the first industrial era. The urban by-products were increasingly replaced by cheaper and more convenient virgin materials.

Especially important was plastic. The first synthetic polymer was invented in 1869 by John Wesley Hyatt to provide a substitute for ivory. This discovery was revolutionary. For the first time humans were not constrained by limited supply of natural materials like wood, metal, stone, bone, tusk, and horn. The new material was considered environmental friendly, a saving for the elephant and the tortoise. Bakelite was discovered in 1907. It was durable, heat resistant, and, unlike celluloid, ideally suited for mechanical mass production. During World War II the plastic industry really took off in the United States, as industrial might prove as important to victory as military success. After the war, use of plastics became widespread in consumer goods. "In product after product, market after market, plastics challenged traditional materials and won, taking the place of steel in cars, paper and glass in packaging, and wood in furniture," Susan Freinkel writes her book *Plastic: A Toxic Love Story* [10].

Still, some of the materials in the new industrially manufactured products held a high value that inspired new recycling activities. Making cardboard from paper and recycling metals and making tar from coal were economically viable. But the general picture was that waste was considered a cost to society. The aim of the evolving waste management systems was to reduce this cost, and the cheapest way to handle it was to remove it from the city streets, out of sight and into, mainly, the surrounding nature.

As clean air regulations reduced the informal incineration, rubbish collection systems were established in most industrialized areas – resulting in big landfills. The changing nature of products is reflected in the fact that the volume of garbage grew faster than the weight. Packaging and newspapers changed the composition of the residual waste. Traditional vehicles were not sufficient, and from the 1930s garbage trucks with compressors became more popular.

Industrial-scale incineration was developed as early as in the 1870s. Burning waste greatly reduced the volume and weight. The "destructors" could be placed near the source of collection, and the ash could be used for mortar. Energy from incineration could be used for energy production and, modern houses, central heating. Advertisements from France in the 1920s promoted central heating as a sign of a modern urban lifestyle.

But energy recovery was not the norm. From the beginning of the nineteenth century on, waste was more than anything considered a burden. "Many sorting plants were equipped with incinerators and only materials with high value were salvaged (in particular metals), the rest was burned," says Sabine Barles in her historical review. Garbage collection was no longer developed with usefulness in mind. The aim was to get rid of it, at lowest possible cost. Landfills were established. In some cities waste was incinerated without energy recovery. Others dumped waste in the sea. In New York barges were used to carry refuse 25 miles from shore from 1872 to 1934, a practical solution – even for the increasing amounts of bulky waste.

Cars, electronic appliances, and gradually plastic created new problems. "Sanitary landfilling" originated first in England, where it emerged in 1912. The waste was put in layers. When heat from fermentation stabilized, it was covered with inert matter such as stone and sand. Then a new layer was added. Gradually this came to be considered the best solution for garbage storage. In between the wars, the number of landfills multiplied.

A linear economy was becoming the norm. New production techniques in combination with the discovery of huge petroleum reserves and sources for raw materials created unprecedented productivity gains and optimism in the 1920s. A range of new chemicals and materials were introduced in the market. The economy was booming.

Then, suddenly, everything collapsed.

On 24 October 1929, the stock markets in the United States crashed due to unregulated speculation. This date marked the beginning of the great depression – the worst decline in modern history. In the United States wages fell 42% as unemployment rose to 25% [11]. How could growth get back on track?

One of the suggestions was the concept of "planned obsolescence": given the idea of unlimited access to resources, it made sense to consciously shorten the lifespan of products. The term was first coined by Bernard London in the pamphlet "Ending the Depression Through Planned Obsolescence" [12]. The idea was originally to use state intervention in the market to increase sales volumes by reducing the time between new purchases. It was suggested as consumer obligation organized by the government to kick-start the economy. In the 1950s the idea was further developed into a design philosophy that has deeply affected business models and waste volumes until our time. The industrial planned obsolescence was, according to Brooks Stevens, "instilling in the buyer the desire to own something a little newer, a little better, a little sooner than is necessary" [13].

This mindset and strategic approach was widely adopted and changed the industrial dynamics. Progress was not a function of technological innovation but of the ability to create and then satisfy market expectations. Given there were no costs associated with the handling of the waste and negative environmental footprint, this strategy was soon proven to be very profitable. Obsolescence strategies in different forms started to shape the patterns – physical design, business strategies, and marketing were all based on the idea of generating growth by increasing production. There was a 1 : 1 relationship between material footprint and the size of the economy.

Many new and creative techniques were used to shorten the life of consumer goods. In many cases products were designed to deteriorate quickly. Some durable products were made in single-use versions, such as disposable cameras. Designs could purposely be made to be impossible or expensive to repair. Manufacturers may make replacement parts unavailable. By using mass marketing the clothing industry invented fashion, a cycle of desirability where change in styling creates a constant demand, even when all the customers are fully equipped with highly functional clothing.

In the 1960s a new form of criticism started to emerge among ecologists, historians, and urban planners. In 1962, *Silent Spring* by American biologist Rachel Carson was published. The book described the impacts of the spraying of DDT. The logic of spraying chemicals into the environment without understanding their effects on human health and ecology was questioned [14].

Engineer Abel Wolman and his famous article "The Metabolism of Cities" [15], published in *Scientific American* in 1965, emphasized how the management of cities had failed. The "urban metabolism" imported the food and most of the resources needed and transformed these materials into solid, liquid, or gaseous waste. The result was damaging to both urban and natural environments.

These critical perspectives gave birth to the modern environmentalism movement. Several local communities experienced high cancer rates and other health effects due to toxic waste. Rusty barrels of chemical waste were discovered in many parts of the world. A new, even more productive, industrial era was on the rise. So was the idea of the biosphere and resource limitations. Were there limits to growth [9]?

2.1.3 The Third Industrial Revolution

If the second shift was run by electrons, the third was fueled by bits. The driver behind the third industrial revolution was to a large part the transformation to digital handling of information. The first trying steps started to evolve after World War II. Inspired by all the inventions of the physical reality, scientists started to explore an interesting question: Can we enhance our brain in the same way that we expanded our mechanical capacity by the use of machines?

The German aircraft engineer Konrad Zuse (1910–1995) was one of the pioneers. The mechanical calculator was invented. But this machine did not save intermediate results to reuse them in the calculation. He found out that you could build a better system by dividing this into three basic components: controller, memory, and calculator. During and after the war, he built the first proto-computer using many of the principles of the digital devices we use today.

This was a time with abundant access to fossil fuels. The green revolution – where industrial fertilizers and new crops greatly increased the world's agricultural output – created a new level of prosperity. A range of new materials and chemicals were developed. Especially important were the new forms of plastics, typically made from by-products of coal or petroleum production. Around 1975 there was an exponential growth in both waste volumes and fossil fuel emissions (Figure 2.1).

After the wake-up call in 1965, environmentalism became an important political consideration in many countries. In the following decade large political bills were introduced concerning waste, both in Europe and the United States. The new perspective in these legislations is that they emphasized the necessity to reduce the production of waste at the source, minimizing waste in the design and reusing materials and energy through recycling, energy recovery, or biological conversion. A polluter pays principle was adopted by the OECD. The first funding mechanism for the collection and treatment was established.

The problem was that productivity grew way faster than the attempts to regulate the negative effects of increasing waste volumes. Until the 1990s the quantity of waste did not decrease, despite regulations. Recycling rates stayed limited. Disposal on land remained widespread and even gained ground in some countries.

What pulled the trigger in the third revolution was the personal computer and global digital networks. The microchip was invented in the 1960s, and ever since the size has decreased, and the processing power increased. In the beginning computers were big and came with a high price tag for corporations and public organizations only. The first personal computer was the Kenbak-1, released in early 1971. It had 256 bytes of memory. Only 40 of these machines were built and sold. But the avalanche of personal computers was triggered. Ten years later the capacity of the machines had developed exponentially. Falling production costs made computers available to both regular business and private homes.

With computation machines available to anyone, a brand new area of innovation opened up in creating operating systems and software tools. The ability to collect and process increased rapidly. Professional users and scientists started connecting the machines, merging and distributing data, and accessing information from places far away. When the World Wide Web was invented in 1991, this created a networked virtual world of data and documents, creating a new form of interaction with profound effects on processes and organizational structures. Global value chains evolved; free trade deals opened up for a more globalized economy. Increasingly the production for richer Western economies in the United States and Europe moved to low-cost societies.

The cargo ships that crossed the oceans with cheap products returned with the garbage of Western consumers. An increasing population and a fast-growing economy created a huge rise in waste volumes in Western countries – especially related to plastics (Figure 2.2). In line with the "out-of-sight philosophy" of the previous decades, the problem was not solved – it was exported. Many developing countries with chronic

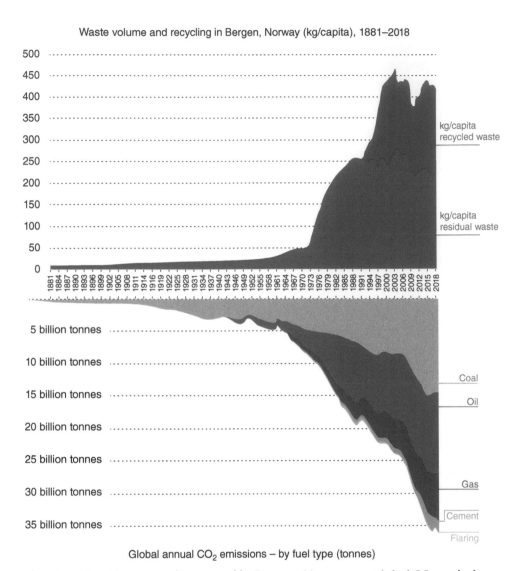

Figure 2.1 Local waste volumes and in Bergen, Norway vs. global CO_2 emissions by fuel type. The inter-municipal waste management company BIR gathered waste statistics dating back to 1881. This is a timeline of waste generated per capita and the recycling rates in the municipal waste management, collected by Toralf Igesund/BIR. The other figure is annual carbon dioxide (CO_2) emissions from different fuel types, measured in tonnes per year.
Source: This graph is based on data from Global Carbon Project (GCP) and Carbon Dioxide Information Analysis Center (CDIAC) [16]. With kind permission from Nick Rigas, D-Waste.

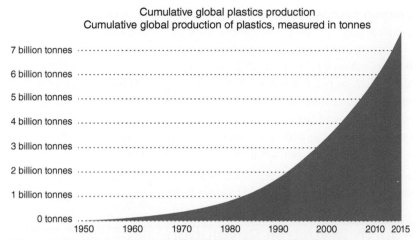

Cumulative global plastics production
Cumulative global production of plastics, measured in tonnes

Figure 2.2 The cumulative production of plastic, globally.
Source: With kind permission from Nick Rigas, D-Waste, based on data from Ref. [17].

poverty and weak regulations became the host of landfills as cities of the Western world developed new collection systems to keep their streets clean and tidy. The services in these waste-importing countries were generally undersized, with lack of equipment, space, and organizational structures to deal with the increasing volumes of hazardous waste. Lack of regulation led to scavenging and informal recycling, similar to the situation in Europe and North America during the first wave of industrialization. The proliferation of garbage in developing countries with chronic poverty, inefficient management structures, and weak states and regulations resulted in sanitary, social, environmental, and economic issues.

At the shift of the century, new and urgent planetary challenges were recognized, and the first scientist started hitting the alarm button on behalf of global ecosystems. Thanks to satellite monitoring and accurate measurement, scientist increasingly recognized that the CO_2 from burning of fossil fuel was changing the atmosphere. Digital climate models showed that anthropogenic emission would rise global temperatures, with serious consequences. The First World Climate Conference was held on 12–23 February 1979 in Geneva, and negotiations have been going on ever since.

At the turn of the century, the problem with ocean plastic was also being recognized. The waste dumped in the ocean was drifting along with global currents, destroying marine habitats, breaking into microscopic particles, and finding the way into the food chain.

This happens in a context of deforestation, habitat destruction, overfishing, and contamination, resulting in a dramatic decrease in biodiversity.

In 2017, 15 364 scientists from 184 countries signed a "Warning to Humanity," stating the urgency of the situation: "We have unleashed a mass extinction event, the sixth in roughly 540 million years, wherein many current life forms could be annihilated or at least committed to extinction by the end of this century," the scientists wrote [18].

The industrial value chains require enormous amounts of fossil energy and virgin natural resources. But it also causes huge problems when all the valuable consumer goods are turned into waste and the energy changes the composition of the atmosphere. So as we turn the page to open a new chapter, with more technological capabilities than ever before, the scientists give us a clear instruction:

> To prevent widespread misery and catastrophic biodiversity loss, humanity must practice a more environmentally sustainable alternative to business as usual.

2.1.4 The Fourth Industrial Revolution

So, here we are. The next big shift is happening, under our feet. We are in the early stages of what many describe as the fourth industrial revolution. The digitization has already transformed a range of industries. Global digital networks, new materials, hefty streams of data, increasing processing power, and alternative energy sources create brand-new generation of possibilities, processes, and solutions.

This time, the challenge is not only making the economy grow but also reducing the resource impact and environmental degradation. We need to fix the root causes of our current planetary crisis.

"It is no longer enough to recycle or to recover excreta simply to limit the quantity of final waste. What matters now is to close the loops and, through recycling and recovery, to limit the extraction of resources at the source. Such a project, driven by industrial ecology and territorial ecology, will succeed only if the levels of consumption fall. It will require a profound reform in society and its way of viewing its waste," Sabine Barles says as a conclusion in her review of the history of waste management.

Basically she says we do not only need a technical revolution related to collection and handling of waste. We need to think broader to redesign the surrounding structures related to extraction, design, manufacturing, and consumption. We need to use business model innovation, technology improvements, regulations, and collaboration as means to create a system shift centered around resource productivity.

There are promising shifts underway, especially in the global energy systems. According to the Frankfurt School report on energy trends [19], global investment in renewable energy capacity hit $272.9 billion in 2018, far outstripping investments in new fossil fuel generation. Due to falling marginal costs in renewable energy, particularly within solar photovoltaics, the capacity increases dramatically year by year. The global energy system has added an estimated 1.2 TW of new renewable energy the last decade. This is more than the entire electricity generating fleet of the United States.

Revolutions tend to feel rather chaotic. We are surrounded by scientific breakthroughs with unpredictable outcomes. Digital platforms are currently the biggest and most profitable companies in the world, replacing the industrial companies of the previous century. The lines between physical, digital, and biological spheres are blurred. Thanks to rapid developments in artificial intelligence (AI), the autonomous machines are already shaping our daily lives. Technical standards enable a new form of interoperability – machines can be combined and connected without human interference. Tools are making decisions themselves, sometimes organized in distributed systems.

All the different industrial revolutions increased the level of complexity (Figure 2.3). Can new and smarter solutions scale faster than our capacity to create problems? The capacity for change increases steadily. What changes do we need? As the exponential graphs are steepening, this is becoming an increasingly important question for business leaders to ask. A new form of industry is about to be born. Leaders will not only have to

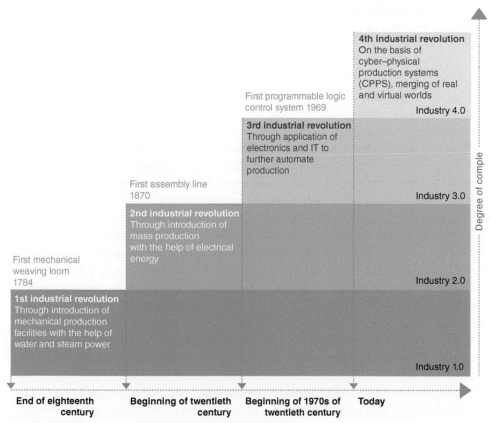

Figure 2.3 The four industrial revolutions build upon each other – creating more complex and sophisticated technological and economical systems.
Source: With kind permission from Deloitte [20].

define how the value chains should be organized but also what kind of value creation they should be optimized for.

2.2 INDUSTRY 4.0

A new generation of smart and partly autonomous production systems is evolving around us. The fourth industrial revolution is a complex transformation with multiple outcomes for every sector and industry. An important part of this revolution is the concept called Industry 4.0.

Industry 4.0 is not a historical fact. It is a possible outcome, an invention in the making, something that is still ahead of us. The term itself was coined in a 2011 research project initiated by the German Federal Ministry of Education and Research. They were asked to identify how new technologies and trends could improve the world and boost productivity. The result was a conceptual invention of a process that connects mechanical and virtual into holistic systems [21].

The key enablers of these "cyber–physical systems" are the improvements in AI, increased machine-to-machine connectivity, and the new systems for real-time storage, processing, and sharing of data. Hardware can now react to real-life situations and collaborate using programmable logic control units and reasoning based on optimization algorithms in connected software (Figure 2.4). Given the right technical frameworks, new generation production systems can provide the unique combination of higher productivity and more flexibility. The price for sensors, processors and bandwidth have declined dramatically in the last decade, and marginal costs for necessary hardware and software are still falling. Sensor prices are half the price compared to ten years ago. The cost of bandwidth has decreased by a factor of nearly 40 times. Processing costs have declined almost 60 times.

The German government found the results of the initial research promising and funded further research to help German businesses become front runners in the next wave of industrialization. The ideas have already spread to a range of research institutions and companies worldwide. The conceptual model of Industry 4.0 kept, and keep on, evolving. Key principles of this approach are increasingly being applied in various industries – manufacturing, energy plants, and medical institutions – and in mobility services.

If Industry 4.0 is igniting revolution still in its active early stage, this means that the outcomes are yet to be seen and that we still have the possibility to discuss what kind of value the technology should create and who should benefit from it. Viewed through an optimistic lens, the concept opens up for a new level of productivity, transparency, and collaboration. It is a door leading into an interesting room that we just recently started exploring.

One of the key differences from previous industrial configurations is that Industry 4.0 is breaking apart the silos. The atoms, the electrons, and the bytes are becoming a part of

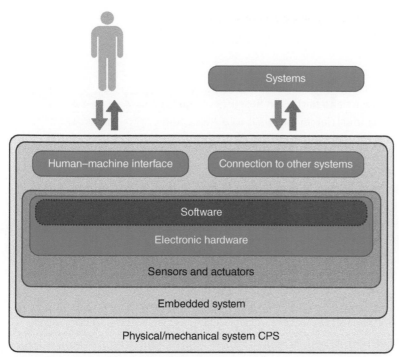

Figure 2.4 The interaction between humans and machines in cyber–physical systems.
Source: With kind permission from Dr. Manfred Broy [22], Technical University Munich.

the same structure. The rise of data-driven networks of physical objects, in some form or another, can be observed in almost every industry. Data is being distributed along whole supply chains in real time – enabling new user experiences and processes (Figure 2.5).

Most digitized industries use both sensors and machines that collect data and communicate and interact with no human interference. The data is being analyzed on the fly, sometimes using AI – where programmatic rules and learning algorithms are encoded into different systems. There are currently many proof of concepts on smart and autonomous production lines. The machines themselves, or the network they are a part of, can increasingly make decisions on their own.

The machines within Industry 4.0 systems can theoretically be placed anywhere geographically. Industrial infrastructure does not have to be limited to a specific geographic place – or even a specific organization. The "factory" can function distributed and decentralized. This concept of automation at higher levels can be applied to a wide range of processes across sectors – including the "production" of mobility and transportation in cities.

Even if there are new and mind-bending innovations launched by the week, we are just starting this exploration. Industry 4.0 is a forward-looking concept, still in the

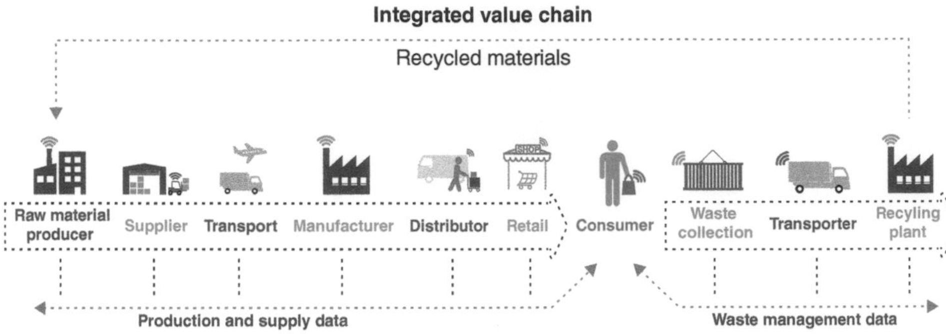

Integrated value chain

Recycled materials

Raw material producer | Supplier | Transport | Manufacturer | Distributor | Retail | Consumer | Waste collection | Transporter | Recyling plant

Production and supply data   Waste management data

Figure 2.5 In an integrated value chain, data can be shared by different stakeholders in real time. Operational models can be optimized either for new customer experiences, more efficient logistics, or on-demand manufacturing. It can also theoretically enable a cradle-to-cradle process where materials are reused in new cycles.

Source: With kind permission from Nick Rigas, D-Waste.

making. The dream? A production system where information is shared in real time in a network of humans, sensors, actuators, robots, and other smart equipment, with superpowers from processors and algorithms delivering insights from the cloud.

2.2.1 The Technologies that Drive the Revolution

Industry 4.0 is not ignited by a specific, sudden invention. It is actually hard to single out one key breakthrough technology. According to Silvija Seres, there are multiple "Gutenberg moments" happening at the same time, maturing in hyper speed [23]. The reference to Gutenberg's printing press in the fifteenth century is illustrative. Before this invention scribes would copy text by hand, page by page. Making books was slow and expensive. Mass production democratized book ownership and borrowing, and its associated knowledge started spreading.

Many of the technical enablers were in fact invented some 20 or 30 years ago. Processors and computer networks have been around since the 1970s. It was developments in electronics and information technology that triggered the third revolution. The transformation from analogue to digital data management led to a whole new level of productivity – as well as the rise of the digital marketplace and social networks.

This is the foundation for the next leap. What is new in Industry 4.0 is that the machines start sensing, talking, and collaborating – with more sophistication and in real time. The devices are being organized in networks with a shared logic. Massive increase in computing power sharply reduced cost per unit, and increase in performance now makes them suitable for industrial use. Different technologies, such as 3D printing, sensor technology, nanotechnology, AI, drones, and robots, are currently being combined. A myriad of devices, intelligent software, and people are connected in new configurations. Thanks to AI and machine learning (ML), both individual machines and whole systems can increasingly operate without human intervention.

Predicting the innovations is impossible. As the revolution is unfolding, new technological breakthroughs will be continuously announced, and marginal costs will fall. Different sources are trying to identify and label the key technologies, but given the exponential development, the landscape is rapidly changing. Some breakthroughs are obviously going to have key roles in the transformation that is about to play out (Figure 2.6).

Figure 2.6 A selection of 10 general-purpose technologies that constitute important drivers in the shift towards Industry 4.0 processes and business models. *Source*: With kind permission from Nick Rigas, D-Waste.

2.2.1.1 Internet of Things

Industry 4.0 means that more devices will be connected digitally and enriched with embedded computing. Sensors, actuators, and processors out in the field communicate and interact both with one another, with humans, and with software. One simple definition of Internet of Things (IoT) described by Sezer et al. [24] is: "IoT allows people and things to be connected anytime, anyplace, with anything and anyone, ideally using any path/network and any service": smart door locks, crash sensors in cars, smartphones, and connected weights. We are surrounded by "things" that now are online and increasingly operate partly autonomous and (in varying degree) intelligent.

2.2.1.2 Artificial Intelligence (AI)

Information from many different sources are necessary for the Industry 4.0 systems to work. This includes data from production systems, enterprise software, and customer solutions. Until recently big data was hard to make sense of due to the complexity and unstructured form of the data. But the last few years AI has become available for everyone through services like Watson (IBM) and is currently being applied to a range of different uses, analytics tools, and applications. It is increasingly easy to see patterns and trends in data. Many such services are becoming available as a service. The analyzing and decision-making capacity of machines are rapidly evolving.

2.2.1.3 Machine Learning (ML)

ML is the learning in which a machine can learn on its own without being explicitly programmed. It is an application of AI that provides system the ability to automatically learn and improve from experience. The goal is to learn from data on a certain task to maximize the performance. By combining ML and simulations, also complex systems can be designed and improved.

2.2.1.4 Autonomous Robots

Both physical and virtual robots are increasingly learning and making decisions on their own. In Industry 4.0 robots are working side by side with humans. Thanks to advancements in robotics, they get a range of new capabilities, and combined with AI and ML, robots are also improving the capability to learn and make decisions on their own.

2.2.1.5 Virtual Simulators

It is now possible to build digital models that mirror the physical world. Many industrial plants are actually built as "virtual twins" before the physical construction. These models can include machines, products, and humans and subscribe to real-world data. They can also be used as sandboxes to test different scenarios – inspired by the gaming industry before making the actual changes in the real world.

2.2.1.6 New Human–Machine Interfaces

Thanks to both improved design and better hardware, machines and humans can now talk fluently, with less frictions and misunderstandings. Touch screens, voice assistants, and wearable technology are among the examples of new forms of interaction. Augmented reality adds a layer of data on top of images from the real world. Live camera streams and smart glasses can also be used to enrich the physical world with binary information (or enriching the data with real-world information).

2.2.1.7 Encryption and Cybersecurity

Trust is key to enable the new and open data-driven business systems in Industry 4.0. Increased connectivity and data sharing create an urgent need to secure critical industrial systems and manufacturing lines. Secure, reliable communication systems are being developed, as well as sophisticated systems for user authentication.

2.2.1.8 Cloud and APIs

Data sharing across companies and machines and real-time interaction with the users and collaborator require instant accessibility of data. The cloud is becoming the norm for data handling, replacing the old local servers. Services are increasingly built on top of application programming interfaces (APIs) that make integrations and programming data-driven services much easier.

2.2.1.9 Additive Manufacturing

Additive manufacturing, such as 3D printing, offers construction advantages, such as complex, lightweight designs. These machines are already integrated in various industries, especially to prototype and produce individual components. With Industry 4.0, these additive-manufacturing methods will more widely be applied in small batch production of customized products.

2.2.1.10 Blockchain

Blockchain, while most known for cyber currency, will possibly have many important applications in a future Industry 4.0 world. Blockchain can be considered an alternative, collaborative way of managing data. Instead of a centralized database, data is stored in a "distributed ledger" – a timestamped chain of blocks that is shared across a network of computers (Figure 2.7). In a blockchain each record is approved by the network, and the records are added to a block. The records are "engraved" and are very difficult to change after they are added to the chain. Blockchain-based record keeping makes compliance, data sharing, and collaboration simpler. The advantage of blockchain is that it enables a trustworthy log of everything that happens related to a product or process. Theoretically you could track the full lives of a component that is being

Blockchain				Database?
A peer-to-peer decentralized ledger for data storage – based on consensus mechanism	(B)	vs.		A centralized ledger for structured data storage managed by an administrator
Distributed network		Architecture		Client–server
Decentralized with consensus		Autorithity		Centralized with administration
Can not be manipulated		Security		Can be manipulated
Read or write data		Data handling		Create, read, update and delete data
Harder to implement and maintain		Cost		Easy to implement and maintain
Can be slowed down by consensus methods		Performance		Fast and scalable
High – due to proof of work		Energy footprint		Low

Figure 2.7 **The characteristics of a blockchain versus the characteristics of a traditional centralized database.**
Source: **With kind permission from Nick Rigas, D-Waste.**

reused within several products during its full lifetime. The system is very robust: if one database fails, one copy fails, and you've got the information stored and cross-checked across multiple nodes. The data is managed in real time – as participants in the system reach consensus, records are encrypted and cannot be tampered with. As blocks of information are chained together, you're creating a perfect audit history – where you can go back through time and see a former state of the database. A blockchain can be organized either as a private, restricted, or open system. Some consider blockchain a key part of future distributed product life cycle management strategies, as it provides data integrity in a real-time environment at value chain level. Others prefer the control and flexibility of traditional database solutions.

2.2.1.11 Advanced Materials

Another possible important general-purpose technology driver (not mentioned in Figure 2.7) is advanced materials. The progress in the design and development of new materials is the "silent enabler" of many of the improvements in products and machinery. Thanks to improving simulation capabilities, it is now possible to calculate how materials behave under different conditions. Thanks to innovations such as nanofiber, brand-new composites and nanomaterials with special characteristics can now be developed on demand. A natural progression has led to the incorporation of AI and the use of complex ML in advanced materials. Algorithms are already being used to figure out various quality molecules and materials and how they may interact in various environments [25]. The plastics in the world are still about 35 times higher than the production of composites, but market shares and production volumes of composites and nanomaterials are steadily increasing. Many new industrial applications can be expected as technologies are successfully developed and commercialized [26].

2.2.2 What Changes Can Industry 4.0 Enable?

It is not the individual innovations, but the reinforcing combination of them, that create the magic of Industry 4.0. We are witnessing the rise of intelligent, flexible, and distributed ecosystems, enabled and enriched by technology. By connecting multiple stakeholders, the old linear value chains are emulated into networked ecosystems that will, in some form or another, deeply transform product development, production, business models, and logistics. We see that different sectors start to overlap and also the birth of sectors that still do not exist on any map.

Based on predictions developed by Deloitte [27] and other sources, we can generally emphasize four main characteristics of an Industry 4.0 business environment (Figure 2.8).

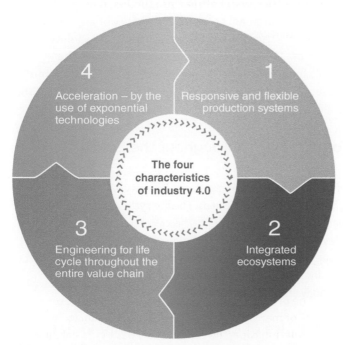

Figure 2.8 Four major characteristics of an Industry 4.0 business environment. *Source*: With kind permission from Deloitte [27], adapted by Nick Rigas, D-Waste.

2.2.2.1 Responsive and Flexible Production Systems

In Industry 4.0 we go from mass production to individualized and customer-specific manufacturing. So-called smart factories organize themselves virtually and enable a more flexible industrial-scale production. They may use many of the mentioned technologies and combine them in vertically integrated systems. Smart sensor technology is used for monitoring production and also enables a condition-based maintenance management. Due to machine-generated data, all stages in the production process are logged. Discrepancies can be registered automatically. As fluctuations in quality or machinery breakdowns can be monitored, maintenance can be organized based on condition, not schedules. Generally, efficient use of materials, energy, and human resources is emphasized.

2.2.2.2 Integrated Ecosystems

Thanks to the open and collaborative design of Industry 4.0 systems, a new generation of value chains and value-creation networks will emerge. The context of real-time data sharing and transparency opens up for a range of innovative business and cooperation models.

These collaborations can be spanning countries and continents – facilitating global value chains – and include inbound logistics, warehousing, production, marketing, and sales to outbound logistics and downstream services. The history of any part or product is logged and can be accessed at any time, ensuring constant traceability. This means that the whole chain, from the supplier through the producer to the customer – or the purchasing through production to sales – can be a part of the same high-level process. Thanks to real-time transparency, factors such as quality, time, risk, price, and environmental sustainability can be handled dynamically.

2.2.2.3 Engineering for Life Cycle Throughout the Entire Value Chain

By breaking apart the silos and creating more holistic systems, Industry 4.0 enables a new form of "through-engineering" across entire value chains. The life cycle value becomes more important both for products and customer management. This calls for better coordination between design, prototypes, development, and manufacturing of new products and services. Adaptations can be made not only in the production but also in the development, ordering, planning, composition, and distribution. In flexible and data-driven production systems, better information is readily at hand at all stages of a product's life cycle.

2.2.2.4 Acceleration by the Use of Exponential Technologies

In Industry 4.0 change happens continuously and fast. The new technologies serve as catalysts to allow for individualized solutions, flexibility, and cost savings. Thanks to the programmable nature of robots and connected devices, the production system can be upgraded. With the help of AI and ML, the machines can themselves improve efficiency and reliability. Increasingly AI can also be used in the design process. Thanks to improved sensor, quality management will be easier, and the robotic capabilities will improve. An example is 3D printing (additive manufacturing). 3D printing allows new products with higher complexity without additional cost. At the same time it enables a total redesign of the supply chain, with faster delivery times, inventory reduction, and even on-location printing of spare parts. New composites and materials will expand the possible applications of this technology. A few very specific inventions can open up for new products and price models and at the same time transform the service ecosystem around the product in profound ways.

2.2.3 Important Concepts in Industry 4.0 Systems

The fourth industrial revolution is a complex transformation with multiple outcomes for every sector and industry. We are still in the early stages of the transformation. On a general level we see centralized and controlled industrial systems we have known for 200 years are now shifting to a decentralized self-controlled systems. The machines in Industry 4.0 systems can be placed anywhere geographically. Industrial infrastructure does not have to be limited to a specific geographic place or even a specific organization. By applying Industry 4.0 principles, organizations increase their capacity to react quickly on the basis of data. Sensors increase the ability to monitor the process. The flexibility opens up for a new and individualized customer relationship.

The idea of creating automated higher-level systems that bind subsystems together can be applied to a wide range of processes across sectors – including the "production" of mobility and transportation in cities. The cyber–physical systems are programmable – both the specific subsystem and the business logic can be upgraded and improved. We go from a tradition of investing heavily in replacing a monolithic process to a system of continuous improvement of independent services.

A range of new concepts, ideas, and principles have been suggested and tested, which are illustrative for what an Industry 4.0 world of products and services might look like.

2.2.3.1 Predictive Maintenance

By using sensors, parameters, and data tools, the remaining service life of machine and plant components can be predicted. This way maintenance can be planned and executed based on actual conditions, not scheduled inspections. Replacements can be done with correct timing. Operating conditions can be analyzed with high accuracy based on patterns of critical operating parameters. The combination of service and digitalization offers tremendous potential, as service phases add substantially to the total operating costs of machines and plants. The concept of predictive maintenance can change pricing models but also affect procurement and organizational structures.

2.2.3.2 Digital Twins

Thanks to real-time sensor data and increasing processing and simulation capabilities, it is possible to create a fully virtual representation of assets, processes, or systems from the physical world. Models like this are often built before the physical construction to test different scenarios and improve the solution. In the digital representation, individual devices can be connected to a wider context, and simulations can reflect the actual complexity. When an industrial plant is operational, the twin reflects the processes happening in the real world – thanks to real-time data using sensors. The twin can sometimes be accurately modeled visually and explored through virtual reality or other kinds of visualization.

2.2.3.3 Smart Factories

One of the associated concepts in Industry 4.0 is "smart factory." This is another word for an Industry 4.0 production system, where individual companies or corporate groups use data management along the whole supply chain – including product development, production, logistics, and interface coordination with customers. The smart factory responds more flexibly to incoming requests. Smart factory is a complex system that integrates the main elements of the Industry 4.0 concept autonomous robots, IoT, and big data. In smart factories intelligent and connected assets collaborate with each other continuously, which results in self-organizing and self-optimizing production. The result can be a more efficient and agile system with less production downtime. It can also lead to a greater ability to predict and adjust to changes – both in the facility and in broader network.

2.2.3.4 Industrial Symbiosis

Industrial symbiosis is a form of resource brokering between local companies together in innovative collaborations using the waste from one as raw materials for another. There are several examples of this happening in different industrial parks – with impressive results in both economic gains and resource savings. The word "symbiosis" is usually associated with relationships in nature, where two or more species exchange materials, energy, or information in a mutually beneficial manner. By using symbiotic connection the need for virgin raw material and waste disposal can be reduced, emissions and energy use can be minimized, and new revenue streams can be created.

2.2.3.5 Lights-Out Manufacturing

Lights-out manufacturing (more often called dark factory) is a manufacturing methodology that aims to remove all human presence. These factories can be considered to run "with the lights off" using robots and autonomous decision making.

The technology necessary for lights-out production is available, and the process may be used in combination with staffed production to meet increasing demand or save money by reduced labor costs.

2.2.3.6 Edge Computing

Edge computing is local processing of data (at the edges of a network, as opposed to central processing) that enables the physical connected assets to function partly autonomously and programmed to only forward key data to a more centralized application. This contributes to efficiency, as bandwidth and computing power are often a constraint, and higher-level systems will often not need to receive important event – not the full schedule of raw data. The embedded AI can process information very near real time, which can be particularly important applications such as self-driving cars or medicine, when a split second is important.

2.2.4 Revolution or Evolution?

Industry 4.0 drives a general focus from tangible to intangible assets – as different industries move further into digital engineering and manufacturing. On a general level the change is about integration – connecting machines, processes, and systems to allow for the creation of intelligent networks. By applying these new technologies, machines can predict their own failure and trigger maintenance processes. Equipment can self-organize and react to unexpected changes in production demand. A new level of interoperability and transparency enables a fundamentally new form of production that works across organizational silos and disciplines.

What will be the consequence? How these trends will play, and how fast, out is not obvious.

To what degree Industry 4.0 will be a revolution or more an evolutionary process depends on how fast and wide the new philosophy is gaining ground and also on the pace of technological and cultural change. If the take-up is fast and the prices of necessary equipment and software keep falling, we will likely see a deep transformational shift in a number of industries – with a socioeconomic and infrastructural transformations as a result.

A German expert group, organized by VDI (the largest German engineering society), looked into the possible effects of Industry 4.0 to reduce the high degree of uncertainty perceived in many companies [28]. One key finding is that established performance indicators are not designed with the long-term perspective in mind and therefore have an inhibitory effect on the investment in new business models. Attractive new markets are not discovered; data and technology are not repurposed and utilized properly. Cooperational potentials are not realized; threats by new and old competitors are recognized too late.

To clarify the long-term effects, these experts discussed both the foreseeable development and possible implications for individual businesses. Their discussions around the subject can be summarized in the following ten theses:

1. *Servicification of products.* Industry 4.0 has the potential to change what is produced, how it is produced, and the way it is used. There is a blurring of boundaries between physical products and services. Sometimes the service adds substantial value or even replaces the primary product. This requires a more holistic life cycle approach.

2. *New levels of efficiency.* In the sectors of industrial production, intensive networking and automation create potential for cost reduction and increase in efficiency, for instance, by self-optimizing production facilities or the completion of routine works through robots. On the other hand, there are new possibilities of realizing a production of individual services with high efficiency and scalability.

3. *Customer adoption.* Customers are online, the physical objects are increasingly connected, and production facilities generate a vast amount of utilization data. If properly utilized, this allows for a significantly better satisfaction of the needs of latent users. Products can increasingly also be upgraded and adapted after the point of purchase and adjusted to changing requirements during the product lifetime.

4. *Faster innovation.* The increasing availability of usage-orientated and contextual data provides new insight. This can be used to speed up the innovation process and fuel the development of new products, services, and business models.

5. *Dynamic pricing.* In a world of connected products, production facilities, and processes, pricing can reflect actual usage intensity ("Pay as you need"). Data and other nonmonetary factors increasingly serve as means of payment tokens ("Pay with X"). This can lower the entry barriers and increase the number of potential customers.

6. *Platformization.* A central aspect of Industry 4.0 business models is the emergence of industrial platforms, or business ecosystems – often with a multisided business model. The platform leadership in a sector can become a very dominant competitive factor. For a long-term competitive advantage, platforms need to ensure a fair balance between added value and the distribution of value among the platform operator and users.

7. *Deeper collaboration.* Thanks to the higher-level systems of Industry 4.0, new value-adding networks can be formed in a decentralized and flexible way. Real-time data can be exchanged and are thereby optimized in the interest of the user. In many cases, this involves ad hoc networks, which are formed and dissolved near real time, based on market trends and fluctuations.

8. *Higher flexibility.* Collaborative added value in Industry 4.0 is based on reliable and transparent cooperation models, which are designed for the possibility of switching partners or modifying the model itself within its life cycle. These open value chains and the associated governance model differ from existing contractual and static models.

9. *Knowledge transfer.* Industry 4.0 platforms lead to a quick exchange of best practices and know-how. This gives rise to potential risk of strategic similarity. Industry 4.0 business models, therefore, must be interested in new forms of differentiation. At the same time, contrary to the classic tendency of strategic isolation, companies must consider the doctrine of typical digital start-ups and speed up the innovation and implementation processes.

10. *Cultural change.* Besides the economic and technical implications, the progressing digitalization will also produce social consequences, including changes in workplaces and culture. To ensure participation and social peace, the development of medium- and long-term strategies must be considered.

Another question is where the revolution will lead us. Will it set the stage for another leap in the material throughput in an economic-technological system that is currently exploiting the scarce resources of the Earth in one end and creating mountains of waste in the other? Or can it be a key enabler of increased resource productivity, or a qualitative leap in how we track materials and utilize resources in the economy? In their report on Intelligent Assets in the circular economy [29], the Ellen MacArthur Foundation talks of how different "circular value drivers" enabled by IoT devices and new technology. Interestingly there are also associated economic drivers that correspond well, creating good reasons for a transition towards new and sustainable operational models (Table 2.1). There are a range of conceivable opportunities. Maybe the most important are yet to be conceived?

2.3 MORE WITH LESS AND THE REBOUND EFFECT

It is easy to be fooled by promises of the future. Technological optimists can easily create a false but convincing story about the future. In 1985 James Krier and Clayton Gillette [30] stated that "their views are rigorously formulated, grounded in an apparent reality, based on knowledge and experience, and artfully defended. There are no crazies among the best of the optimists; they are conservative, respected experts who command enormous authority. They have a very specific position – namely,"that exponential technological growth will allow us to expand resources ahead of exponentially increasing demands. "This is the precise meaning of technological optimism as a term of art."

Will any technological revolution lead to a more efficient use of resources? Will Industry 4.0 in its nature create a more sustainable world? The short answer: no.

Table 2.1 The relationship between circular value drivers, technology, and economic value creation in different parts of value chain.

Level in value chain	Circular value driver	Improvement	Technical enablers (examples)	Economic value drivers
Production	Increase resource utilization	Negative externalities	Industrial symbiosis, resource stream monitoring, smart contracts	Reduced handling costs and new revenue
Manufacturing	Extended use cycle	Product life length	Condition sensors, lot integration, sharing platforms	Recurring payments and lasting customer relationship
	Increased utilization	Users per product	Availability monitoring, sharing platforms, access control systems, mobile phones	Higher revenue for each item produced
Service	Predictive maintenance	Uptime	Condition sensors, IoT monitoring, maintenance planning	Increased customer satisfaction
	Increased reparability	Product material footprint	Ecodesign, modular components	Revenue from service and spare parts
Retail	Sustainable consumption	Product preferability	Value chain transparency, standards, footprint calculations	Premium price due to sustainability
Logistics	Reduced logistical footprint	Transport footprint	Geo sensors, level sensors, route optimization	Lower-cost association with distribution/collection
	Closed loop cycles	Resource return rates	Reverse vending machines, customer loyalty apps, voucher systems, serialization	Customer loyalty and materials related to cost
Waste handling	Increased source sorting	Quality waste fractions	Smart containers, pay-as-you-throw systems, customer feedback systems	Increased value of waste fraction
	Increased central sorting quality	Quality recycled granulate	Material scanners, digital screeners, industrial robots, AI optimization	Increased sorting rates and value of granulate
Ecosystem	Regeneration of natural capital	Ecosystem condition	Ecosystem monitoring sensors	Thriving ecosystem services

Source: Inspired by Ellen MacArthur Foundation [29], adapted by Anders Waage Nilsen.

At a macro and global level, greater efficiency can significantly increase overall energy consumption. Sometimes this increase can be more than over 100%; the innovation actually creates a "backfire."

In energy economics this paradoxical effect is a well-known and documented fact. The behavioral response in society will reduce and sometimes cancel out the expected gains from new technology. This phenomenon, which can be applied to both natural resources and labor, is called the rebound effect (or take-back effect).

The rebound effect is generally expressed as a ratio of the lost benefit to the expected environmental benefit when holding consumption constant [31].

If you improve the fuel efficiency in cars by 5%, the results may be only 2% drop in fuel consumption, if this causes people to drive more than they did before.

Even if the rebound effect is undisputed among economists, outcomes in complex system dynamics can be hard to predict. With cars, for example, the efficiency savings made by robotic production result in more people buying vehicles, increasing the number of cars and the emissions. Similar effects can be seen in farming – where automated processes and huge industrial farms have meant more food can be produced more efficiently. This productivity gain could have left larger areas for rich and diverse ecosystems, but what happens is the opposite: the consumption of high-impact foods such as red meat increases, with negative effects for biodiversity and CO_2 emissions. Several reports on home energy use are also saying that energy-efficient buildings don't perform as expected. The reason? People. The occupants behave in more complex ways than designers account for. They purchase vacuum cleaning robots, watch TV on big plasma screens, open windows to compensate for central heating and read books on smartphones in need of regular charging.

What systemic effects will play out in Industry 4.0 society? Will the gains in productivity and transparency result in a lower environmental footprint or increase consumption? This is closely related to how these changes will affect the economy – both for the individual consumer and for businesses within different industries. According to theory, there are three economic reactions to technological changes [32]:

1. Direct rebound effect, caused by substitution: The lower cost of use results in an increase of consumption.

2. Indirect rebound effect, caused by higher purchasing power: The lower cost of use results in consumption of other goods and services. Less spent on fuel means more spent on luxury goods.

3. Economy-wide effect: A general fall in service cost reduces the price of other goods, creates new production possibilities, and increases economic growth. Lower fuel prices lower the cost base of every business in the economy.

The American think tank the Breakthrough Institute concluded in a literature review that at an economy-wide level the rebound effect could have a serious impact. "For every two steps forward we take with below-cost efficiency, rebound effects mean we take one or more steps backwards, sometimes enough to completely erode the initial gains made," said the report's lead author, Jesse Jenkins [33].

The highest rebound in energy use from efficiency occurs "not at the consumer level but in the productive sectors of the economy (industry and commerce). Improving the efficiency of a steel plant may result in lower cost of steel, greater demand for steel, and also create greater economic growth – all of which will drive significant rebound in energy use following efficiency improvements."

The discussion on rebound effects associated with climate change policy is interesting, but yet unresolved. As the rebound is so closely associated with cost base and purchasing power, some claim that the effect can be mitigated by raising the price of energy or resources using measures like energy taxes or a carbon tax. Advocates of carbon pricing point to energy rebound as a reason to put in place carbon taxes.

Given the urgency of reducing the resource footprint, this discussion is interesting also in an Industry 4.0 perspective. Will we need to regulate the material throughput in new ways?

Maybe the next industrial revolution will be enabled by possibilities but also by limitations and constraints?

2.4 RADICAL SOLUTIONS TO DIFFICULT PROBLEMS

The linear economy is no solution to the problem of waste. And a linear strategy is no solution to the challenge of shaping the outcome of the upcoming industrial revolution. We are facing a systemic challenge, with clusters of interrelated or interdependent challenges. This kind of challenge is what is described as "messy problems." They are, unlike tame simple-to-solve problems, more like puzzles: we solve them by resolving complexities, not by breaking them down into constituent parts [34].

Germany Trade and Invest (GTAI) defines Industry 4.0 as:

> A paradigm shift ... made possible by technological advances which constitute a reversal of conventional production process logic. Simply put, this means that industrial production machinery no longer simply "processes" the product, but that the product communicates with the machinery to tell it exactly what to do. [35]

But process innovation in the actual production phase is only one part of the story about how this will unfold. Technology is a tool, not an end in itself. It must be used, by someone, actively, to shape new solutions and generate value – not just within individual businesses but also within wider loops of production and resource management. And as history tells us, technology is not a one-way route to a better society. To shape the outcome of this revolution, we need specific innovations and new regulatory incentives, but we also need to explore the combinatory

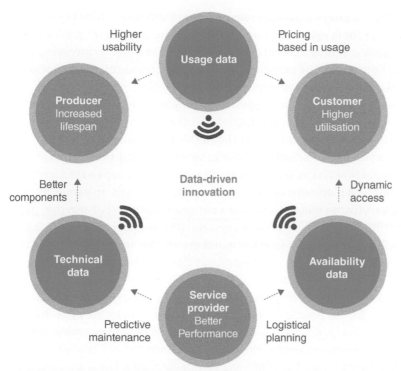

Figure 2.9 By using new technologies a range of new business models and operational processes can be enabled.
Source: With kind permission from Nick Rigas, D-Waste.

effects – how one part of the system affects the other. In complex systems change tends to play out through cascading effects and evolve over time. Industry 4.0 has left the station. How can business leaders take part in giving, not only taking, directions? The discussion on how to use digital assets to create new value has just started, but there are obvious advantages for maintenance, monitoring, life cycle, and design process (Figure 2.9).

2.4.1 Transforming Business Models

Innovation is not the same thing as invention. Changes in customer behavior, globalization, and technological innovations are currently creating a window of opportunity for new business models and processes. By recombining existing solutions it is possible to create and capture value in new ways and transform the relationship with both clients and partners. The collaborative business models enabled by Industry 4.0 technologies will probably contribute to deep transformations in many value chains in the years to come.

Generally Industry 4.0 drives business from a closed to an open business model. In the integrated value chain externals, both individuals and businesses are involved in both design and product development. There are many examples of involving a large number of people instead of only hired experts. The car manufacturer Local Motors, for example, announces challenges for car engineering on its innovation platform, where members can hand in suggestions. They use open-source designs and a network of "microfactories" for production. In other examples the designer in the process may actually be the customers themselves, choosing from a range of materials and components to create a unique product. This way the customer becomes a business partner just as much as a traditional client. Shapeways.com, a spinoff from Philips, is a platform for 3D printed consumer goods on demand. A product printing service, an online shop, and a designer community form the prototype of a possible future manufacturing system. Designers upload their 3D design, select materials, and offer it to other members via the online shop. After an order is received, Shapeways.com builds the product close to the final destination and ships it.

Thanks to IoT and remote monitoring, manufacturers can go from selling a product to taking responsibility for the function throughout the whole product life cycle. By staying connected with their products, businesses can not only repair the broken items but also prevent failures from happening. This means that firms can add continuous revenue streams with subscription-based lifelong service contracts. There are currently many examples of this. AVL List is a leading provider for powertrain development. This firm offers remote usage and condition monitoring in addition to product sales. This way they can proactively exchange weak parts to avoid breakdowns and optimize product lifetime.

Any product or service is a means to achieve something for someone, not an end in itself. Industry 4.0 opens up for making the actual output or result a service, replacing the product in itself. This kind offering can take the form of full-service packages and takes responsibility for safe operations and compliance. Kaeser, for example, innovates from selling compressors to selling compressed air per cubic meter. Kaeser takes full responsibility and operates compressors at the customer's site. This is also a business model that enables continuous revenue streams.

2.4.2 Creating Collaborative Ecosystems

In the old industrial economy, individual companies produced specific products or services they believed consumers needed or simply accepted. In a connected world this static supply-side approach is replaced by a new and more responsive model, giving consumers an extended role and pushing the whole value chain to collaborate and coordinate resources more efficiently. Industry 4.0 gives rise to a new generation of data-driven business ecosystems. These are networks of organizations – including suppliers, distributors, customers, competitors, government agencies, and so on – involved in the delivery of a specific product or service through a combination of competition and cooperation. Each entity in the ecosystem affects and is affected by the others, creating a constantly evolving relationship in which each entity must be flexible and adaptable in order to survive, as in a biological ecosystem. Industry 4.0 ecosystems can be connected through platforms or base their collaboration on standards or smart contracts. They can be globalized networks or located in the same area.

One of the examples of the latter is the industrial symbiosis project at Kalundborg, Denmark, which is a model of environmental sustainability. In this park, the different businesses collaborate to increase utilization of surplus energy and waste across different industrial processes. The symbiosis project, however, is not the result of a careful environmental planning process. It is rather the result of a gradual cooperative evolution of four neighboring industries and the Kalundborg municipality. Although it begun by chance, the project has now developed into a high level of environmental consciousness. The participants are constantly exploring new avenues of environmental cooperation.

With the new abilities of Industry 4.0, projects like this can enhance their collaboration further using real-time data exchange and possibly also "smart contracts" – computer codes that have the ability to auto execute specific functions. It works by using a simple "if x then y" operating logic. Smart contract code can be written to receive data confirming the consignment of cargo to a buyer, verify the data, and then automatically release payment that had been held in escrow to the seller.

2.4.3 Killer Apps of the Industry 4.0

These are only some examples on how the early adopters have started using the interconnected digital technologies to enable new services and new ways of generating revenue. What will happen the day when the majority of companies embrace decentralized, individualized, and on-demand production? And how will the value chains evolve when customers prefer service-based models to tangible products?

We are still waiting for the big breakthrough. A killer application (shortened to killer app) is any computer program that is so desirable that it proves the core value of some larger technology trend. A killer app can increase sales of the platform on which it runs. Industry 4.0 is still in its infancy. Even if we have seen many innovations such as software platforms, we still haven't seen the killer app making the revolution available to every business or end consumer.

What we need more than anything is solutions that contribute to closing the gap of a wasteful economy and regenerating depleted global ecosystems. Hopefully the killer app doesn't kill.

Key Take-Outs of Chapter 2

Key take-outs	Why it is important
There are many technological revolutions throughout human history that, often through unpredictable cascading effects, transforms societal and economic structures in surprising ways. Sometimes the effects can be transformational within the span of a few decades	Technology can transform a society fast, often faster than most people expect
In the beginning of industrialization, there was a scarcity of resources. Many value chains that reused and repurposed different materials. Circularity and closed loops for resources were actually an ideal in the hygiene movement	Circular economy and technological innovations for resource optimization are not new concepts. There is a lot to learn from previous chapters in industrial history
Access to energy and extracted resources has been a key driver in all the different steps of the industrial revolution. The driver behind industrialization has to a large extent been the development of technologies for increased extraction and consumption	The innovation in the first industrial revolutions was centered around "more from more" and economical principles that do not calculate in the real costs of negative externalities
There is a close connection between the handling of information in society and how the production and economy is organized. The digitalization of the third industrial revolution increased productivity and created a more globalized economy. It also globalized the handling of waste	We need a broad and conscious reflection on system effects of new innovations – how new digital enablers will affect value chains and the material throughput in society in a sustainability and resource perspective
The fourth industrial revolution is happening right now. Many of the possible concepts are described by scientists based on technology projections and anticipations about the future	Living within a time of dramatic technology shifts is a possibility to use technology as a lever for new solutions and help enable regulations and innovations that will change the course in a new and more sustainable direction
The idea of "cyber–physical systems," a network of connected machines that collaborate in smart ways, is a key concept. This is anticipated to create deep several changes on a value chain level that deeply affects business strategy	Many traditional concepts of business will be challenged by the collaborative and integrated nature of Industry 4.0. This is an important organizational challenge that requires leadership and curiosity

Key take-outs	Why it is important
Many different technological drivers are playing out simultaneously – ranging from connected hardware to artificial intelligence and new software. They can be combined to create new processes and organizational structures	The changes ahead are not given, but many of the enablers are already technologically mature
On a value chain level, faster pace of innovation, end-to-end integration, collaboration, and more dynamic business relationships are expected. A lot of new concepts can possibly change of challenge existing business models	The relationship between companies is becoming more important, and we will see the rise of collaborative industrial ecosystems
Earlier industrial revolutions have proved that there are systemic rebound effects: gains in efficiency are canceled out by increase in consumption. These effects can be unpredictable	It is not given that smarter systems and technological improvements will create a more sustainable and circular economy
There is an ongoing transformation, but we are still waiting for the big breakthrough of Industry 4.0. Managers can now play a major role by redesigning business models and creating new partnerships on a value chain level	There is a big opportunity for businesses that want to win new positions in value chains that are about to change

REFERENCES

1. Bar-Yosef, O. (2002). The upper Paleolithic revolution. *Annual Review of Anthropology 31*: 363–393.
2. Floud, R., Humphries, J., and Johnson, P. (2014). *The Cambridge Economic History of Modern Britain: Volume 1, Industrialisation*, 1700–1870. Cambridge University Press.
3. Coren, M.J. (2018). The speed of Europe's sailing ships is revamping historian's view of the industrial revolution. *Quartz*; 2018 [cited 3 February 2020]. https://qz.com/1193455/the-speed-of-europes-18th-century-sailing-ships-is-revamping-historians-view-of-the-industrial-revolution (accessed 13 April 2020).
4. Crouzet, F., Teich, M., and Porter, P. (1997). The industrial revolution in national context: Europe and the USA. *Choice Reviews Online 34*: 346–386.
5. Barles, S. (2014). History of waste management and the social and cultural representations of waste. In: *The Basic World Environmental History*, vol. *4* (eds. M. Agnoletti and S.N. Serneri), 199–226. Cham: Springer.

6. Herbert, L. (2007). *Centenary History of Waste and Waste Managers in London and South East England.* Northampton: Chartered Institution of Wastes Management, London and Southern Counties Centre.

7. Kumar, S. (2016). Municipal solid waste management in developing countries. http://dx.doi.org/10.1201/9781315369457.

8. Crafts, N. (2004). Productivity growth in the industrial revolution: a new growth accounting perspective. *The Journal of Economic History 64*: 521–535.

9. David, P.A., Stanford University. Center for Economic Policy Research, University of Warwick. Department of Economics (1989). *Computer and Dynamo: The Modern Productivity Paradox in a Not Too Distant Mirror.* Paris: OECD. 45 p.

10. Freinkel, S. (2011). *Plastic: A Toxic Love Story.* HMH. 336 p.

11. Amadeo, K. (1929). Worst stock market crash in U.S. history. *The Balance.* [cited 23 February 2020]. https://www.thebalance.com/stock-market-crash-of-1929-causes-effects-and-facts-3305891 (accessed 13 April 2020).

12. Circular Glasgow (2018). Ending the depression of planned obsolescence. [cited 23 February 2020]. https://circularglasgow.com/ending-the-depression-of-planned-obsolescence (accessed 17 April 2020).

13. GDCA (2012). What does Brooks Stevens have to do with "Planned Obsolescence"? [cited 23 February 2020]. https://www.gdca.com/brooks-stevens-and-planned-obsolescence (accessed 17 April 2020).

14. Ferguson, D.E. and Carson, R. (1963). Silent spring. *Copeia 1963*: 207.

15. Wolman, A. (1965). The metabolism of cities. *Scientific American 213*: 178–190.

16. Ritchie, H. and Roser, M. (2017). CO_2 and greenhouse gas emissions. Our World in data. [cited 11 February 2020]. https://ourworldindata.org/co2-and-other-greenhouse-gas-emissions (accessed 17 April 2020).

17. Geyer, R., Jambeck, J.R., and Law, K.L. (2017). Production, use, and fate of all plastics ever made. *Science Advances 3* (7): e1700782.

18. SkubałA, P. (2018). World scientists' second warning to humanity: the time for change is now. *BioScience 68*: 238–239.

19. UNEP. GLOBAL TRENDS IN RENEWABLE ENERGY INVESTMENT 2019. [cited 29 January 2020]. https://wedocs.unep.org/bitstream/handle/20.500.11822/29752/GTR2019.pdf?sequence=1&isAllowed=y (accessed 17 April 2020).

20. Deloitte. Industry 4.0 – challenges and solutions for the digital transformation and use of exponential technologies. [cited 11 February 2020]. https://www2.deloitte.com/content/dam/Deloitte/ch/Documents/manufacturing/ch-en-manufacturing-industry-4-0-24102014.pdf (accessed 17 April 2020).

21. (2013). Zukunftsprojekt Industrie 4.0. *ZWF Zeitschrift für wirtschaftlichen Fabrikbetrieb 108*: 275–275. https://doi.org/10.3139/104.013051.

22. Broy, M. (2011). *Cyber-Physical Systems: Innovation durch softwareintensive eingebettete Systeme.* Springer-Verlag. 141 p.

23. Beattie, A., Shaw, F., Moss, L., and Hurley, P. (2018). Moments that will change the world – Ethos. [cited 30 January 2020]. https://ethos-magazine.com/2018/09/moments-that-will-change-the-world (accessed 17 April 2020).

24. Alcácer, V. and Cruz-Machado, V. (2019). Scanning the Industry 4.0: a literature review on technologies for manufacturing systems. *Engineering Science and Technology, an International Journal 22* (3): 899–919.

25. AI and Nanotechnology (2019). How do they work together? AZoNano.com. [cited 12 February 2020]. https://www.azonano.com/article.aspx?ArticleID=5116 (accessed 17 April 2020).

26. Anonymous (2020). Thematic paper: Industry 4.0, advanced materials (nanotechnology) – internal market, industry, entrepreneurship and SMEs – European Commission. [cited 3 February 2020]. https://ec.europa.eu/growth/tools-databases/regional-innovation-monitor/report/thematic/thematic-paper-industry-40-advanced-materials-nanotechnology (accessed 17 April 2020).

27. Deloitte (2017). *Industry 4.0 Challenges and Solutions for the Digital Transformation and Use of Exponential Technologies*. Deloitte.

28. VDI (2018). VDI Status Report. Seamless and dynamic engineering of plants. https://www.vdi.de/presse/publikationen/publikationen-details/pubid/vdi-status-report-seamless-and-dynamic-engineering-of-plants (access 17 April 2020).

29. World Economic Forum (2015). Intelligent assets: unlocking the circular economy potential, by Ellen MacArthur Foundation and World Economic Forum as part of Project MainStream. [cited 7 February 2020]. https://www.ellenmacarthurfoundation.org/publications/intelligent-assets (accessed 17 April 2020).

30. Krier, J.E. and Gillette, C.P. (1985). The un-easy case for technological optimism. *Michigan Law Review 84*: 405.

31. Grubb, M.J. (1990). Communication energy efficiency and economic fallacies. *Energy Policy 18*: 783–785.

32. Sorrell, S. and Dimitropoulos, J. (2008). The rebound effect: microeconomic definitions, limitations and extensions. *Ecological Economics 65*: 636–649.

33. Jenkins, J., Shellenberger, M., and Nordhaus, T. (2011). *Energy Emergence: Backfire and Rebound as Emergent Phenomena*. Breakthrough Institute.

34. Hancock, D. (2017). Tame, messy and wicked risk leadership. http://dx.doi.org/10.4324/9781315241838.

35. Kagermann, H., Helbig, J., Hellinger, A., and Wahlster, W. (2013). Recommendations for Implementing the Strategic Initiative Industrie 4.0: Securing the Future of German Manufacturing Industry. Final Report of the Industrie 4.0 Working Group. 112 p.

Chapter 3
Un(mis)understanding Circular Economy

The closed economy of the future might similarly be called the "spaceman" economy, in which the earth has become a single spaceship, without unlimited reservoirs of anything, either for extraction or for pollution, and in which, therefore, man must find his place in a cyclical ecological system which is capable of continuous reproduction of material form even though it cannot escape having inputs of energy.

— Kenneth E. Boulding

Recommended Listening
The Godfather (Soundtrack), **Nino Rota**
Because it is epic, lyric, romantic, full of good intentions and hopes, but finally it's tragic too.

Industry 4.0 and Circular Economy: Towards a Wasteless Future or a Wasteful Planet?, First Edition.
Antonis Mavropoulos and Anders Waage Nilsen.
© 2020 John Wiley & Sons Ltd. Published 2020 by John Wiley & Sons Ltd.

3.1 A GLOBAL TREND

Circular economy has become a global trend almost rapidly. It is promoted by UN and European Union (EU), it is becoming a cornerstone for environmental policies by many governments, it has been adopted as a policy framework by hundreds of cities, and it is widely supported by most of the world's biggest companies and the World Economic Forum and, of course, by almost all the NGOs active in the field of environment. It has also its own global institution, the Ellen MacArthur Foundation (EMF), an organization established with the sole purpose of promoting and facilitating moves towards a circular economy. In this paragraph, we will briefly present the circular economy policies or initiatives in different countries, just to highlight the global coverage of circular economy.

3.1.1 China and EU Are Leading the Way

Starting from the EU, in March 2019, the European Commission issued a comprehensive report on the implementation of the Circular Economy Action Plan [1], which was adopted in 2015. The report presents the main achievements under the action plan and sketches out future challenges to shaping EU economy and paving the way towards a climate-neutral circular economy where pressure on natural and freshwater resources as well as ecosystems is minimized. In 2018, EU adopted its Circular Economy Package [2] that includes a strategy for plastics in circular economy; a proposal for port reception facilities; a communication on how to address the interface between chemical, product, and waste legislation; a monitoring frame; and a report on critical raw materials and the circular economy. In addition, EU has created the Circular Economy Stakeholder Platform [3], a virtual open space that aims at promoting Europe's transition to a circular economy by facilitating policy dialogue among stakeholders. There is no doubt that the most important impacts by the adoption of circular economy have been to the waste sector. A serious revision of the waste legislation has entered into force by July 2018. The main elements are the following:

- A common EU target for recycling 65% of municipal waste by 2035.
- A common EU target for recycling 70% of packaging waste by 2030 (plus targets for each material).
- A binding landfill target to reduce landfill to maximum of 10% of municipal waste by 2035.
- Separate collection obligations are strengthened and extended to hazardous household waste (by end of 2022), biowaste (by end of 2023), and textiles (by end of 2025).
- Minimum requirements are established for extended producer responsibility (EPR) schemes to improve their governance and cost efficiency.
- Prevention objectives are significantly reinforced, in particular, requiring member states to take specific measures to tackle food waste and marine litter as a contribution to achieve EU commitments to the UN SDGs.

It is important to notice that according to the agenda set by the new European Commission President Ursula von der Leyen, EU aims to become the world leader in circular economy. A new circular economy action plan focusing on resource-intensive and high-impact sectors such as textiles and construction is expected. Austria, Germany, France, Netherlands, and Belgium are playing a central role to push the whole EU towards circular economy.

China is the only country that has developed the concept of CE and has practiced it as a development strategy on a large scale [4]. China has developed a circular economy strategy from 2002, and it has tested its applicability in several pilot cities around the country. Actually, in China circular economy is not considered a tool to reduce waste or improve environmental conditions and performance, but it is rather faced as a new model for economic growth that will drive both economic and environmental sustainability [5]. If the size of its economy is taken into consideration, then the potential global benefits from China's shift to circular economy are going to be very important. Enacted on 29 August 2008, the relevant law defines the circular economy as "a general term for the activities of reducing, recycling and recovering in production, distribution and consumption." It specifies that [6]:

- reducing means the decrease of resource consumption and waste generation in production, distribution, and consumption;
- recycling means the full or partial reuse, repair, or remanufacturing of waste;
- recovering means the reuse or regeneration of waste as raw materials.

The law included a comprehensive set of measures aiming to stimulate the transition to circular economy, indicators, and quantitative goals. At the same time, emphasis is given on packaging material recycling and the standards for product packaging design to avoid excess waste. The 5-year plan for 2010–2015 sets a goal of increasing resource productivity by 15%, which was then adapted and adopted on a regional level. The framework law covered various actors (the state, local authorities, businesses, associations, etc.) with different roles and coordinated the required cooperation in different levels. Local governments were obliged to include the principles of the framework law in their local planning documents. Another major law concerns the EPR framework, as part of the 13th Five-Year plan. This framework will be implemented in 2020, and the aim is that all relevant laws and regulations will be finalized by 2025. The policy objective [7] is to integrate environmental impacts related to the distribution, use, and disposal of products into producers' design and production decisions. Similar to the remanufacturing policy, the EPR policy regulates the behavior of producers, but it is also consumption oriented. Finally, China has set ambitious goals for waste prevention and recycling. By 2020, at least half of all used packaging material must be biodegradable. The recycling rate should reach 50% for all waste types by 2025, while all the new products should contain at least 20% recycled materials.

Interestingly, EU and China, in April 2019, announced [8] that they will step up their cooperation on circular economy based on the memorandum of understanding that was signed in 2018 and describes strategic exchanges and dialogue on best practices of circular economy, industrial parks, chemicals, plastic and waste, eco-design, eco-labeling, green supply chains, and EPR schemes.

3.1.2 Other Government Initiatives

Japan started its efforts on circular economy around 1990, and it has already developed a framework law governing all actions related to the circular economy. The main drivers seem to be the geographical limitations as well as the restricted natural resources (only 20% of the land area is habitable). The legislation developed involves three different levels [9]. First, the framework law for a "Sound Material Cycle Society" that manifests the waste hierarchy and the stakeholder engagement. Second, the resource efficiency law that promotes the 3Rs (reduce, reuse, recycle) and sets the rules for waste management (recycling and incineration). Third, there are several sector-specific laws that deal with the sectoral recycling methods, the quantities and practices of waste management, etc. Eco-design has a special role in Japan, and energy efficiency is also of special interest, while the whole approach is based on the principle of continuous improvement, incorporating regular revision of targets, annual progress reports, and 5-year plans.

In the United States, the federal government does not deal with resource efficiency and circular economy; however several states adopt relevant approaches, including 3Rs and zero waste. A recent report [10] registered more than 200 circular economy initiatives in the United States but concluded that they are not called "circular economy." In addition, it was found out that less than 8% of the companies that were included in the report offered take back options and that the main barrier of the existing initiatives was the effort to stay within the dominant economic and production model.

In Russia, although several laws and regulations for specific waste streams were introduced in early 1970s, during the USSR period, the collapse of post-Soviet Russia created a huge gap in a legal and administrative aspects that resulted in a relevant collapse of the waste management sector too [11]. However, the Russian government has started several "green" initiatives in line with the circular economy approach aiming to control and eliminate waste mainly from the big state business and follow the EU best practices wherever this is possible. In addition, a serious reform is on the way regarding municipal solid waste infrastructure, recycling, and institutional development.

Canada is also considering circular economy initiatives in different levels of the economy. There is a bottom-up approach with several companies setting and testing out new circular business models, municipalities are shifting to zero waste programs, and policy initiatives are underway at all levels of government. Till now the most emblematic national initiative is the Canada-wide Action Plan on Zero Plastic Waste [12], which was approved in November 2018 by the Canadian Council of Ministers of the Environment and aims to reduce the harmful environmental impacts of plastic waste through greater prevention, collection, and value recovery to achieve a more circular plastics economy.

The overall picture is that in China circular economy is promoted as a top-down national political objective, while in other areas and countries as EU, Japan, and the United States, it is a tool to redesign bottom-up supply chains. However, it seems that waste management policies are a core element of all the circular economy approaches, and especially in EU, China, and Japan they create serious changes to municipal waste management and recycling.

Regarding the Global South, although until recently circular economy was considered as an approach suitable only for wealthy countries, it seems that several governments are at least examining possibilities to involve circular economy in policies and strategies. Nigeria, Rwanda, South Africa, Turkey, India, and Colombia have been mentioned. The key issue here is that in developing countries there are several activities that are resource efficient and can be considered as circular economy initiatives, but most of them are implemented by the informal sector. A recent report [13] by Chatham House describes the situation like this: "Although developing countries are often more 'circular' than wealthier countries – in the sense that few things left on the street are not retrieved for recycling or repairs – this is largely out of economic necessity. A key question is how the Circular Economy will affect people employed in informal sectors who play significant roles in waste-management processes; whether or how to 'formalize' such jobs is a well-known development challenge. Looking to the future, developing countries also need to address rapidly rising consumption among the middle classes."

3.1.3 Private Sector Initiatives

Last but not least, besides the government initiatives, what makes the shift to circular economy so rapidly evolving as a global trend is the huge support it receives from the biggest companies of the world. An indicative example was the setup of the EMF's Circular Economy 100 (CE 100) initiative that according to the official announcement [14] came as a result of "the exceptional response from business to the publication of two reports by the Foundation, launched during the World Economic Forum in Davos, with analysis by McKinsey, which demonstrate over $1trillion of opportunities posed by circular economy innovation." The creation of the new alliance was supported by the Foundation's founding partners, Cisco, BT, National Grid, B&Q, and Renault, while the first 20 members included also the Coca-Cola Company, M&S, IKEA Group, Morrisons, Tarkett, FLOOW2, Heights UK, iFixit, Ricoh, Vestas, WRAP, Turntoo, and Desso. Today the program is going on with great success, and besides the members it has announced several partnerships with big companies including Google, Danone, Philips, Unilever, BlackRock, H&M Group, and SC Johnson. The UN Global Contact study on the views of CEOs on Sustainability 2013 [15], in the mining and metals sector, after surveying 1000 CEOs, from 103 countries and 27 industries, remarks that "Absent almost entirely from our conversations in 2010, the concept of the circular economy has taken quick hold among CEOs focused on innovation and the potential of new business models. Already, a third of CEOs in this year's Study – and fully 46% in the mining & metals sector – report that they are actively seeking to employ circular economy models."

Figure 3.1 is presenting a selection of the most important circular economy initiatives worldwide categorized per countries, cities – regions and business [13].

3.1.4 Why Now?

As the concept of circular economy is rather old, it is useful to answer the question: why it became so popular now? There must be several drivers that contributed to make circular economy a global trend. Obviously, global warming, loss of biodiversity,

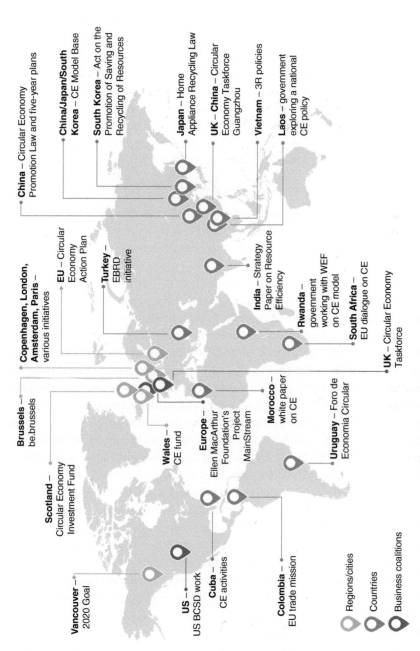

Figure 3.1 Circular economy initiatives and their typology worldwide.

Source: From Ref. [13]. with kind permission from Chatham House.

urbanization, pollution, the rise of the new middle class, and the rising resource demands, as we discussed them in Chapter 1, have played a crucial role setting the scene and the need for better policy responses. The Task Force on Resource Management of the International Solid Waste Association (ISWA), in its report "Circular Economy: Trends and Emerging Ideas," [16] has suggested three broad key drivers that push for such a change in resource management, namely, the commodity prices and raw material supply, environmental legislation, and green taxation. In ISWA's report, the accelerating pace of change (regarding the adoption of circular economy) was attributed to the involvement of business leaders, the new landscape of partnerships between the NGOs and the business sector, and the ongoing technological shift. The final conclusion was that "In a global world it is clear that the circular economy will accelerate when global economic pressures force change. The pace of that future change is going to be variable and will require economic, legislative and fiscal measures to combine to become mainstream. A race has started between global nations to secure quality secondary raw materials and gain the economic advantage that flows from putting in place effective policy and fiscal change that will drive innovation to recover and use these materials."

An interesting finding from a detailed review study [17] was that the circular economy is not an appropriate tool for growth-oriented economic systems, but it fits better to steady-state oriented economies as well as in (potentially) declining ones because circular economy, resource efficiency, and environmental protection are crucial factors for a smooth transition to new production and consumption models capable to delay the decline. This fits a lot to EU, a declining economy that suffers from austerity and low growth rates, that imports 57% of fossil fuels and 40% of metals used [18]. Measured in raw material equivalents, about 40% of material inputs into the EU economy are imported, making a more circular system a condition for the long-term viability of the EU.

Especially regarding the business involvement in circular economy, it is obvious that their primary interest is to secure their supply chains and develop a long-term plan to control the supply of the most important materials. It is well recognized [19] that where companies control the full cycle of a material or product, they choose circular models to offset the need to hedge for the price volatility of new materials. This also avoids the (normally uninsured) risk that lack of availability of resources will constrain production. The popularity of circular economy in the business world has also been linked with the perceived similarity of the concept (for the broad public audience) to 14 different approaches (including recycling, zero emissions, corporate social responsibility, blue economy, sharing economy, and industrial ecology) that are considered as "business sustainability movements." [20] Concepts like these ones usually stay for a max of 10 years in the center of discussions and marketing practices. Afterwards, they remain part of the discussions and some times of the practices, but they are substituted by new concepts that build on them and create synergies for the profile of the business sector.

As circular economy becomes a global trend, its geopolitical impacts are starting to be outlined and discussed, especially the ones about the developing world and global trade. We will come back on them, but right now, we will focus on circular economy as a forgotten practice of our not so distant past, because this will set the scene for the rest of the chapter.

3.2 CIRCULAR ECONOMY COMES FROM OUR PAST

The history of solid waste management is a real mirror of the social and economic history of human societies. What is considered waste is straightforward linked with our cultural context, the raw materials used, and the way objects are produced, the abundance or scarcity of specific materials, and the way human symbiosis is organized in settlements and cities. This is more than clear when someone considers the different words used to describe what we call waste. S. Barles has written that [21] "In the first category, terms are associated with the themes of loss and uselessness: *déchet* in French from the verb *choir* (to fall), *refuse* and also *garbage* in English (which primarily refers to animal offal), *rifiuti* in Italian, *residuo* in Spanish, *Abfall* in German. In the second category, terms emphasize the dirty or repulsive nature of these particular materials: *immondice* in French, *immondizia* in Italian, from the Latin *mundus* which means clean; *ordure* in French from the Latin *horridus*, meaning horrible. Finally, terms in the third category describe the materials that make up the waste: *boues* in French, *spazzatura* in Italian, *Müll* and *Schmutz* in German, *rubbish* in English derived from *rubble*." The careful reader will easily spot that the first category refers to the lost value (economy in broader terms), the second one is related to health and environmental impacts, and the third category describes the materials that are found in waste, which actually determine both the lost economic value or the potential value of recovery and the environmental and health impacts. Well, this is an impressive bottom-up understanding of the key principles of modern waste management shaped by folk wisdom.

For thousands of years, the quantities of waste produced by human societies were rather small, the collection and disposal of waste was done in primitive ways, or it was not existing at all as organized activity, as it is still the case for 35–40% of the population worldwide. Two driving forces, urbanization and industrialization, converged and finally shaped the origins of what we call modern waste management. First, urbanization created the conditions that made waste- and pollution-related diseases a serious threat for the whole society, something that started to be felt from the early medieval period. The rise of neo-Hippocratism in medicine [21] and the importance it put in the relationship between health and environment shifted a lot of doctors in Europe to look at the environmental conditions for potential explanations of morbidity and mortality. Second, it was the first and the second industrial revolution, with the factories as their emblems, that besides increasing waste and pollution, they also stimulated the resource value of waste [22]. In Paris, the value of urban excreta [23] was increased due to its industrial and agricultural use; thus organized collection and removal systems became profitable and subject of business. In London, as David Wilson explains [24], "The industrial revolution and rapid urban expansion led to an excess in demand for bricks and 'breeze' for building, for which municipal waste became an important raw material. So, the London parishes began to let contracts, effectively granting an exclusive franchise to a private contractor to collect the waste in their area." In his emblematic work *The Capital*, written in 1867, Karl Marx mentioned [25] that turning waste products into something valuable reduced the cost of the raw material to the extent to which it is again saleable and that these savings increased profitability. Marx even went so far as to say that after economies of scale, waste recovery was the second big source of economy in industrial production.

In brief, urbanization demonstrated the importance of public health in waste management, and industrialization highlighted the resource value of waste. However, during all the years before waste collection and removal from cities became an organized daily activity run by municipalities and businesses, there was another activity that was permanently implemented in a kind of natural way. Susan Strasser describes this activity in Chapter 1 of her famous book *Waste and Want: A Social History of Trash* [26]: "Everyone was a bricoleur in the preindustrial household of the American colonies and, later, on the frontier; saving and reusing scraps was a matter of course. Cloth, wood, and food could only be obtained by arduous spinning, weaving, chopping, sawing, digging, and hoeing, by bartering with other products of strenuous work, or by spending scarce cash. Whether things were purchased at stores or crafted on farms and plantations, the value of the time, labor, and money expended on materials and their potential value as useful scraps were evident to the people who worked with them."

Reuse of items, systematic repair, storage of equipment (materials and spare parts for future use), creative rearrangements, and fit-for-purpose applications for specific needs were the rule and not the exception in the daily lives of the vast majority of people. Especially people in rural areas were using all food leftovers to feed animals, and since they had more space available, they used, as our grandfathers and grandmothers did, to store everything that could be used in the future for repair or as a spare part for a customized creation. This social practice was dominant and resulted in minimum amount of waste generated per capita. It is still a dominant practice in many developing countries where informal recyclers, in poor communities, consider wasted materials as a potential income source through different practices that involve bricolage [27]: waste materials, either picked from bins or dumpsites or bought at a low price, are used to create new items that can be sold for a price like toy cars or decorative articles made from drinks cans, necklaces and bracelets from paper, and recycled metals or bags made from plastics and recycled clothes.

3.2.1 The Post-World War II Acceleration

In a way, it can be said that the main practices that we are trying to introduce with the circular economy like the extension of the life cycle of products, reuse, creative repair, and resource recovery were followed for hundreds of years as a natural practice by the vast majority of people. In a similar way, a research project [28] that examined the development of industrial recycling networks concluded that "industrial symbiosis" is not, as is usually believed, a break with past practices, but rather a widespread phenomenon that has been neglected by contemporary researchers and provided historical evidence about it. So, how did this change, and the modern Western societies became so much wasteful? A view in the period that started after the end of World War II helps to answer this question. Although some single-use and easily disposable products were existing for long, like the paper collars or the cone-shaped soda cups, after 1945 easy disposability and single use were advertised as a serious advantage. The benefits of single use started, rightly, to be advertised for napkins and towels. Then, aluminum foil became popular for several kitchen uses, food packaging became the norm, precooked frozen food (with its special packaging) was important for people that were tired from their works or simply they did not have time or attitude to cook, and cleaning products in plastic packaging became very popular. As Susan Strasser points out, what is important is to understand that single use and easy disposability were heavily advertised as modern, convenient, and timesaving. Notably, the advertisement practices generated a huge amount of single-use waste papers. Probably nothing describes better the business atmosphere of this period than two quotes mentioned in an article [29] about plastics in *The Guardian*. The first is a quote of the economist Victor Lebow who said in 1955 that "Our enormously productive economy demands that we make consumption our way of life...We need things consumed, burned up, worn out, replaced and discarded at an ever-increasing pace." In a similar way, in 1963, Lloyd Stouffer, the editor of *Modern Plastic* journal, said "You are filling the trash cans, the rubbish dumps and the incinerators with literally billions of plastics bottles, plastics jugs, plastics tubes, blisters and skin packs, plastics bags and films and sheet packages. The happy day has arrived when nobody any longer considers the plastic package too good to throw away."

Coming back to the question: how Western societies abandoned the circular economy practices and became so wasteful or, in a different context, how we created so linear economies? The answer seems to be related to the industrial revolutions. It is the tremendous resource, energy and labor productivity gains, the rapid development of coal and of the petroleum industry, the advances in chemistry, and the continuous search for better, lighter, stronger, and fit-for-purpose materials that made massive industrial production possible and stimulated massive consumption.

3.2.2 Industrialization Stimulated the Linearization of the Economy

As Mario Giampietro explains, the great linearization [30] of the global economy happened due to the industrial revolutions. This deserves some more details. Assuming that we have the technosphere (human societies and economies, governed by processes under human control) and the biosphere (environment, primary resources, and sinks of the waste and emissions of the technosphere, governed by processes outside human control). Figure 3.2 shows the flows between the biosphere and the technosphere.

In Figure 3.2 we can distinguish three types of flows:

1. Primary flows – Flows crossing the interface between the technosphere and biosphere, either entering into the technosphere (extracted from primary

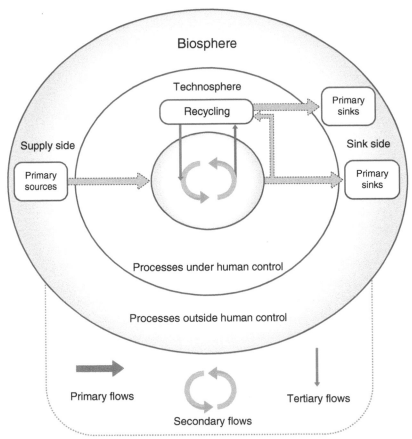

Figure 3.2 Flows inside the metabolic pattern of human society connecting the technosphere and the biosphere.
Source: **From Ref. [30], © 2019 The Author, published by Elsevier B.V. under the CC BY license, adapted by Nick Rigas, D-Waste.**

sources, such as coal mines and aquifers) or exiting from the technosphere into the biosphere (waste and emissions going into primary sinks, such as the atmosphere, water table, and dumpsites).

2. Secondary flows – Flows derived from the exploitation of primary flows (e.g. fossil fuels are primary flows; gasoline is a secondary flow). Secondary flows are at the same time inputs and outputs produced and consumed within the technosphere.

3. Tertiary flows – Flows derived from the recycling of secondary flows.

For millions of years, the available primary flows from biosphere and the available final sinks were much larger than the requirements and the outputs of the technosphere. In these conditions, the biosphere was able to offer (primary resources and energy) and receive (emissions and waste) without undermining the continuous flow. In these conditions, according to Giampietro, the biosphere takes easily care of recycling, and it is capable to absorb and metabolize the outcomes of the technosphere. Circularity occurs as the natural rule for the flows between biosphere and technosphere.

This changed with the industrial revolutions and the systematic exploitation of fossil fuels, which is directly linked with the rise of human population and the urbanization process. Due to its increasing size and rate of growth, technosphere now requires a much larger and faster supply of primary flows, while the emissions and waste going into the biosphere require much larger capacity of final sinks. In this situation, the biosphere can no longer produce all that is required and absorb and recycle all that is produced by the technosphere; consequently some of the recycling has to be implemented inside the technosphere. Technically speaking, the metabolic pattern [31] of the whole system shows that inside the technosphere, both the densities and paces of flows per unit (e.g. crop yields per hectare) are much larger than the natural flows per unit in the biosphere. So, the biosphere is not able to metabolize the emissions and waste of the technosphere, and circularity is no longer possible. This is where we are today, and this is called linearization of flows.

Giampietro explains [30] that, for example, the yield of grain per hectare from a crop field is at least an order of magnitude larger than the available quantity of biomass from unmanaged land and continues: "The pace and density of the natural deposition of nitrogen in soil … does not permit yields of 7–10 t/ha of grain typical of modern agriculture. Maintaining such yields require heavy doses of artificial fertilizer…Indeed, with the event of the industrial revolution the agricultural sector moved from low external input to high external input agriculture… While the former relied on nutrient recycling through a complex network of interactions among ecological fund elements (thus guaranteeing soil health, biodiversity, healthy aquifers, etc.), the latter is based on linearization of flows through the use of fossil energy (stressing ecological fund elements). This continuous human struggle to boost the pace and density of natural flows has resulted in a tremendous increase in agricultural productivity inside the technosphere: from less than 1 t/ha of grain in pre-industrial agriculture to more than

10 t/ha in industrial agriculture. An even more impressive improvement has been achieved in labor productivity – from about 1 kg of grain per hour of labor in pre-industrial agriculture to around 1000 kg/h in industrial agriculture. The price to pay for this increased agricultural productivity has been a progressive liquidation of ecological funds (which would slow down productivity because of their low flow/fund ratio)."

The same is true for energy demands. The energy supply of modern society predominantly consists of a linear exploitation of nonrenewable stocks of fossil energy allowing a density and pace of flows that are orders of magnitude higher than those of circular renewable fund flows, such as biomass. Thus, the last 200 years' transformation of the energy and agriculture sector demonstrates the impact of the industrial revolutions. The mode of energy and food production changed dramatically from being almost entirely based on circular fund flows (inputs produced and wastes absorbed by ecological funds) to almost complete dependence on linear stock flows (inputs extracted from stocks and wastes overwhelming environmental sink capacity) [30].

The concept of the Great Acceleration [32], developed by the International Geosphere-Biosphere Programme and Stockholm Resilience Centre, presents the same picture in 24 indicators that depict the dramatic acceleration in human enterprise and the impacts on the Earth system over the last two centuries, but more specifically from 1950s to today. In the time framework of a single human life, the human footprint on Earth has become similar to geological footprints, something that has been described as the geological era of Anthropocene [33].

The straightforward linkages between industrial revolutions and the development of linear economy are clear. As we are experiencing IND4.0, a further linearization of the economy should be expected, unless humans will be able to shift to circular economies that will rebalance our working relationship with the biosphere and the supporting ecosystems. But now, we are ready to answer the question, "what is a circular economy?"

3.3 WHAT IS A CIRCULAR ECONOMY?

Although there are a lot of political agendas, many conferences and reports, and hundreds of scientific papers speaking about circular economy we are still far away from crystallizing a definition that will allow the concept to be transformed to a concrete scientific and measurable project. In contrast, the term "circular economy" is used in many different ways and meanings by people and organizations with completely different political and business interests. It seems that it has become another buzzword. According to the Wikipedia, a buzzword is a word or phrase, new or already existing, that becomes very popular for a period of time. Buzzwords often derive from technical terms yet often have much of the original technical meaning removed through fashionable use, being simply used to impress others. We have seen the same thing happening with sustainability and green economy few years ago, when Robert Engelman precisely described that "we live today in an age of 'sustainababble', a cacophonous profusion of uses of the world 'sustainable [development]' to mean anything from environmentally better to cool." [34] As Martin Geissdoerfer and his colleagues already wrote [35], "While the terms Circular Economy and sustainability are increasingly gaining traction with academia, industry, and policymakers, the similarities and differences between both concepts remain ambiguous. The relationship between the concepts is not made explicit in literature, which is blurring their conceptual contours and constrains the efficacy of using the approaches in research and practice." So what is exactly circular economy?

Usually, the concept of circular economy, as we discuss it today, is related to the economist Kenneth E. Boulding. Boulding wrote the study "The Meaning of the Twentieth Century," which is considered a very important contribution to the foundation of sustainable development principles [36]. However, his essay "The Economics of the Coming Spaceship Earth," which was published in 1966, is related more to the current version of circular economy. Probably the most well-known excerpt of this essay is this one:

"I am tempted to call the open economy the 'cowboy economy,' the cowboy being symbolic of the illimitable plains and also associated with reckless, exploitative, romantic, and violent behavior, which is characteristic of open societies. The closed economy of the future might similarly be called the 'spaceman' economy, in which the earth has become a single spaceship, without unlimited reservoirs of anything, either for extraction or for pollution, and in which, therefore, man must find his place in a cyclical ecological system which is capable of continuous reproduction of material form even though it cannot escape having inputs of energy." It is interesting to notice that Boulding's essay became so famous because (i) he focused on the need to rethink our relationship with nature and (ii) he emphasized on the need to change the economic principles in order to shift to new production and consumption patterns.

Other researchers consider the work of David Pearce and Kerry Turner, *Economics of Natural Resources and the Environment*, [37] as the origin of circular economy as they were the first that used the term explaining that in a circular economy "everything is an input to everything else." The role of the environment is central in their approach because it involves three economic functions: resource supplier, waste assimilator, and source of utility. McDonough and Braungart with their "cradle-to-cradle" concept [38] and Benyus with his work on biomimicry [39] are also usually mentioned as major contributors to the circular economy concept. Schröder [40] believes that the roots of the concept of "circular economy" go back to classical political economists (e.g. Ricardo, Smith, Quesnay) who saw the system of production and consumption as a circular process. In any case, the concept of circular economy is in the same family with other sustainability concepts as it is shown in Figure 3.3 [41]. All these concepts were developed as a response to the growing pressures that industrialization and the resulted pollution created to ecosystems and public health. Despite of the fact that

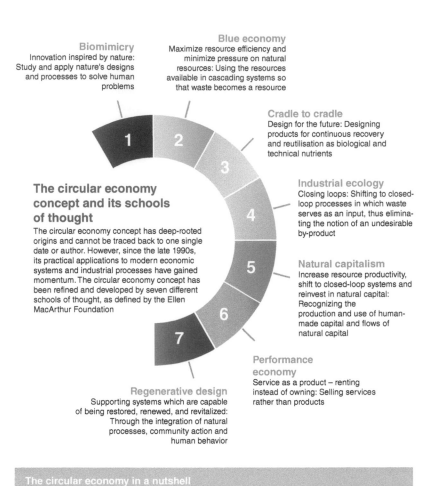

Biomimicry
Innovation inspired by nature:
Study and apply nature's designs
and processes to solve human
problems

Blue economy
Maximize resource efficiency and
minimize pressure on natural
resources: Using the resources
available in cascading systems so
that waste becomes a resource

Cradle to cradle
Design for the future: Designing
products for continuous recovery
and reutilisation as biological and
technical nutrients

**The circular economy
concept and its schools
of thought**

The circular economy concept has deep-rooted
origins and cannot be traced back to one single
date or author. However, since the late 1990s,
its practical applications to modern economic
systems and industrial processes have gained
momentum. The circular economy concept has
been refined and developed by seven different
schools of thought, as defined by the Ellen
MacArthur Foundation

Industrial ecology
Closing loops: Shifting to closed-
loop processes in which waste
serves as an input, thus eliminating the notion of an undesirable
by-product

Natural capitalism
Increase resource productivity,
shift to closed-loop systems and
reinvest in natural capital:
Recognizing the
production and use of human-
made capital and flows of
natural capital

**Performance
economy**
Service as a product – renting
instead of owning: Selling services
rather than products

Regenerative design
Supporting systems which are capable
of being restored, renewed, and revitalized:
Through the integration of natural
processes, community action and
human behavior

The circular economy in a nutshell
In contrast to the traditional, linear, throwaway economy, in the circular economy, we design
and create products that are easy to share, lease, reuse, repair, refurbish and recycle while using
regenerative resources and renewable energy. The goal is to minimize waste and to keep products
and resources in the economy for as long as possible. Ideally, this win-win approach benefits both
the economy and the environment

Figure 3.3 Circular economy and the schools of thought that shaped it.
Source: From Ref. [41], with kind permission from MOVECO.

all these concepts became trendy for several years, the results achieved are at least
questionable and, in many cases, not measurable.

According to one of the most usual definitions (based on EMF studies [42]), the
concept "circular economy" describes "an industrial system that is restorative or
regenerative by intention and design. It replaces the 'end-of-life' concept with
restoration, shifts towards the use of renewable energy, eliminates the use of toxic
chemicals, which impair reuse, and aims for the elimination of waste through the
superior design of materials, products, systems, and, within this, business models."

Based on cradle-to-cradle principles and systems thinking, this interpretation of the concept involves the distinction of two different types of materials: materials of biological origin that can return to the biosphere as feedstock (e.g. forest products) and technical materials that cannot biodegrade and enter the biosphere (e.g. plastics and metals). The concept aims to deliver a "transformation of products and their associated material flows such that they form a supportive relationship with ecological systems and future economic growth." An obvious remark is that this rather technocratic definition lacks any link to broader sustainability and especially its social components, like employment, social equity, fight against poverty, etc.

According to the EU definition [43], "The transition to a more circular economy, where the value of products, materials and resources is maintained in the economy for as long as possible, and the generation of waste minimized, is an essential contribution to the EU's efforts to develop a sustainable, low carbon, resource efficient and competitive economy. Such transition is the opportunity to transform our economy and generate new and sustainable competitive advantages for Europe." A research report by CEPS [44] refers to three problems in the EU definition, namely, (i) the lack of importance in energy resources, (ii) the absence of the social dimension of sustainability, and (iii) the missing emphasis to the quality of materials and resources.

ISWA presents a definition [45] that involves the waste hierarchy, the energy recovery, and the need for final sinks too: "The general aim of the circular economy is to utilize the maximum value of resources and keep them in use for as long as possible. In a circular economy, materials and products are designed with waste prevention in mind, and are re-used, recycled or recovered. When the raw materials they contain can no longer be re-used the energy they contain is extracted to displace virgin fuels and residues are safely disposed to landfill. In order for the circular economy to function effectively, enabling policy and market conditions are needed across all sectors."

3.3.1 Hundred Fourteen Answers for One Question

CEPS researchers [44] found out that there is big diversity in circular economy definitions that reflects the different priorities and views of the different stakeholders involved. Definitions start from focusing exclusively to material flows and resources and go up to major transformation of the economic system that goes far beyond waste and resource management. They concluded that "The circular economy is a complex concept and it is unlikely that in the short term there can be an international consensus on its meaning." Another interesting research paper [46], after analyzing 327 papers, concluded that there is a lack of consensus on the use of different terms and definitions, and the literature analyzed shows two different clusters. One cluster focuses on eco-parks and industrial symbiosis, mostly in the context of China. The second cluster is concerned with supply chains, material closed loops, and business models.

In a recent blog post [47] at ISWA's website, with the title "114+1 Definitions for a Circular Economy – Finding a Common Language," Julian Kirchherr and Ralf van Santen presented the results of their research [48] on the definitions of circular economy. As they wrote, "…when we investigated its core principles, aims and enablers, we found 114 circular economy definitions set forth by scholars and practitioners over the last years. We uncovered two core principles. First, a serious dependence on the use of the Rs (reduce, reuse, recycle) framework, and second an emphasis on systems perspective." They found out a serious diversion of the definitions regarding the waste hierarchy. Seven percent of the definitions portray circular economy simply as recycling. Recycling is the most common element of the definitions (79%), followed by reuse and reduce (75 and 55%, respectively). Unfortunately, waste hierarchy is only mentioned in only 30% of the 114 definitions. Interestingly, only 42% of the definitions consider the systemic perspective, yet with little agreement on what levels of the economy are included. The majority of definitions, 21 and 24%, respectively, focus on the macro (national) and meso (regional) level, with only 19% including the micro (product) level. This indicates that a transition is considered to require especially efforts on the regional level, rather than simultaneous change on all three levels. Surprisingly, sustainable development was linked with circular economy only in 12% of the definitions, and in most of the cases it was related only with economic development, ignoring environmental and social aspects.

Interestingly, the authors, besides recognizing the conceptual blurriness, developed an additional definition trying to cover the gaps they noticed. According to this definition [48], "circular economy is an economic system that replaces the 'end-of-life' concept with reducing, alternatively reusing, recycling and recovering materials in production/distribution and consumption processes. It operates at the micro level (products, companies, consumers), meso level (eco-industrial parks) and macro level (city, region, nation and beyond), with the aim to accomplish sustainable development, thus simultaneously creating environmental quality, economic prosperity and social equity, to the benefit of current and future generations. It is enabled by novel business models and responsible consumers." A similar definition has been proposed by J. Korhonen [49]: "Circular economy is an economy constructed from societal production-consumption systems that maximizes the service produced from the linear nature-society-nature material and energy throughput flow. This is done by using cyclical materials flows, renewable energy sources and cascading-type energy flows. Successful circular economy contributes to all the three dimensions of sustainable development. Circular economy limits the throughput flow to a level that nature tolerates and utilizes ecosystem cycles in economic cycles by respecting their natural reproduction rates."

A study [17] that analyzed 1031 research papers about circular economy concluded that "At both theoretical and practical levels circular economy is mainly rooted in environmental economics and industrial ecology, with a high emphasis on technological innovation in the form of cleaner technologies as well as on recycling rather than reuse. The latter is a key principle in circular economy and should be prioritized with adequate policies. Moreover, the high emphasis on increasing resource efficiency is not fully consistent with the often-claimed need for decreasing resource use as well as the high reliance on non-renewable resources." According to an OECD working paper [50], we can categorize three types of circular economy views – definitions. The first defines circular economy relative to a traditional linear economic system, i.e. one that focuses on closing resource loops. The second views the circular economy putting emphasis on the importance of slower material flows, either with some material circularity or even in a linear concept. The third, and broadest, view of the circular economy is that it involves a more efficient use of natural resources, materials, and products within an existing linear system. This broad view of the circular economy affects potentially all economic activities, not only those that have a high material use profile but also the one applied in most cases.

To conclude there is no single commonly accepted definition of the term "circular economy," but different definitions share the basic concept of decoupling of natural resource extraction and use from economic output, having increased resource efficiency as a major outcome. A lot of scientists believe that there is an urgent need to develop a commonly accepted and internationally recognized definition about circular economy that will allow academics, practitioners, and policy makers to communicate in a common language, as it is required to make the implementation of circular economy more pragmatic. It seems that the definitions presented above by J. Kirchherr and J. Korhonen are more suitable as they incorporate all the dimensions of sustainable development (economy, environment, society). However, it has to be stressed that it is possible, if not necessary, to have a number of coexisting narratives [40], each reflecting different and sometimes competing social, economic, and environmental framings and therefore different versions of circular economy. This will drive diversified political strategies and different country-specific pathways, especially for the developing world, since for circular economy also, it is true there is no one-size-fits-all approach [51].

The ultimate goal of promoting the circular economy is, in all the cases, the decoupling of environmental pressure from economic growth. A common element in the mainstream narrative is that instead of the traditional waste management practices, circular economy promotes product, component and material reuse, remanufacturing, refurbishment, repair, cascading, and upgrading as well as solar, wind, biomass, and waste-derived energy utilization throughout the product value chain and cradle-to-cradle life cycle. There is a broad consensus that the circular economy has the potential to stimulate existing efforts to advance environmental sustainability and solving global environmental problems. Now it's time to discuss about some common problems in the mainstream circular economy narratives and the butterfly effect.

3.4 FROM GOOD INTENTIONS TO SCIENCE

Making circular economy a global practice that will deliver the expected results requires a deeper understanding of material flows and their interactions with the biosphere. This is a very difficult task because we are talking about the kingdom of complexity. It is not easy, fast, nor simple to explain, quantify, project and visualize material flows and interactions without scientific tools capable to deal and manage complex phenomena. In this paragraph, we will outline some basic conceptual problems in the mainstream narrative of circular economy. Our view is not to undermine the concept of circular economy but to stimulate a deeper scientific research that will allow its transformation to a concrete, measurable, and scientifically defined approach, capable to meet, at least partially, the expectations it has created. There are millions of people with good intentions that are inspired by circular economy and try to contribute in their own ways, so it is the duty of scientists to build on these good intentions for the benefit of our world.

Having this in mind, it will be interesting to go deeper and examine in more details the mainstream narrative of circular economy, as it is mainly expressed by the EU and the EMF. Figure 3.4 presents the EMF original butterfly diagram of circular economy [52]. Figure 3.5 presents the narrative of circular economy simplified in three arguments and the underlying assumptions and some issues that need further discussion.

To frame the relevant discussion, there are three important concepts that will be briefly explained. The first is about system dynamics and planetary limits [53] (as described in Chapter 1). The second is about thermodynamics and its application on circular economy, and the third is the distinction between natural and engineered systems.

In general terms, systems can be described in terms of stocks, flows, and feedback. Stocks are accumulations of things (not necessarily physical) that change over time through the actions of inflows and outflows. Feedback occurs when changes in the size or composition of a stock affect the rates of inflow and/or outflow. Feedback can be either balancing or reinforcing, in other terms negative or positive. Another important characteristic is the throughput, the amount of material or energy passing through a system or process. In our case, we have a continuously growing global economy (the system) that is fueled by a continuous increase of consumption (positive feedback) and relies on continuous increasing inflows of energy and matter and outflows of emissions and waste (throughput). According to the system dynamics view, the system will eventually run into some kind of physical constraint, in the form of a rebalancing feedback. Some scientists believe that global warming is exactly this rebalancing feedback. This can be better illustrated using the metaphor of the "Empty" and "Full" world developed by Daly [54] and Goodland [55] as shown in Figure 3.6.

The main idea is that in an "Empty" world, the supporting ecosystem (biosphere) is or must be capable (i) to provide the energy and the matter required to sustain

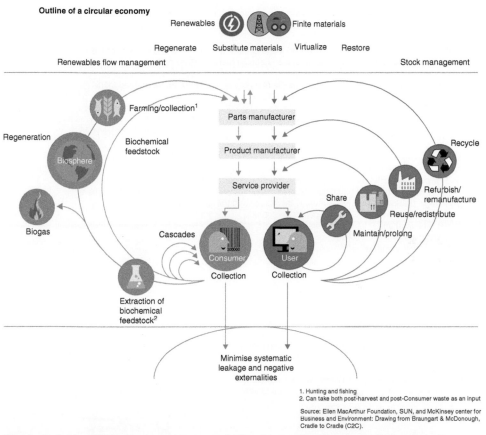

Outline of a circular economy

Renewables · Finite materials

Regenerate · Substitute materials · Virtualize · Restore

Renewables flow management · Stock management

Farming/collection[1]

Regeneration

Biosphere

Biochemical feedstock

Parts manufacturer

Product manufacturer

Service provider

Recycle

Refurbish/remanufacture

Reuse/redistribute

Share

Maintain/prolong

Biogas

Cascades

Consumer · User

Collection · Collection

Extraction of biochemical feedstock[2]

Minimise systematic leakage and negative externalities

1. Hunting and fishing
2. Can take both post-harvest and post-Consumer waste as an input

Source: Ellen MacArthur Foundation, SUN, and McKinsey center for Business and Environment: Drawing from Braungart & McDonough, Cradle to Cradle (C2C).

Figure 3.4 The Ellen MacArthur Foundation butterfly diagram of circular economy in which technical and biological materials have their own closed loops. *Source*: From Ref. [52], with kind permission from the Ellen MacArthur Foundation.

the global economy and (ii) to have available high capacity in final sinks to receive waste and emissions. While the first condition is usually part of all the relevant discussions, the second condition (the availability of final sinks) is usually ignored or seriously underestimated. So, meeting those two conditions, the biosphere is capable to receive and metabolize–circulate (or if you prefer to recycle naturally) the outflowing emissions and waste without serious disorders of the global biogeochemical processes. In contrast, in a "Full" world, the expansion of the economy, after a certain point, requires energy and matter that are not (easily or at all) available, and the final sinks are not capable to receive and metabolize the outflowing waste and emissions. In this case, the biosphere is not able to metabolize–circulate the waste and emissions coming out of the economy. Let's see some examples that materialize the concepts.

Circular economy narrative	Underlying assumptions	Issues to discuss
1 Waste will be eliminated through developing closed material loops and redesign of materials, products, systems, and business models	3Rs and close loops demand less resources and energy than the linear system. Circular loops are natural. Technology will advance resource efficiency and boost sustainability	• Are closed loops possible? • More or less primary resources? • How about anthropogenic stocks? • Energy demands to develop closed loops • Rebound effects and Jevons paradox
2 There are two different types of materials: materials of biological origin that can return to the biosphere as feedstock (e.g. forest products) and technical materials, which cannot biodegrade and enter the biosphere (e.g. plastics and metals)	The distinction of two types of materials is valid. Biosphere is the source and the final receptor of organic materials	• Types of waste are covered • Biosphere and technosphere interaction • System boundaries • Mixed biological and technical materials
3 Circular economy aims to deliver a transformation of products and their associated material flows such that they form a supportive relationship with ecological systems and future economic growth	The decouple between economic growth and resource extraction–use and their environmental impact is possible. Business have interest to promote circular economy	• Social footprint of circular economy • Economic impacts of circular economy? • Developing countries • Geopolitical impacts • Decouple or degrowth?

Figure 3.5 The main points of the circular economy narrative and the underlying assumptions and issues that should be discussed.

Source: With kind permission from Nick Rigas, D-Waste.

3.4.1 We Live in a "Full" World

Our world is becoming more and more "Full" due to the continuous expansion of the global economy. The global GDP, in constant 2010 US $ prices, has grown from 11.36 trillion in 1960 to 82.46 trillion in 2018, according to the World Bank statistics [56], a rise of 725% in 58 years. The global economy is expected to more than double in size by 2050, far outstripping population growth, thanks to continued technology-driven productivity improvements [57]. The result is that in order to find new energy resources, we are looking for fossil fuels in the most difficult and remote places in the world. We pump oil from 3 km depth below the sea, as it was the case of Deepwater Horizon in the Gulf of Mexico. When you drill in such depths and conditions, then accidents that discharge more than 780 000 m³ of oil (or 4.9 million

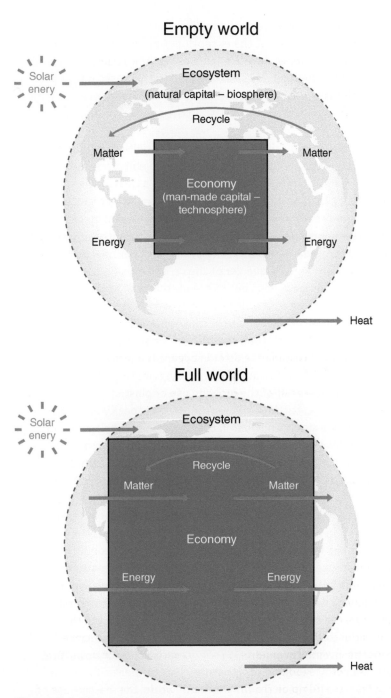

Figure 3.6 The metaphor of "Empty" and "Full" world as developed by Daly and Goodland.
Source: From Ref. [30], © 2019 The Author, published by Elsevier B.V. under the CC BY license, adapted by Nick Rigas, D-Waste.

barrels) [58] destroying the local economies, marine life, beaches, wetlands, and estuaries are going to become the rule, not the exception. Back in 1949, the average depth of oil wells was almost 1108 m, while in 2008 it was 1830 m. Today, the world's deepest drill is in the Chayvo oil field on the Sakhalin Shelf in the Russian Far East, going 12.2 km deep [59]! This is exactly the process of the "Full" world: our economy expands deeper and deeper in the biosphere to sustain the continuous increasing energy demands.

The case of global warming is a typical case where human activities overcame the capacity of final sinks. The continuous rising carbon dioxide emissions resulted in increasingly high concentration in atmosphere, which acts as a final sink for carbon dioxide. What we call global warming is the consequence of the fact that the atmosphere, as a final sink, is already beyond its capacity to receive and metabolize the outflows of the global economy. The rise in carbon dioxide concentration parallels closely the rise in primary energy use and in GDP. The ocean, another final sink for carbon dioxide, currently absorbs about a quarter of the carbon dioxide added to the atmosphere from human activities each year, but this leads to ocean acidification, which tracks the carbon dioxide curve closely. While the ocean uptake of carbon dioxide significantly reduces its impact on climate, it causes marine ecosystems and biodiversity to change. For example, corals and shellfish are finding it more difficult to build their shells. The case of plastic marine litter in oceans is another case that illustrates the importance of final sinks for the global ecosystem. The concentration of plastics in the oceans is a direct result of the growing use of plastics in a continuous growing economy that results in restricting the ocean's natural capacity to absorb plastic pollution. The results are already obvious in the food chain, the damages in biodiversity, the losses in fishing and tourism industry, etc. What is important to consider is that violating the limits of final sinks creates serious long-term health and environmental impacts, as both the cases of global warming and marine litter demonstrate.

So, what really matters when considering the interaction of the economic process (technosphere) with its environmental context (biosphere)? The answer is the size of the required input flows and the waste outflows generated by the economy (throughput of the technosphere) compared with the size of the primary sources and primary sinks made available by ecological-natural processes (biosphere). In a "Full" world, some of the outflowing waste needs to be internalized into the economic process, and this is what we call closed loops or circular economy. However, this internalization increases the cost for the economy and reduces its performance. Production factors must be invested in generating services (recycling of flows) that would otherwise have been provided for free by nature.

As Turner [60] and Meadows [61] have shown, in a "Full" world, the massive use of renewable energy can only delay but not reverse the end of the growth due to physical restrictions. So, this brings us to the energy aspects. Geoffrey West, in his emblematic book *Scale: The Universal Laws of Life, Growth, and Death in Organisms, Cities, and Companies* [62], gives a nice example of how big is the energy consumption per capita today. "In the United States it is … 11,000 watts, which is more than one hundred times larger than its 'natural' biological value (100 watts). With this amount of power…the effective human population of the planet accordingly operates as if it were much larger than the 7.3 billion people who actually inhabit it. In a very real sense, we are operating as if our population were at least thirty times larger, equivalent to a global population in excess of 200 billion people." After few paragraphs, West highlights that the real revolutionary character of the industrial revolutions "was the dramatic change from an open system where energy is supplied externally by the sun to a closed system where energy is supplied internally by fossil fuel. This is a fundamental systemic change with huge thermodynamic consequences, because in a closed system the Second Law of Thermodynamics and its requirement that entropy always increases strictly applies. We 'progressed' from an external, reliable, and constant source of energy to one that is internal, unreliable, and variable."

3.4.2 Thermodynamics Are Fundamental

The first law of thermodynamics says that energy can neither be created nor destroyed; it can only change form. The second law of thermodynamics says energy is always tending towards more disorganized forms. The overall result is an increase in the degree of disorder or randomness, which is called entropy. Thermodynamics says that no energy transfer is 100% effective because of losses. It is useful to make here the distinction between exergy and energy. Exergy represents the work potential, i.e. the useful portion of the energy used by a fridge to freeze water in the ice tray. While energy cannot be destroyed, exergy can. In other words, the fridge degrades some of the useful electricity into useless disorganized heat dissipated in the room. Due to the laws of thermodynamics, all economic activities as well as circular economy activities consume energy, increase entropy, and decrease exergy. As Giampietro put it [30], "According to the second principle of thermodynamics irreversible processes alter the qualitative characteristics of material flows. Recycling can be done, but only to a certain extent and at a certain cost, and only if the corresponding primary resources are available. Hence, the amount of primary waste outflows of an economy can be reduced by recycling (provided the inputs required by the recycling process itself do not exceed the waste outflow recycled), but a continuous production of waste is unavoidable."

Regarding the description between natural and engineered (man-made) systems, as Rammelt mentions [63], ecosystems (biosphere) are driven by high-exergy and low-entropy resources and generate almost no waste. In contrast, engineered systems are driven by the extraction of low-exergy resources. At the other end, engineered systems produce, accumulate, and dispose high-entropy emissions and waste. In the same school of thought, Nielsen and Müller [64] explained that in natural systems, the cycles are local, decentralized, and develop towards being increasingly closed with decreasing emissions and waste as a consequence. In engineered systems, the cycles are becoming increasingly global; they are transport intensive and have evolved to be open with increasing emissions and waste as a consequence.

But these differences do not mean that all the natural systems are circular and all the engineered ones wasteful and linear. In a great piece [65] titled "Cycles, spirals and linear flows," Paul Brunner puts things in a more historical-geological perspective: "Seen from a natural history point of view, the development of today's world was a linear process, with no cycles. Where then does the notion 'cycles of nature' come from? If smaller time scales are taken into account, regularly occurring phenomena can be observed: The daily cycles, the seasonal pattern during a year, the corresponding hydrological cycle, the cycling of carbon and nitrogen, and others. These cycles are important for nature as well as mankind. At a second examination, most of the natural cycles are not closed, they deviate from a cycle. During the beginning of the photosynthetic period, oxygen was a useless waste product that slowly increased the oxygen concentration in the atmosphere. It may be more appropriate to use the metaphor of a spiral: At first sight, it looks like a cycle, but when you come closer, you realize, that the natural system does not come back to exactly the same point. From a larger distance, the movement follows the form of a spiral, which – in contrast to a cycle – is headed towards a certain direction. While a cycle defines a static situation returning to the same state, a spiral allows progress, resulting in new developmental possibilities. Cycles, spirals, or linear flows – what has all this got to do with waste management? The discussion about the direction of waste management has always been influenced by the underlying world views. In our times, the notion of a cycling economy is the predominant paradigm. There is no doubt that reuse and recycling conserve energy and resources, and that they contribute significantly to reduce pollution. But it is also a fact that waste management is a key element for controlling linear flows, too."

Now, after these introductory comments, we can move to the discussion on the subjects mentioned on Figure 3.5, based on the butterfly diagram presented in Figure 3.4. The discussion will be organized following the categorization presented in Figure 3.5. For all the three points of the circular economy narrative (left column of Figure 3.5), we will discuss the underlying assumptions (middle column) and the relevant issues for further discussion (right column). The first part of the narrative (item number 1 of the left column) is discussed in Section 3.5. The second part of the narrative (item number 2 of the left column) is discussed in Section 3.6. Finally, the third part (item number 3 of the left column) is discussed in two paragraphs. In Section 3.7 we will discuss economic growth, the hopes for decouple, and the role of business, and in Section 3.8 we will discuss about the social footprint of circular economy and its economic and geopolitical impacts.

3.5 CIRCULARITY IS NOT SUSTAINABILITY

According to the first part of the narrative in Figure 3.5, waste will be eliminated through developing closed material loops and redesign of materials, products, systems, and business models. The main underlying assumption is that product reuse, remanufacturing, and refurbishment demand less resources and energy and are more economic than conventional recycling of materials as low-grade secondary materials. Materials should first be recovered for reuse, refurbishment, and repair, then for remanufacturing, and only later for raw material utilization, similar to traditional recycling. According to the circular economy narrative, combustion for energy recovery should be the second to last option, while landfill disposal is the last option. In this way, the product value chain and life cycle retain the highest possible value and quality as long as possible and are also as energy efficient as it can be.

In a recent study about the limitations of circular economy, Jouni Korhonen and his colleagues put straightforward the first and most important problem: "This ideal of a single cyclic system, while desirable, is not realistic. In the world, approximately 75% of the energy production is based on non-renewable sources extracted from the lithosphere that are combusted. The combustion releases emissions to biosphere in forms and concentrations that nature cannot tolerate or assimilate. This is the most obvious example of the linear throughout economy and the best example of the limits of the current circular economy visions." Several authors have mentioned that developing closed loops for a wide array of products and materials does not automatically generate environmental benefits because in many cases the closed loops would require additional energy quantities for logistics and reprocessing. As an example, a study on the greenhouse gas (GHG) emission factors for recyclables [66] found out that recycling soil, plasterboard, and paint contributes to increasing GHG accumulation, while recycling carpets has a negligible contribution to GHG savings. So, the message is clear: the closed loops should be developed on a case-by-case analysis based on specific costs and benefits. In any case, it is obvious that a narrative of circular economy focusing explicitly on materials and ignoring the increasing energy demands for the development of closed loops is not complete and can't form the basis for proper decision making.

A second problem is related to the fact that thermodynamically speaking, every closed loop will have leakages. Andreas Bartl, in ISWA's report "Circular Economy: Cycles loops and Cascades" [67], wrote that "Recycling can help to close cycles and to feed materials back into the production process. However, cycles are never perfect, and leakages are an inevitable reality. Materials will, intentionally as well as unintentionally, be mix up. Molecules will undergo a degradation and exhibit reduced intrinsic properties. To a certain extent, substances will always be released into the environment in such concentrations that makes recovery impossible." In the same report, the phenomenon of cascading is explained as another barrier to recycling:

a repeated use of the resource over time takes place. It is obvious that the quality is decreasing over time and for each additional utilization a quality drop has to be accepted. However, as the resource is passing through several phases, the overall use of resources can be significantly reduced.

In another ISWA report titled "Circular Economy: Closing the Loops" [68], Costas Velis revisited the evidence regarding the feasibility of recycling and focused on two key engineered materials: polypropylene (PP) of fossil nonrenewable origin and paper and board of renewable biogenic materials. The study found out that "a critically important contribution of the waste management and reprocessing industries lies in keeping the environment clean while closing the loop. In many cases, legacy issues have to be addressed during the recycling process, such as in the presence of certain brominated flame retardant additives in plastics that proved carcinogenic, or compounds associated with inks in paper, such as bisphenol-A (BFA): it is very important to depollute the material flows and close the loop while preventing dispersion of polluting substances. Such depollution function, results in some inevitable losses and sets another limit to what can be sustainably recycled. ... there is insufficient understanding of the environmental and wider sustainability performance of the closing of the loop." Another finding was that the (perpetual) closed-loop model may not be feasible in many occasions, because of small but sufficient deterioration of fundamental material properties during its use, collection for recycling, and reprocessing: e.g. the length of fibers is shortened for paper/board in each reprocessing cycle. As a result, extra virgin raw material still needs to be added at some percentage. It was also concluded that another limit to the closed-loop model stems from the losses that inevitably occur during the multiple stages of closing the cycle: starting with ability to collect sufficient quantities and continuing with the rejection of unsuitable material (damaged, contaminated) and the limited separation efficiencies of the sorting equipment.

The discussion on circular economy makes clear that we need new tools, more complex and robust than the usual cost–benefit analysis and LCA modeling, that will help us to assess and evaluate the feasibility of developing closed loops for selected materials in a multidimensional way that should cover the social, environmental, economic, and technical domains. Researchers from the University of Leeds proposed [69] a new conceptual approach that combines scientific and engineering methods with a sociopolitical narrative providing an analytical framework for making the transition to a resource-efficient future. Figure 3.7 presents the proposed concept for complex value optimization of resource recovery and demonstrates its multidimensional character. There is a need for further research on this field that will allow us to develop tools suitable to the complexity of circular economy instead of using the same tools we are using to make decisions for the linear economy.

Another point regards the expectations about resource efficiency and the hopes that it will advance sustainability. As we have already seen in Chapter 1, the

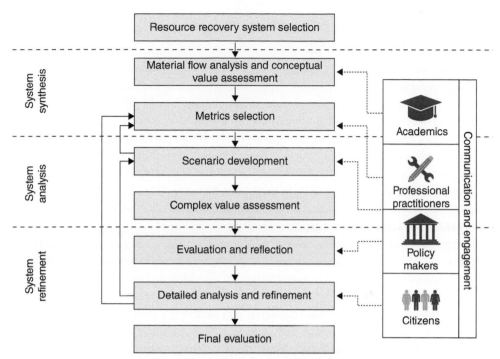

Figure 3.7 Framework for complex value optimization for resource recovery.
Source: From Ref. [69], Crown Copyright © 2017, published by Elsevier Ltd. under the CC BY license.

advances of resource efficiency do not contribute automatically to sustainability due to the rebound effect and the Jevons paradox. Daly argued that eco-efficiency is not sustainability [54]. The crucial parameter is to keep the growth of the physical scale of the economy (which is not the GDP growth but the material and energy flow footprint) under control and within the biosphere limits. If those limits are crossed, even if the whole system works with optimum eco-efficiency, Daly pointed out that "optimally loaded boats sink though they would sink optimally." In a more general way, as Rammelt put the whole issue [63] in a rather humorous way, "capitalism operates on the assumption that the output of production is a function of capital, labor and natural resources; shortages in the third factor elicit development of substitutes and higher efficiency in the first two factors. The suggestion in economic textbooks is generally that natural resources are not a limiting factor. If we follow the logic of the production function, we could ultimately bake a cake with only the cook and his kitchen; we do not need flour, eggs and sugar. We could also make our cake a thousand times bigger with no extra ingredients, if we stir faster and use bigger bowls and ovens." In reality, of course, there are biophysical limits that simply can't be crossed, as we have already explained.

3.5.1 Anthropogenic Stocks Are Ignored but Rapidly Growing

Anthropogenic stocks are also an important and rather underestimated issue in circular economy discussions. In a recent study [70], Fridolin Krausmann (Institute of Social Ecology of the Department of Economics and Social Sciences of BOKU University in Vienna) and his team examined the socioeconomic metabolism of the global economy between 1900 and 2015. They found out that for the whole period global material extraction rose by a factor of twelve, reaching to 89 billion tonnes per year in 2015. A shift from materials for dissipative use to stock building materials resulted in a massive increase of in-use stocks of materials to 961 billion tonnes in 2015. Around 1900, merely 0.5 billion tonnes of materials were added to the stock of manufactured capital each year. By 2015 this flow had grown more than 69-fold to 31 billion tonnes/year. Most of these accumulated materials were nonmetallic minerals used in construction (concrete, asphalt, bricks, sand, and gravel). In addition, 33 billion tonnes of metals, 15 billion tonnes of wood, 3 billion tonnes of plastics, and 3 billion tonnes of glass were employed in in-use stocks. For a better illustration of the importance of the problem of anthropogenic stocks, the annual growth [71] in Germany's anthropogenic material stock is about 10 tonnes/capita/year. Similarly, in Austria, as illustrated in Figure 3.8, the annual municipal solid waste generation is about 500 kg/capita, the total waste generated is 5 tonnes/cap, and the annual additional stocks are 8–10 tonnes/capita [72].

All these anthropogenic stocks (roads, buildings, infrastructure, durable equipment) are a valuable accumulation of future secondary materials and future waste too, when the stocks reach the end of their life cycle. To put it in another way, the future waste is already here, so a real circular economy approach should take into consideration how we will deal with massive stocks and the involved secondary materials. However, it seems that this is not an issue for the EU circular economy package that focuses too much on municipal solid waste, although it recognizes that it is a rather tiny fraction of the total quantities.

The huge rise of stocks creates an important collateral damage: it relatively downgrades the importance of closed cycles and recycling in the current waste flows. This is demonstrated by Johann Fellner and his co-authors at the paper "Present Potentials and Limitations of a Circular Economy with Respect to Primary Raw Material Demand" [73]. The paper examines what will be the benefits if 100% of the waste stream can be recycled, despite the second law of thermodynamics and the problems relevant to quality of materials. Then it calculates the savings in GHGs for 100 years in the range of 140 kg of carbon dioxide equivalents per capita per year (CO_2-eq./capita/year). Comparing these savings with the average EU GHG emissions that are around 9000 of carbon dioxide equivalents per capita per year (CO_2-eq./capita/year), it comes out that the emissions per capita will be reduced by a tiny 1.6% and energy savings would at maximum contribute to a reduction of 1.8%. The authors concluded that

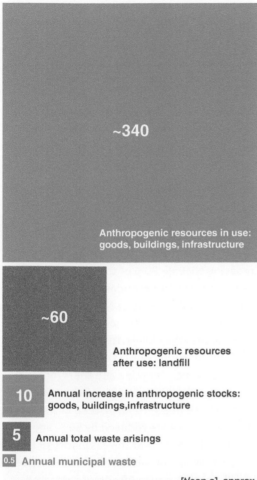

~340

Anthropogenic resources in use:
goods, buildings, infrastructure

~60

Anthropogenic resources
after use: landfill

10   Annual increase in anthropogenic stocks:
goods, buildings,infrastructure

5   Annual total waste arisings

0.5 Annual municipal waste

[t/cap.a], approx.

Figure 3.8 Municipal solid waste, total waste, and anthropogenic stocks per capita (upper part) and comparison between total landfilled quantities and quantities in use (lower part) in Austria.
Source: With kind permission from Nick Rigas, D-Waste, based on data from Ref. [72].

"because a significant share of commodities is still used to build up our infrastructure and thus accumulates in societies' material stock, the overall potential for reducing primary raw material consumption and accompanying impacts (including costs for primary raw materials) is limited at present, even for a highly developed economy like the EU."

The final comment in this paragraph regards the overall scope and aim of circular economy. To model the resource extraction till 2050 [70], Fridolin Krausmann and his team assumed, in a business as usual scenario, that by 2050 there will be a convergence of diet patterns and of per capita stocks at the 2010 level of industrialized countries. The results are astonishing. The result is a 2.4-fold increase of material extraction between 2015 and 2050, reaching 218 billion tonnes resource extraction in 2050. The global metabolic rate doubles to 22 tonnes/capita/year, which is more than currently observed in most industrialized countries and far beyond the global target corridor of 6–8 tonnes/capita/year, which has been proposed by the International Resource Panel as a goal for 2050 in order to remain within a safe operating space. In such a scenario, the total solid waste generated will be close to 40 billion tonnes/year. So, what can we finally achieve if we adopt circular economy immediately, and how it will reshape our future? How it will reduce resource extraction using secondary materials instead of virgin ones?

3.5.2 What Can We Really Achieve Advancing Circular Economy?

To answer this question, we have to start by saying that 15–20% of the resources extracted are energy resources that can't be substituted by secondary materials, but they can only be reduced through advances in renewable energy use. The second we have to think is that roughly 50% of all the resource extracted goes to build new anthropogenic stocks. In an important effort to assess the circularity of the global economy in 2005, Haas, Krausmann, and Wiedenhofer found out [74] that that only 7% of all materials entering the global economy are currently recycled and therefore used in closed loops. Then assuming (i) that renewables will substitute 50% of the extracted energy resources, (ii) that recycling for metals will rise to 80–100% instead of 71% that was assessed in 2005, (iii) that the net additions to stocks are reduced by 50%, and (iv) that food waste is reduced by 50–100%, they calculated that the circularity of the global economy can rise to 20–34%, instead of 7%. What can we learn by this exercise?

First, the energy transition to renewable will directly reduce resources extracted and the relevant emissions and waste due to fossil fuel utilization. However, reaching to 50% renewables worldwide is a quite ambitious target. Haas noted that "The energy transition, which is urgently required to stay within the planetary boundary for climate change, is mutually interlinked with the Circular Economy. No energy transition means a tight limit for circularity, and no Circular Economy means a stumbling block for climate mitigation." This is one more reason for which we need a circular economy that integrates the energy challenge.

Second, the importance of anthropogenic stocks is very high, and we need urgently to find a way to reduce net additions and increase recycling of demolition waste. A special problem here is that the circularity of construction–demolition waste is very sensitive to transportation distances.

Third, if we will be able to meet all the four assumptions–conditions that were described before, we will have one-third of the materials in circular loops. This is a realistic measure of what can be delivered by adopting circular economy globally, and of course it will create substantial environmental benefits. However, any reader can easily assess if the four assumptions–conditions are easy to be met in the current global landscape.

Haas closed his paper by saying that all the previous are far from realistic if we continue to have an annual increase of resource consumption in the range of 2–3% per year. The paper closes with the phrase "Achieving a reversal of the global growth trend in resource consumption into a no-growth or even shrinking dynamic remains the greatest challenge." Now it's time to deal with the butterfly effect.

3.6 THE BUTTERFLY EFFECT

Almost 50 years ago, Edward Lorenz – an American mathematician, meteorologist, and one of the founders of the chaos theory – posed a question during the meeting of the American Association for the Advancement of Science: "Does the flap of a butterfly's wings in Brazil set off a tornado in Texas?" Lorenz has explained that putting this question he intended to demonstrate the idea that some complex dynamical systems exhibit unpredictable behaviors such that small or negligible changes in the initial conditions could have profound and widely divergent effects on the system's outcomes. Because of the sensitivity of these systems, outcomes are unpredictable. This idea became the basis for a branch of mathematics known as *chaos theory*, which has been applied in countless scenarios since its introduction, including modeling of the global economy. Some scientists believe that the shift to circular economy will increase the already high complexity of the economic system to levels that are similar to a chaotic system; others believe that the increased complexity can be managed, but we need new tools for that [75]. However, in this paragraph, when we talk about the butterfly effect, we are dealing with something completely different and certainly less complex. We are dealing with the butterfly diagram of the EMF because the flap of its wings has created business and policy turbulences. More specifically, we will deal with item number 2 of the narrative in Figure 3.4 and the distinction between two types of materials.

The butterfly diagram, presented in Figure 3.4, has become an emblem of the circular economy, and it dominates the relevant discussions and presentations. The powerful visualization has served as a basis for a better understanding of the circular economy approach, and certainly it inspired decision makers, business leaders, and researchers to dive deeper into the concept of circularity. There is also another famous diagram regarding circular economy, the one that was used by the EU in its relevant documents, as shown on the upper side of Figure 3.9. Both diagrams, the one of Figure 3.4 and the one on the upper side of Figure 3.9, despite being completely different in shape and level of details, have a common characteristic. They both show that circularities and residual waste are an issue for the products that are

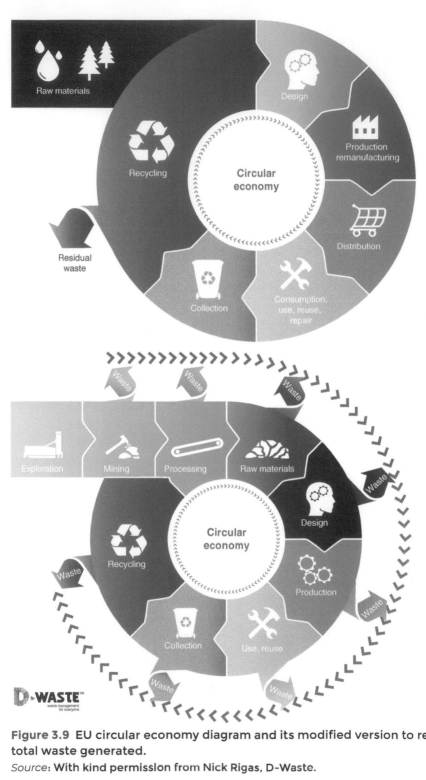

Figure 3.9 EU circular economy diagram and its modified version to reflect the total waste generated.

Source: With kind permission from Nick Rigas, D-Waste.

manufactured, distributed, consumed, and arrived at the end of their life cycle. This is rather misleading. Let us explain why.

Take the example of aluminum, a basic metal for the global economy. Its production involves three stages: mining bauxite ore, refining bauxite to alumina, and then smelting alumina to produce aluminum. To produce 1 tonne of aluminum, the mining process generates 10 tonnes of waste rock and 3 tonnes of toxic red mud (from the alumina extraction). For copper, another basic component of our economy, things are even worst: for 1 tonne of copper, we generate 140–200 tonnes of solid waste. Mining waste is one of the most important streams, worldwide, in terms of quantities. According to EUROSTAT [76] in 2016, the total waste generated in the EU-28 by all economic activities and households amounted to 2538 million tonnes. 25.3% of it, or almost 643 million tonnes, was mining waste. For a comparison, household waste was just 8.5% or just 216 million tonnes, almost one-third of mining waste. Still, it seems that mining waste is not included in both diagrams. In a more general way, the primary sector, covering the extraction of raw materials including mining, mineral extraction, agriculture, fishing, and forestry, probably the most wasteful sector in our economy is, visually at least, excluded from both diagrams. This is completely misleading because any potential circularity or resource recovery improvement in the primary sector will have huge effects in the overall system by activating the domino effects. In a 2016 report [77] it was mentioned that worldwide almost 50% of the GHGs are related to primary materials; thus if we were able to reduce the emissions related to materials about 20–30% with circular economy strategies, we would be able closing half of the emission gap between existing commitments and the 1.5 °C pathway in 2030.

A second obvious problem is that both diagrams do not visualize the fact that in each stage of the product management like manufacturing, distribution, repair or refurbishment, and recycling, there are waste streams generated. In Figure 3.8 on the lower side, a modified diagram shows exactly these waste streams. A third problem is that both diagrams demonstrate circular flows, but they miss to visualize and thus to give the proper importance on the anthropogenic stocks that are almost half of the extracted resources, as we have already discussed.

Of course, it is pretty clear that any diagram is a visual simplification of the complex concept of circular economy, so there will be things that are not or cannot be illustrated. Still, the high importance of these diagrams in communication and awareness about circular economy makes some improvements necessary.

Both the diagrams have also been criticized for creating a misleading understanding regarding the relationship between technosphere and biosphere. Giampietro [30] underlines that "in the EMF definition of the circular economy, the concept of circularity refers only to 'products, components and materials', that is, flows under human control inside the technosphere. However, it does not explain how these flows can be recycled without using ecological processes from the biosphere (energy, water, land, biomass, minerals). Nor does it mention the natural resources embodied in the goods and services imported from abroad." It has also been mentioned that [78] "The diagram suggests the belief of a perfect natural environment where waste does not exist (biological wing) and that people are separate from nature and live in a wasteful technosphere (technical wing); rather than fully embracing the notion of ecosystem stewardship in which people are an integral part of the environment."

3.6.1 How About Composite Materials?

Researchers have also questioned the separation of "biological" and "technical" material flows. According to the narrative that supports the diagram on Figure 3.4, "technical" materials are finite materials, used in a closed-loop system through sharing, maintaining, reusing, remanufacturing, and recycling of products. Conversely, biological materials are renewable and organized in an open-loop system of resources cascading through subsequent steps of extraction, production of bio-based materials, energy recovery, and returning nutrients to the biosphere to feed the next cycle of primary produce. Velenturf and her colleagues in their paper "Circular economy and the matter of integrated resources" [78] highlight that "Large proportions of material flows contain composites and mixtures of organic (such as agricultural produce) and inorganic (such as metals) elements that are technically difficult and costly to separate. Composites and mixtures occur in the environment naturally, for example in the form of most sedimentary rocks (conglomerate and aggregate), mineral and metal ores, soils and living organisms. Materials can also be designed, consciously or subconsciously, to become integrated during extraction, production, consumption and disposal. Examples are acid/metal mine drainage, precious metal wastes (e.g. road dust and furnace linings), steel slag, car components, paint, sewage water and bioenergy residues (e.g. ashes and digestate)." They also propose a new circular economy diagram that according to their approach demonstrates better both the relationship with the supporting biosphere and the waste streams generated in the different phases, as shown in Figure 3.10.

3.6.2 The Importance of the System's Boundaries

The critiques on the distinction between technical and biological materials and the relationship between technosphere and biosphere bring also another interest point for discussion, the system's boundaries. The system's boundaries can be further analyzed in spatial–geographical and temporal boundaries. The spatial and geographical boundaries relevant to circular economy are directly linked with the globalization of production and the global supply chains that characterize many multinational products. Increased international trade in the last few decades has reduced poverty in many developing countries and raised living standards and purchasing power. At the same time, it has radically changed the footprint of waste management around the world.

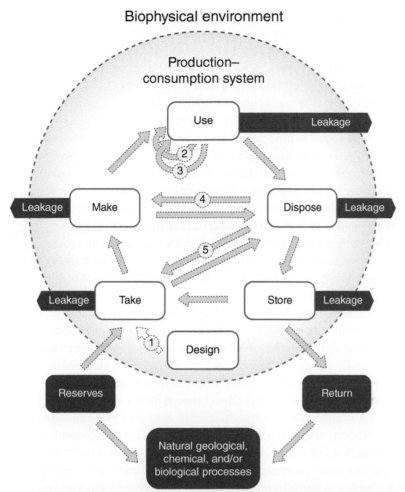

Biophysical environment

Production–consumption system

Use

Leakage

Make

Leakage

Leakage

Dispose

Leakage

Take

Leakage

Store

Leakage

Design

Reserves

Return

Natural geological, chemical, and/or biological processes

Figure 3.10 Integrated resource flow diagram for the circular economy (legend: thick arrows are natural materials, thin arrows are industrial materials, dotted arrow is immaterial; [1] prevention by designing out all avoidable wastes, [2] shared consumption, [3] reuse and repair, [4] remanufacturing, [5] recycling). *Source:* From Ref. [78], © 2019 The Authors, published by Elsevier B.V. under the CC BY license, adapted by Nick Rigas, D-Waste.

Until a couple of decades ago, products were dominantly manufactured near their areas of consumption, and wastes were managed near their source of production. Nowadays, vast amounts of mass-produced consumer products are being distributed worldwide. Solid waste management is linked increasingly to resource management, and so it has evolved into a complicated global network of material and recyclable waste flows, affecting various aspects of the environment and everyday life. In ISWA's report "Globalization and waste management" [79], it was clearly mentioned that there is a globalization footprint in every local waste management system and this footprint affects the composition of solid waste, the recycling markets, the waste trafficking networks, the resource value of waste in developing countries, and the governance patterns. It was concluded that this "global footprint should be taken into consideration when local efforts are made for a more sustainable waste management system. There is a need for further research to identify more specifically the globalization footprint and create appropriate ways to qualify it." The same is true for circular economy too; there is a globalization footprint that should be considered in each and every circular supply chain. Thus, it becomes clear that for certain products the circular economy either will become global or will not exist. So when the supply chains and the physical material flows cross multiple geographical and administrative boundaries, there is a serious question: who is going to take the responsibility to coordinate the interventions required to develop circularities? How we will avoid another round of problem displacement and shifting as it was the case with the exports of plastics to China that was counted as recycling in EU and the United States while most of them were burnt as a cheap fuel?

Korhonen raises the issue of temporal system's boundaries [49]. As he explains, the physical flows of materials and energy create both short-term and long-term environmental impacts, and this should be taken into account when designing reuse, remanufacturing, and recycling projects. The product durability might also create problems because as he mentioned "When the value and service is utilized many times, need for resource extraction for new products should be reduced. However, due to the fact that many of the impacts human mobilized material flows generate in nature are currently unknown, extending product lifetime might create economic and organizational structures that risk unsustainability in the long-term. This is if those products turn out to create negative impacts, impacts currently unknown. In such a situation, short lifetimes and continuous innovation and market penetration of new products might have an environmental advantage. Here is a conflict between product reuse and traditional recycling, which does not try and prolong the product life, rather utilizes its waste materials for the low-quality raw material value only."

Costas Velis has raised some relevant issues about global secondary material markets in his article "Circular economy and global secondary material supply chains" [80]. After explaining that for plastics recycled in Western countries the global secondary material supply chains are actually materializing recycling away from the place recyclables are collected, he concludes that in fact we are not able to know the environmental, occupational, and public health conditions under which plastics are reprocessed, and of course the recycling rate achieved. The second issue regards the complexity and volatility of global recycling markets and how much we really know to intervene successfully to them. He puts the question: "we anticipate benefits from circular economy, but for whom?" And he explains: "Waste electrical and electronic equipment (WEEE – or e-waste) often are illegally exported or legally exported as used functional equipment, just to end up in some of the most polluted places in the world... Academics have put forward the hypothesis that for such recovery of resources, the least environmental standards and lowest wages global pathway may materialize." His last phrase raises an important question that will be discussed in Chapter 4. "Such sub-standard global supply chains are a reason why the waste hierarchy is possibly an outdated, or at the very least insufficient, concept when it comes to today's global systems of resource recovery. Secondary and energy material savings, actual or perceived, cannot be at the cost of certain communities. Alignment of local and global sustainability can and should be perused."

3.7 THE END OF GROWTH AS WE KNOW IT

Circular economy aims to deliver a transformation of products and their associated material flows such that they form a supportive relationship with ecological systems and future economic growth. The mainstream narrative of circular economy is based on a main underlying assumption that a decouple between economic growth and resource extraction – use and their environmental impact – is possible. Circular economy, in general, seems to be embedded in a green growth narrative putting emphasis in new business development and market opportunities, more efficient ways of producing and consuming, and creation of local jobs. Economic growth, in GDP terms, is argued to be driven by "increased consumption due to correcting market and regulatory lock-ins that prevent many inherently profitable circular opportunities from materializing" [81]. Economic growth in terms of GDP means more consumption, and guess what: all the big companies, including the fanatic supporters of circular economy, make scenarios for their future based on the fact that consumption will continue to grow continuously (and they will do what is required to further stimulate it). A report that was issued by the World Resources Institute in March 2017 has the interesting title "The Elephant in the Boardroom: Why Unchecked Consumption is not an Option in Tomorrow's Markets" [82]. Here is an excerpt that explains who the elephant in the room is.

"Many CEOs speak 'sustainability', and many multinational companies have invested resources to build internal capacity on sustainability. It has become common for these companies to establish greenhouse gas emissions reduction targets and renewable energy goals and to address water risk and deforestation. However, underneath this welcome progress lies an uncomfortable truth: Most businesses' growth is still predicated on more people buying more goods. The world will have more than 9 billion people by 2050, and the middle class will have swelled by 3 billion by 2030. On top of this, consumer expectations for yet more are being stoked by trends such as fast fashion. The rapid expansion of consumption-driven markets in the coming decades is the anticipated engine for continued business growth. Without a change to current business models in which growth is predicated on selling more goods to more people, environmental stresses will pose increasing business risks and costs. Ultimately, these stresses will be a brake on business growth. Whether we look at consumer durables, fast-moving consumer goods, or consumables (this paper looks at all three), the pattern and risk of selling more stuff to more people is the same, and we see that improved practices are not sufficient to counteract anticipated global growth. Business growth predicated on consumption is the new elephant in the corporate boardroom. It is uncomfortable and unmentioned, both because the model has worked so well financially in the past and because addressing it challenges the traditional business model. Analysis of sustainability reports cited in this paper uncovers an alarming lack of attention to natural resource limits. The few quotes in this paper attributable to corporate spokespeople boldly referencing resource limitations are notable for their rarity." As rightly is pointed in this report, "Without a change to current business models in which growth is predicsated on selling more goods to more people, environmental stresses will pose increasing business risks and costs." So it is clear that from a resource perspective, the business models based on continuous increase of consumption are coming to an end, unless we will be able to find a way to decouple economic growth and increasing consumption from environmental problems.

For several years now, decoupling (breaking the link between economic growth and environmental impacts) is the new catchword in environmental debates. We can distinguish between relative decoupling (environmental impacts growing slower than the economy) and absolute decoupling (environmental impacts are declining, while the economy is growing). As a result, we can say that relative decoupling implies a gain in efficiency rather than removal of the link between impact and GDP. Analyzing the trends between 1990 and 2012, different countries demonstrate different results that support a relative decoupling between GDP growth and energy and/or material consumption [83]. OECD countries and Germany even create hopes for absolute decoupling. OECD countries GDP increased by almost 250% with a rather flat energy and material consumption. Germany almost doubled its GDP in real prices while reducing energy consumption by 10% and total material use by 40%. At the global level we can also observe only relative decoupling with the world GDP increased by almost 350% between 1990 and 2012, while energy and material use increased by 54 and 66% over the 22 years, respectively.

James Ward and his colleagues, in the research article "Is Decoupling GDP Growth from Environmental Impact Possible" [83], comment on these findings: "However, there are several limitations to the inference of decoupling from national or regional data. There are three distinct mechanisms by which the illusion of decoupling may be presented as a reality when in fact it is not actually taking place at all: 1) substitution of one resource for another; 2) the financialization of one or more components of GDP that involves increasing monetary flows without a concomitant rise in material and/or energy throughput, and 3) the exporting of environmental impact to another nation or region of the world (i.e. the separation of production and consumption…An additional mechanism of decoupling is associated with growing inequality of income and wealth, which can allow GDP to grow overall…without a proportional increase in material and energy flow (i.e. relative decoupling) when a wealthy minority of the population derives the largest fraction of GDP growth but does not necessarily increase their level of consumption with as much demand for energy and materials."

3.7.1 The End of "Green Growth"

The relative decoupling is of course more than welcomed, and certainly it helps to reduce pollution and health impacts providing more time to humanity to further mitigate the anthropogenic environmental impacts and up to a certain point rebalance its relationship with the supporting biosphere. However, as we have already discussed, in a world with finite resources and limited capacity in final sinks, the relative decoupling has only one impact: it delays the time required for crossing the existing physical limits. Going back to the metaphor (Figure 3.6) of the "Empty" and "Full" world, the relative decoupling makes the transition from "Empty" to "Full" longer in time, but of course it can't reverse it. Rammelt comments [63] that "Within a growth economy, the adoption of these engineering practices …at best… merely delays the time it takes to reach the boundaries of the biophysical envelope. At worst, the resource and energy savings generate profits that are reinvested in growth, which doesn't delay, but speeds up depletion and pollution." The ecological economist Herman Daly has famously said, commenting on the relative decoupling, that "It is true that in 1969 a dollar's worth of Gross National Product was produced with one-half the materials used to produce a dollar's worth of Gross National Product in 1900, in constant dollars. Nevertheless, over the same period total materials by consumption increased by 400 percent."

In July 2019, the European Environmental Bureau published its report [84] "Decoupling Debunked – Evidence and arguments against green growth as a sole strategy for sustainability." The report made a careful assessment to determine whether the scientific foundations behind this "decoupling hypothesis" are robust or not. The conclusion is that there is no empirical evidence supporting the existence of a decoupling of economic growth from environmental pressures on anywhere near the scale needed to deal with environmental breakdown. This is the case for materials, energy, water, GHGs, land, water pollutants, and biodiversity loss for which decoupling is either only relative, or observed only temporarily, or only locally. The report also lists seven reasons for which such decoupling appears unlikely to happen in the future. These reasons include the rising energy expenditure, rebound effects, problem shifting, the underestimated impact of services, limited potential of recycling, insufficient and inappropriate technological change, and cost shifting. By the way, this report has also a serious political consequence: it is the official announcement of the death of the so-called green growth. Green growth narrative was exclusively based on the fact that with gradual and radical improvements in environmental efficiencies and advanced technological solutions, we will be able to control and mitigate the environmental impacts of human activities without changing the systemic causes rooted in the economic system.

We already had a lot of examples that demonstrated that in several cases green growth created more and worst environmental impacts [85]. The green fuel drive has led to large areas in Borneo being cleared of forest in order to plant oil palm. This has led to devastation of crucial habitat for clouded leopards and orangutans, among many other species. Another issue is the reliance of much green technology upon rare earth metals, such as neodymium, which is mined at considerable environmental cost. Ethanol production requires more fossil fuel than it produces, while biofuels use the equivalent of ten acres for every car per year. Moreover, demand for biofuel has contributed directly to the loss of millions of acres of tropical forest that are replaced by soy fields for biofuel production. Even without forest destruction, essential farmland is being displaced for green energy production, putting huge pressure on food production in poor countries.

3.7.2 IND4.0 and the Religion of Continuous Growth

A world of continuous growth, as it is measured by the GDP, is a world that can evolve only in two ways: either by a catastrophe that will substantially reduce the population and/or the living standards (as it might be the case of global warming) or by a magic technological breakthrough that will allow unlimited economic growth with finite resources. This is where the narrative of IND4.0 comes. The truth is that as every industrial revolution, IND4.0 inspires new hopes and a kind of techno-optimism that we will be able to resolve the serious problems that our economic model, capitalism, has accumulated by revolutionary technology advances that will even allow abundance of resources for everyone. Giampietro comments [30] that "In the 1950s, we were promised that nuclear energy would produce electricity 'too cheap to meter'. In the 1970s, genetically modified crops were supposed to eradicate hunger from our vocabulary. In the 1980s, the hydrogen economy was going to solve our dependence on fossil energy. Having failed to do so, the same result was promised for the first generation of agro-biofuels in the 90s." Rammelt puts the issue in a more general way: "Environmental engineering has so far failed to bring about the level of absolute decoupling that is required to sustain the current economic system. Present expectations of dematerialization, recycling and loop-closing should be tempered by the fact that these engineering principles have been around for a very long time and that their environmental gains have been overwhelmed by economic growth. Several practices and concepts with strong engineering content nevertheless promise an absolute reduction in the environmental impacts of production and consumption systems in growth-based economies. For several reasons, this is a false promise."

If we are not able to separate economic growth from the growing environmental impacts, then there is another very important question. How about a world without continuous economic growth? 250 years of almost permanent GDP growth have created an impact: it seems almost impossible to imagine any other way than the one of continuous growth. After all, a rising per capita GDP, in this view, is equivalent to increasing prosperity. This is undoubtedly one of the reasons why GDP growth has been the single most important policy goal across the world for most of the last century. Such a response clearly has an appealing logic for the world's poorest nations that deserve to get out of the poverty traps and achieve better living standards. Tim Jackson, author of the book *Prosperity without Growth: Economics for a Finite Planet,* believes that the specific type of economic growth we experience is also responsible for the growing inequality and that a meaningful approach to prosperity should involve the end of poverty. But he puts the questions [86]: "But does the same logic really hold for the richer nations, where subsistence needs are largely met, and further proliferation of consumer goods adds little to material comfort? How is it that with so much stuff already we still hunger for more? Might it not be better to halt the relentless pursuit of growth in the advanced economies and concentrate instead on sharing out the available resources more equitably? In a world of finite resources, constrained by strict environmental limits, still characterized by 'islands of prosperity' within 'oceans of poverty', are ever-increasing incomes for the already-rich really a legitimate focus for our continued hopes and expectations? Or is there perhaps some other path towards a more sustainable, a more equitable form of prosperity?"

For some researchers, the answer to the questions above is "sustainable degrowth." Van den Bergh and Kallis mention [87] that "Sustainable degrowth goes also beyond decoupling material and energy use from growth (also referred to as 'dematerialization'), postulating that efficiency improvements alone are not sufficient and might be counterproductive. Limits and reductions in the scale of production and consumption are the key to achieving a future of low material use. Technological and knowledge progress is not to be arrested under sustainable degrowth but redirected from more to better. Denying the imperative of growth is not synonymous with turning back the clock to a fictitious pre-industrial, communal past. Sustainable degrowth is about constructing an alternative sustainable future. Research and technological innovations in a degrowth trajectory would involve innovations for consuming less through lifestyles, political measures and technologies which embody appropriate and chosen limits, rather than continuous innovation to spur consumption. Finally, degrowth is offered as a social choice, not imposed as an external imperative for environmental or other reasons." Other researchers believe that degrowth is also questionable as a concept and that the solution should combine growth in some sectors with degrowth in others. Janicke commented [88] that "It is also questionable, however, to say that giving up on economic growth as a paradigm is the necessary condition to tackle the environmental crisis. In actuality, solving such problems is about radical growth in environmental and resource-saving technologies. It is also about radical 'de-growth' in products and processes that undermine long-term living and production conditions."

No matter what the final answer is, the truth is till now we have only seen relative decoupling in rates that are not capable to prevent humanity from crossing the planetary limits. And although we can have many questions about the concept of degrowth, it's time to accept that the political narrative of the continuous economic growth supports and sustains the business as usual approaches that accelerate the speed towards the world's biophysical limits. Daly has described perfectly the religion of the continuous growth [89]:

> Growth is widely thought to be the panacea for all the major economic ills of the modern world. Poverty? Just grow the economy (that is, increase the production of goods and services and spur consumer spending) and watch wealth trickle down. Don't try to redistribute wealth from rich to poor, because that slows growth. Unemployment? Increase demand for goods and services by lowering interest rates on loans and stimulating investment, which leads to more jobs as well as growth. Overpopulation? Just push economic growth and rely on the resulting demographic transition to reduce birth rates, as it did in the industrial nations during the 20th century. Environmental degradation? Trust in the environmental Kuznets curve, an empirical relation purporting to show that with ongoing growth in gross domestic product (GDP), pollution at first increases but then reaches a maximum and declines.

Giampietro [31] refers to a famous quote of Boulding that nicely exposes this complete absence of awareness about the impact of the economy to biosphere between economists: "Anyone who believes exponential growth can go on forever in a finite world is either a madman or an economist."

It seems that we are already at the end of the economic growth as we know it. What comes next depends a lot on the answer to the following question: who will shape the content and enjoy the benefits of circular economy? This is the subject of the next paragraph.

3.8 CIRCULAR ECONOMY FOR WHOM?

On 21 June 2016, I posted a blog titled "Shaping the Social Footprint of Circular Economy." [90] In the opening paragraph I wrote: "the mainstream thinking about circular economy seems to ignore a well-established fact, that environmental, economic and social problems are interwoven in each and every country, region and community of our world... As a consequence, each and every intervention in resource management, and especially worldwide interventions related with global supply chains, will result in very specific social and economic impacts. If we want to boost circular economy, we need to dig deeper and discuss more the economic and social impacts." Circular economy is directly linked with improvements in resource efficiency. We know well that resource efficiency is not at all a neutral value. Advanced resource efficiency can be achieved either through technological and organizational improvements or through slavery, trafficking, and child labor.

King Leopold of Belgium had developed an interesting way to advance the rubber productivity in Congo by the end of nineteenth century. Leopold persuaded the world that he was acting for humanitarian reasons (actually he covered his activities under the scientific–philanthropic International African Society) and managed to become the owner of nearly a million square miles of unmapped jungle, 75 times the size of Belgium itself. Unfortunately for the people living in Congo, this was the time that Dunlop discovered the pneumatic tire for bicycles, driving a global boom of rubber use and prices. Leopold named his colony "The Congo Free State" and applied quotas for rubber for each and every village: each community was told to provide 10% of their number as full-time forced labor and another 25% part-time. The routine penalty for failing to bring in enough rubber was the severing of a hand. A Baptist missionary wrote a letter to *The Times* concluding "This rubber traffic is steeped in blood" [91]. One of the captains of the riverboats that were carrying rubber and ivory, Joseph Conrad, wrote the famous book *Heart of Darkness* describing the daily horror and the atrocities that decades of millions of people were suffering. In 1904, a report for the UK House of Commons suggested that at least three million people had died during the last 20 years due to the atrocities related the rubber quotas.

What happened with the rubber in Congo by the end of the nineteenth century happens again today, although without cutting hands, with cobalt. Cobalt is a critical metal for electric car batteries and mobile phones, and 70% of the global reserves are in Congo. Although the majority of Congo's cobalt comes from large mining sites, one-third of it is coming from 150 000 informal miners who dig by hand [92]. Recently, a landmark legal case was launched against Apple, Google, Tesla, Dell, and Microsoft [93]. In a lawsuit filed in Washington, DC, Congolese families describe how their children were driven by extreme poverty to seek work in large mining sites, where they were paid as little as $2 a day for backbreaking and dangerous work digging for cobalt rocks with primitive tools in dark underground tunnels. The lawsuit accuses the companies of aiding and abetting in the death and serious injury of children who they claim were working in cobalt mines in their supply chain.

Another famous example of cruel resource efficiency concerns the resource recovery policies applied in Nazi concentration camps. Clothes, hair, shoes, jewelry, and golden teeth were systematically recovered and utilized, while forced labor was used to build feedlots for pigs that were fed on kitchen trash as well as in the reutilization of all sorts of extracted secondary material. Anne Berg provides more details in her article "The Nazi rag-pickers and their wine: the politics of waste and recycling in Nazi Germany" [94]. Describing scrap collectors and salvaging practices inside Germany and in Nazi-occupied Europe, Berg's article argues that waste management and recycling were integral to the Nazi racial order and crystallized as central strategies to administer the chaos of war. Concentration camps were central in this effort as it is described here: "In the face of the Red Army's advance, the Reich's Raw Materials office requested that all textiles be transported to save them from destruction. Out of the 570 wagons stuffed with secondary materials extracted from the camps in Lublin and Auschwitz by February 1943, thirty-four wagons were filled with men's and women's clothing, 400 wagons with rags, 130 with bed feathers, five with mixed secondary materials and one wagon with women's hair. These figures, chilling as they are, only represent the material collected by the Reich's economic ministry. Other collections by the Volksdeutsche Mittelstelle [ethnic German centre], the Reich's Youth Leadership, the concentration camps and others added another 825 wagons of clothing of various kinds."

Heike Weber argues that, both during World War I and World War II, the German state resorted to "total recycling" programs to mobilize both people and resources and contribute to the preparation of the national economy as well as home front morale for "total war." At his article "Towards Total Recycling: Women, Waste and Food Waste Recovery in Germany, 1914–1939" [95], Weber explains that "National Socialist waste policies stand out as unique for the time. Not only did it systematically revalue waste work as condign service to the Volksgemeinschaft (national community), but it included utterly cruel and perverse aspects such as the Aryanisation of the scrap trade, the reutilization of personal belongings or even hair of concentration camp victims or a systematic exploitation of raw as well as waste resources in occupied countries." In another article [96] describing the development of a bone recycling network in Germany and France during World War II, Denton and Weber explain that "The Nazi recycling system relied heavily on the labor of women, children and forced labor and functioned along the lines of National Socialist racial ideology, leading to many inhuman, even perverse, consequences, such as reusing the belongings of concentration camp victims and even their hair or gold teeth." Berg is concluding her analysis on the waste policies of the Nazi regime with a remarkably tragic but true comment [94]: "Millions of volunteers, workers, soldiers and ordinary citizens rallied together to turn trash into 'treasure' with the single goal of sustaining a system that perpetuated grossly distorted hierarchies of waste and value, epitomized by the economy of Auschwitz, where human hair was woven into blankets for German soldiers as memories, lives and the future of millions were recklessly hoisted on to the trash heap of history."

3.8.1 The Social Footprint of Circular Economy Is Shaped by Corporate Interests

A circular economy based (at least partially) on slavery practices might be already a reality. The International Labour Organization estimates that 20.9 million people are victims of forced labor globally for the period 2002–2011. The Global Slavery

Index estimates for 2016 that there are 45.8 million people in some form of modern slavery, a population 5 million more than the population of California, the largest state in the United States. In Congo, as the report "Modern slavery in the Democratic Republic of Congo" [97] describes, "The 2013 Free the Slaves study of South Kivu finds that 866 of the total sample of 931 persons interviewed across all three mining sites (93 percent) were enslaved in one of more types of slavery. The 2011 Free the Slaves study of North Kivu finds that 40 percent of respondents interviewed in Bisie were found to be in confirmed situations of slavery, with an additional 10 per cent showing strong indicators of enslavement." You might think that this is an exaggeration but maybe you can search at the Web and you will easily find titles like "Recycling slavery gang jailed for 32 years" [98], "A worker at a vast Romanian dump smuggles in a Sky News camera to show how thousands of people are forced to scavenge at the site" [99], "Modern slavery at UK recycling company" [100], and other similar. Actually, the issue became so important in the United Kingdom that MRW launched a campaign to raise awareness of how forced labor and modern-day slavery can damage businesses in the UK waste management and recycling sector. Corin Williams, MRW editor, wrote [101] that the campaign "Root it out" is necessary after, based on information gathered and published articles, it was assessed that around two-thirds of victims of human trafficking are placed in a recycling facility at some point during their ordeal.

Alexandre Lemille, in a great post in Medium [102], addresses the social footprint of circular economy and writes that if we assume that we do want a circular economy that will bring better life for all, "…we might have to think beyond just a circular economy as it is designed today: with the same corporate powerful actors, in the same financial paradigm, replicating current human interactions and power relation. In a sea of challenges, building a circular economy with 'profit maximization' as — again — the same narrow-minded corporate objective and, without putting the people at its core first, might not deliver the intended gigantesque intentions that we say it will have on our planet and its inhabitants." He continues explaining that there are at least two missing dimensions in the way circular economy is designed today: (i) optimization of all resources, including humans, which means integrating end of inequality and unemployment in the circular economy narrative and considering poverty as an externality of the linear model, and (ii) distributed powers, which means "ensuring that 'success' encompasses all values that are created in a world of abundance where each decision has multiple ripple effects, thus, if rewarded well, that could be benefiting us all in symbiosis."

The truth is that the fanatic support provided to circular economy approaches by many big companies is somehow encouraging but it has stamped a footprint on the mainstream circular economy narrative in three ways:

First, by restricting its content almost exclusively on material flows and trying to make it compatible with the endless economic growth and the stimulation of further consumption. We already discussed about that in Section 3.7. In addition, in some cases, the support to circular economy can be very easily considered as another form of greenwashing. A famous company, one of the biggest plastic bottle producers in the world, commits $5 million/year to promote recycling of plastic bottles in the United States and $800 million/year for advertisement, again in the United States, to boost further consumption of its products. What do you think the result will be? More recycling or more consumption? The report "Stakeholder Views Report Enablers and Barriers to a Circular Economy" [103] stresses the fact that currently, operating on linear business models seems cheaper and more financially rewarding than adopting circular solutions. Virgin raw materials often cost less than secondary ones, and current taxation patterns are not supportive of the transition, "…meaning circular business models sometimes represent a less profitable choice for businesses. During the stakeholder exchanges… an interesting point emerged: for businesses, circular economy is not an end per se, but rather a means to an end, as it allows to gain business value through innovative and sustainable solutions."

Take also the case of Volkswagen, which, despite a huge marketing campaign touting environmental benefits and portraying the company as a green pioneer, shocked the world when it was revealed that the company fitted its vehicles with software that allowed the car's system to identify when it was being emissions tested and perform with lower engine power. As Eric Lane mentions [104], "…the Volkswagen scandal is much more than just another example of greenwashing. That is, the German automaker's use of software to deceive brings a novel technological aspect to greenwashing." Actually, this is where IND4.0 meets circular economy not to advance it but to fake it according to the most linear driver of our world: profitability and corporate power of big multinational companies.

Second, by disconnecting circular economy from its social impacts and the broader context of sustainability, even by hiding or graying that there will be winners and losers. As we already shown circular economy is not a neutral system, it will be materialized through a broader social–political framework, and there is no guarantee that the final results will be positive for the societies. The following questions [90] are not yet addressed, and usually they are ignored, in the relevant discussions:

- As an example, according to a McKinsey study, in EU almost 2 million new jobs are expected by the advances of circular economy and €600 billion in savings until 2030 [81]. These two million new jobs will be part-time activities paid with peanuts or full-time well-paid jobs? These €600 billion in savings will make the big and rich bigger and richer, or they will provide social benefits for the poorest part of the societies? Is circular economy going to create new local supply chains that will provide job opportunities and help communities to manage unemployment? Or it will be realized by corporate robots?

- Big multinational companies have already developed global supply chains. They have also the power, the know-how, and the economic interest to control (at least some of) the crucial resources globally, and they will do it in their own way. So, will the circular economy be utilized as one more brick in the wall that restricts access to crucial resources? Will the countries, especially the poorer ones, keep control of their crucial resources and ensure that citizens will have access to them?

- The new business model of sharing instead of owning goods, or the so-called sharing economy, is another breakthrough related to circular economy. Obviously, there are substantial environmental benefits when you share a car, but, finally, the owner that provides the specific service has full control of your mobility and, in case of monopolies, the owner can terminate the service whenever required according the owner's interest. How we will ensure that everyone will have access in these services and that they will be affordable? How we will frame the sharing economy with a proper legal framework to avoid overexploitation of the individual, decentralized labor force, like the Uber drivers?

- What exactly will be the role of civil society in the road towards circular economy? Is it going to be an active stakeholder as the final beneficiary or simply a consumer of "Big Ideas for Big Money"? Will the circular economy be shaped increasing the cohesion and the inclusivity of our societies, or it will create more fragmented and polarized continents, regions, cities, and neighborhoods? Will circular economy be implemented reducing the extreme inequality or increasing it? Will it result in a more balanced and better society, where democracy and human rights, especially the ones of the poorest part, will be fully respected or in an economic dictatorship of the ones that will control the crucial resources?

Third, by ignoring or underestimating the fact that circular economy will have serious geopolitical impacts. Beginning with trade, OECD warns [105] that the circular economy transition will likely introduce structural changes to the economy and may have potential impacts on trade flows. Import and export demand for primary materials, secondary materials, and waste may decrease in certain economies. At the same time, the circular economy transition may bring new opportunities for trade in services. According to the OECD report "International Trade and the Transition to a More Resource Efficient and Circular Economy: A Concept Paper," trade can provide potential opportunities towards a global circular economy by channeling waste and materials to destinations where there is comparative advantage in sorting and processing these materials. Unnecessary trade barriers such as import and export restrictions on waste and scrap should be avoided to the extent possible. The report states clearly the risk that, unless proper global cooperation takes place, this might be another form of waste trafficking, following the minimum standards – minimum cost pathway. The report concludes that "there are a number of critical areas at the interface between trade and circular economy that merit further data collection and investigation on definitions and classifications. These include trade in waste, secondary materials, secondhand goods, and goods for refurbishment and remanufacturing. International co-operation on circular economy value chains could be explored for possible harmonization of quality standards of materials, promoting demand for second-hand goods and secondary raw materials, to remove unnecessary regulatory barriers, and to avoid environmentally harmful activities." A report by UNCTAD concludes that "Regions exporting scrap material need to ensure compliance with health protection standards and that such material is optimally prepared for recycling. Countries importing recyclable materials may consider stimulating high-quality recycling and remanufacturing that are safe for workers and the environment. It is also important to differentiate between waste and used and remanufactured goods, creating clear protocols for their acceptability at national borders."

Preston noticed [106] that "A transition to a circular economy brings certain trade-offs that require careful management. In the absence of a coordinated and strategic approach to the circular economy at national or international level, there is the risk that companies will adopt tokenistic – or, at worst, harmful – activities under the umbrella of the circular economy which preclude more sustainable or higher-value material use. Waste-to-energy initiatives using sub-standard incineration practices, for example, may bring environmental and human health risks and may also be drawing on waste streams better suited to second-life products. Trade-offs may also arise where circular solutions imply significant shifts in industrial policy: in resource-intensive economies, for example, circular approaches can support value addition but may also risk job losses among those employed in resource extraction and primary processing."

The applicability of circular economy in developing countries is also another important aspect. It is true that developing countries are facing a growing waste crisis, which has major consequences for environmental and health outcomes. In this view, a circular economy strategy could help developing countries to follow a more sustainable pathway for their development, avoiding the resource-intensive Western paradigm. As Preston and Lehne [13] noticed, "Lower-income countries are in many ways more 'circular' than their developed- economy counterparts – the question is how to turn this into a development opportunity. Much economic activity in lower-income countries revolves around sorting and reusing waste. However, higher-value, employment-generating opportunities for reuse and remanufacturing are yet to be captured."

This is clearly a window of opportunity where development aid can stimulate circular strategies for low-income countries. This window of opportunity has to be considered in accordance with the shifting geopolitical balance in favor of the Global South. Schroder describes [40] the situation like this: "While historically, countries in the global north seemed to hold all the power in the domain of transnational waste dumping and trade, the situation is rapidly evolving with several countries in the global south instituting domestic policies to protect local environments. In 2018, China banned waste imports as the country is waging 'war on pollution.' Before that, in July 2016, the east African countries like Rwanda, Kenya, Tanzania and Uganda decided to increase tariffs on imported second-hand textiles as cheap clothes from abroad were threatening local industries...As the situation of transnational waste dumping and trade demonstrates, a circular economy cannot be achieved in a vacuum, but only through collaboration between countries. Transferring waste (or responsibility) to other countries is not possible in the long run as this would simply mean moving pollutants to 'other corners' of our world, an unjust solution in the short-term, and an unsustainable one in the longer term."

On the other side, the Global North, through its globalized multinational companies, controls almost all the important global value chains. This means that higher-value, employment-generating opportunities like product design, marketing, and retail are sited in the North, while activity with lower economic value revolving around sorting, reusing, and recycling waste is relegated to lower-income countries. Wilson, in a TEARFUND report, highlights [107] that the design and implementation of eco-design policies in Europe could "determine the success or failure of the circular economy in Low Income Countries. Many of these circular industries are fed by imports of waste and used products from European and other high-income nations. For example, estimates suggest that more than 90 per cent of discarded computers in high-income countries are exported to low- or middle- income countries."

3.8.2 Time to Rethink Governance

It is clear that circular economy is challenging not only the business models worldwide but also the governance patterns. The implementation of the circular economy cannot be limited to simple procedures of reduction, reuse, and recycling. And the urgency to implement circular economy as the primary way to reduce carbon dioxide emissions and mitigate global warming does not allow us to wait until business interests and markets will (if) decide to develop a bottom-up circular economy system that will gradually become global after 200 years. Potting and his colleagues [108] have noticed that the actual circular economy transition should lead to closing cycles at the level of individual products, i.e. in the related product chains. The transition process may differ across products and between circularity strategies, where lower circularity strategies are still closer to a linear economy and higher circularity strategies are closer to the circular economy. Technological innovation, they note, is mainly relevant for lower circularity strategies, whereas "socio-institutional changes become more important for higher circularity strategies increasingly involve transforming the whole product chain (i.e. systemic changes). Socio-institutional changes refer to differences in how consumers relate to products, how all actors in a product chain cooperate to achieve circularity, and all institutional arrangements needed to facilitate this."

This brings the aspect of governance as a major challenge. The implementation of a circular economy requires serious changes on institutional levels [109] that will allow us to overcome barriers like the internalization of externalities such as carbon emissions, regulations to overcome imperfect information on circular economy alternatives, and education and incentives to overcome cultural lock-in to linear economy business models and user practices. Attention to social dimensions is also important to avoid potential unintended consequences and to ensure that the social benefits of circular economy solutions are spread widely.

The report "Governance for the Circular Economy" [110] describes the building blocks of governance for a circular economy. It stresses the central role of governance for circular economy, and it highlights the importance of building relationships with value and supply chains and between producers and customers. In addition, it demonstrates the need to follow circularities in every step of the supply chain and to utilize authentic experiences and in situ observations for a circular economy governance. Figure 3.11 shows the governance framework for regenerative systems. And it's a good example of the serious, complex, and nonlinear changes required to address the challenges of circular economy.

Before closing this paragraph, we feel it is important to say few more words about the role of labor and the importance of institutional structures. A report by the European Commission regarding the impacts of circular economy to employment [111] concluded that by moving towards a more circular economy, GDP in the EU increases by almost 0.5% by 2030 compared with the baseline case. The net increase in jobs is approximately 700 000 compared with the baseline through additional labor demand from recycling plants, repair services, and rebounds in consumer demand from savings generated through collaborative actions. While the net impacts on EU employment are positive, the sectoral composition of employment will change. Sectors that produce and process raw materials will decline in size, while the recycling and repair sectors will experience additional growth. Services sectors and electricity are also expected to grow in employment. Sectors that could lose out include those that produce durable goods such as agriculture, forestry and wood products, extraction (energy) and manufactured fuels, extraction (non-energy), chemical food manufacturing, metals, plastics, electronics, nonmetallic minerals, motor vehicles (including sales), electronics, machinery, cars, and accommodation. Construction sector employment is expected to fall from productivity gains as a result of new building techniques. Although the figures above are positive, they do not represent a huge opportunity that will change the conditions of labor or the

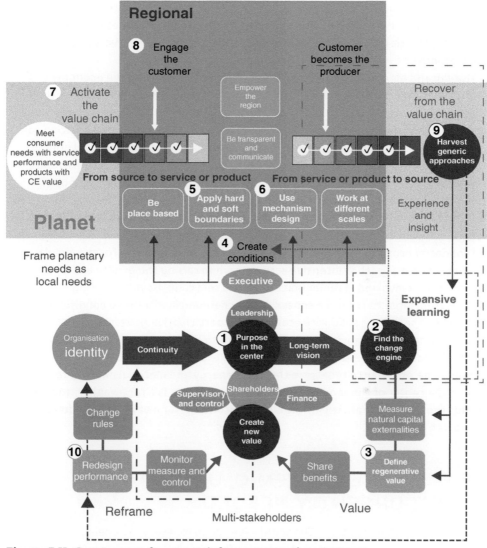

Figure 3.11 Governance framework for regenerative systems.
Source: From Ref. [110], with kind permission from Origame.

total economic system – they are rather a minor improvement that will happen in 10 years. A review of the assumptions of the study confirms that this approach to circular economy downgrades it to a minor business as usual improvement and lacks the systemic change required. As it has been noticed, efficiency and profitability can be achieved without necessarily reducing energy use in absolute terms. Similarly, maintaining products and materials at the highest potential value through reuse, remanufacture, or recycling means that cost-effectiveness underlies these circular economic activities, possibly at the expenses of lower energy intensity and higher labor intensity.

Moreau and his colleagues commented [112] that circularizing the economy through material recycling would need to go beyond economically viable efforts alone and the externalization of costs to future generations. In fact, circular economy may require a significant increase in material throughput and addition to stocks before closing material cycles, in particular within the current institutional conditions. Greater throughput, however, runs against the biophysical constraints of natural sources and sinks and the tacit hypothesis that smaller throughput of primary resources means lower environmental and possibly social impacts. To tackle this contradiction, several economists have emphasized the need to reestablish labor at the core of the economy given its renewable nature. As such "the social and solidarity economy is an instructive and constructive example for the circular economy, increasing labor-intensive activities while raising the quality and diversity of human work involved in remanufacturing and recycling. Beyond shifting the bulk of taxes from labor to resource consumption, changing institutional conditions in support of social and solidarity economy and participative governance would ensure a more suitable environment for cultivating both biophysical resources and human labor" [112]. So let me rephrase it: in the transition to the circular economy, the only way to rebalance the extra energy and material needs is to put labor in the center of the required change, because labor is a renewable resource. But this requires another view of the circular economy, far away from the dominant technocratic approach.

3.9 IT'S HUGE, SYSTEMIC, UNCERTAIN BUT URGENTLY NEEDED

Mahatma Gandhi famously said, "The world has enough for everyone's need, but not enough for everyone's greed." Karl Marx has written that "All progress in increasing the fertility of the soil for a given time, is a progress towards ruining the lasting sources of that fertility." Both of those phrases demonstrate the challenges of circular economy and the barriers we have to overcome if we want to give the right answer to the trillion-dollar question, as we presented it in Chapter 1.

We hope that now it becomes clear that circular economy is a huge systemic change that will not happen by wishful thinking nor by big events in five-star hotels. It is a fundamental change that requires a serious rearrangement of the foundations of our economic system, and this is probably the biggest barrier:

profits are still linear and the more linear profits we have, the stronger the business interests that continue the linear business model. Actually, one of the key messages of this chapter is that market itself will never promote a circular economy as it is required to avoid crossing the biophysical limits. In the best case, market itself will deliver a kind of circular economy that will allow the global players to control material flows of special interest, in a form of circular privatization of critical materials. In the worst case, circular economy will be used for greenwashing, and after some years it will be substituted by another term, so a new round of greenwashing will start.

Sir Nicholas Stern, the author of one of the most influential reports about global warming, said that [113] "The problem of climate change involves a fundamental failure of markets: those who damage others by emitting greenhouse gases generally do not pay...Climate change is a result of the greatest market failure the world has seen." In the same way of thinking, we can already say that marine litter is the result of one of the greatest market failures the world has seen. We can already say that the loss of 60–70% of biodiversity is another result of one of the greatest market failures the world has seen. As the years go by, without changing the dominant economic model, the list of market failures will only get bigger.

The problem is that this is a failure only if you expect the market to resolve the problem, as unfortunately most of the political elites believe. Maybe it's time to move away from this belief and try to see what the global community and each country can do, considering the markets but without bowing to them. This means that the implementation of a circular economy is a matter of political struggle, both regarding the content of it (for whom?) and the speed and the depth of the interventions. The logical consequence is that the project circular economy, besides huge and systemic, is also uncertain. We can speak not only for the political uncertainty involved but, as we tried to show in this chapter, for the great efforts required to transform good intentions to a concrete scientific paradigm, capable to be implemented with measurable results.

Still, circular economy is urgently needed to avoid the crossing of biophysical limits and the risks of a gradual but broad collapse of human societies. But we have to be careful. The risks related to the impacts of the continuously increasing global resource extraction are more like a slope in which we slide down rather than a cliff that we approach fast. The recent misleading discussion about the 12-year window to prevent climate change has a lot to teach us about the way we phase and perceive the risks. As Kate Marvel, a NASA climate scientist, said [114], "12 years isn't a deadline, and climate change isn't a cliff we fall off – it's a slope we slide down. We don't have 12 years to prevent climate change – we have no time. It's already here. And even under a business-as-usual scenario, the world isn't going to end in exactly twelve years." But it will become much worst, I would add....

Key Take-Outs of Chapter 3

Key take-outs	Why is it important?
Circular economy is a global trend that is promoted mostly by EU and China, but it is also supported by other governments and by some of the biggest companies in the world. Recently, even developing countries are focusing on the concept and develop relevant initiatives	As circular economy becomes a subject on global agendas, its geopolitical impacts should be further studied, and its suitability for the developing world must be examined in depth
Reuse of items, systematic repair, storage of equipment, materials or spare parts for future use, creative rearrangements, and fit-for-purpose applications for specific needs was the rule and not the exception in the daily lives of the vast majority of people for hundreds of years	Single-use and easy disposability practices became mainstream after World War II. Besides the massive production that was made possible due to the technological advances, massive consumption was stimulated by massive advertising campaigns that promoted single use and easy disposability as modern, convenient, and timesaving solutions
Industrial revolutions skyrocketed resource, labor, and energy productivity and created the linear economy, as we know it today. Its main characteristic is the dependence on nonrenewable resource flows, like fossil fuels	As we are experiencing IND4.0, a further linearization of the economy should be expected, unless humans will be able to shift to circular economies that will rebalance our working relationship with the biosphere and the supporting ecosystems
There are 114 different definitions of circular economy, and there is a lack of consensus regarding the definitions and the terminology used by the various actors. That creates blurriness and makes the different initiatives noncomparable. All definitions aim to decouple economic growth from resource consumption. A lot of the definitions define circular economy like a more advanced recycling practice and lack any link with the broader context of sustainability	Many believe that there is a need to develop a commonly accepted and internationally recognized definition about circular economy that will allow academics, practitioners, and policy makers to communicate in a common language, required to make the implementation of circular economy more pragmatic. However, different and even competitive narratives will still reflect the need to adapt and adopt circular economy in local contexts
In any case, the laws of thermodynamics and the planetary biophysical limits are the boundaries that determine what can be achieved by circular economy	Circular economy has to address several scientific challenges in order to become a scientific paradigm capable to be implemented and deliver measurable results

Key take-outs	Why is it important?
Circularity is not sustainability, and circular flows are not necessarily always preferable to linear ones. The energy requirements, the barriers to recycling, and the rising anthropogenic stocks limit the potential gains from circular flows	There is an urgent need to develop new tools capable to evaluate the feasibility of circularities in a multidimensional way. These tools should incorporate the science of complexity
Circular economy should be integrated in different geographical and spatial levels, while in some industries it should be also of global coverage	We lack mechanisms of global and regional coordination and governance, and we also lack intervention tools to global–international markets
Relative decoupling is evident, but there is no sign that absolute decoupling happens, or that it can happen in the future – technological advances can't pursue absolute decoupling	It is urgent to rethink the dominant paradigm of continuous economic growth and find alternative ways to advance and measure prosperity
The mainstream definitions refer to circular economy as an ambitious new economic paradigm, on global, national, or company level that combines economic growth with environmental protection, ignoring the social footprint of this transformation	This is the impact of the business sector on the shape of circular economy, and it represents a serious risk to disconnect circular economy from the social dimensions of sustainability
Circular economy is directly linked with improvements in resource efficiency. Resource efficiency is not neutral. Advanced resource efficiency can be achieved either through technological and organizational improvements or through slavery, trafficking, and child labor	There is a need to put the question circular economy for whom? The social and geopolitical dimensions of circular economy require further research in order to be understood and integrated in the concept
A real circular economy requires systemic, global, and fundamental changes to the roots of capitalism. Markets themselves can't promote circular economy because the profits remain still linear	The implementation of a circular economy is a matter of political struggle, both regarding the content of it (for whom?) and the speed and the depth of the interventions

REFERENCES

1. European Commission (2019). Report from the Commission to the European Parliament, The Council, The European Economic and Social Committee and the Committee of the Regions on the Implementation of the Circular Economy Action Plan; SWD(2019) 90 final; Brussels, 2019.
2. European Commission (2016). Circular economy strategy: environment. https://ec.europa.eu/environment/circular-economy (accessed 28 December 2019).
3. European Union (2018). European Circular Economy Stakeholder Platform: A Joint Initiative by the European Commission and the European Economic and Social Committee https://circulareconomy.europa.eu/platform (accessed 28 December 2019).
4. Heshmati, A. (2015). *A Review of the Circular Economy and Its Implementation*, DP No. 9611, 63. IZA.
5. Lehman, M., de Leeuw, B., and Fehr, E. (2014). *Circular Economy Improving the Management of Natural Resources*. Swiss Academy of Arts and Sciences.
6. Institut Montaigne (2016). *The Circular Economy: Reconciling Economic Growth with the Environment*. Paris: Institut Montaigne.
7. Zhu, J., Fan, C., Shi, H., and Shi, L. (2019). Efforts for a circular economy in China: a comprehensive review of policies: China's Circular Economy Policy. *Journal of Industrial Ecology 23* (1): 110–118.
8. EC (2019). EU and China Step up Their Cooperation on Environment, Water and Circular Economy; Press release; Brussels https://ec.europa.eu/info/news/eu-and-china-step-their-cooperation-environment-water-and-circular-economy-2019-apr-01_en (access 1 April 2019).
9. Wuyts, W., Miatto, A., Sedlitzky, R., and Tanikawa, H. (2019). Extending or ending the life of residential buildings in Japan: a social circular economy approach to the problem of short-lived constructions. *Journal of Cleaner Production 231*: 660–670.
10. Circular Colab. *The State of the Circular Economy in America Trends, Opportunities, and Challenges*. Circular Colab.
11. Fedotkina, O., Gorbashko, E., and Vatolkina, N. (2019). Circular economy in Russia: drivers and barriers for waste management development. *Sustainability 11* (20): 5837.
12. Canadian Council of Ministers of the Environment (2019). Canada-Wide Action Plan on Zero Plastic Waste Phase 1. Action Plan PN 12892019, p. 9.
13. Preston, F. and Lehne, J. (2017). *A Wider Circle? The Circular Economy in Developing Countries; Energy, Environment and Resources; Briefing*, 24. London: Chatham House.
14. FLOOW2 (2013). Ellen MacArthur Foundation launches new program (CE100). https://www.tilburgdeelt.nl/news-detail/~/items/ellen-macarthur-foundation-launches-new-program-ce100-141.html (accessed 28 December 2019).
15. Accenture (2014). *The UN Global Compact & Accenture, CEO Study on Sustainability: Industry Insight: Mining and Metals*. New York: UN Global Compact.

16. Martin, B. (2015). *Circular Economy: Trends and Emerging Ideas*. Vienna, Austria: ISWA Task Force on Resources; International Solid Waste Association ISWA.

17. Ghisellini, P., Cialani, C., and Ulgiati, S. (2016). A review on circular economy: the expected transition to a balanced interplay of environmental and economic systems. *Journal of Cleaner Production 114*: 11–32.

18. Bourguignon, D. and Orenius, O. (2018). *Material Use in the European Union Towards a Circular Approach*, European Parliamentary Research Service; Briefing PE 625.180. European Parliament.

19. House of Commons Environmental Audit Committee (2014). *Growing a Circular Economy: Ending the Throwaway Society*, Third Report of Session 2014–15. London: House of Commons.

20. Tóth, G. (2019). Circular economy and its comparison with 14 other business sustainability movements. *Resources 8* (4): 159.

21. Barles, S. (2014). History of waste management and the social and cultural representations of waste. In: *The Basic Environmental History*, vol. *4* (eds. M. Agnoletti and S. Neri Serneri), 199–226. Cham: Springer International Publishing.

22. Melosi, M.V. (2005). *Garbage in the Cities: Refuse Reform and the Environment*. University of Pittsburgh Press.

23. Barles, S. (2017). *A Metabolic Approach to the City: Nineteenth and Twentieth Century Paris. In Resources of the City*, 300. Taylor & Francis.

24. Wilson, D.C. (2007). Development drivers for waste management. *Waste Management and Research 25* (3): 198–207.

25. Circular Academy. Circular economy: origins of the concept. http://www.circular. academy/circular-economy-some-definitions-2 (access 17 April 2020).

26. Susan, S. (1999). *Waste and Want: A Social History of Trash*. New York: Macmillan.

27. Holt, D. and Littlewood, D. (2017). Waste livelihoods amongst the poor: through the lens of bricolage waste livelihoods. *Business Strategy and the Environment 26* (2): 253–264.

28. Desrochers, P. (2002). Regional development and inter-industry recycling linkages: some historical perspectives. *Entrepreneurship and Regional Development 14* (1): 49–65.

29. Buranyi, S. (2018). The plastic backlash: What's behind our sudden rage – and will it make a difference? *The Guardian* (13 November 2018).

30. Giampietro, M. (2019). On the circular bioeconomy and decoupling: implications for sustainable growth. *Ecological Economics 162*: 143–156.

31. Giampietro, M. (2011). *The Metabolic Pattern of Societies: Where Economists Fall Short*, 1e. Routledge.

32. Steffen, W., Broadgate, W., Deutsch, L. et al. (2015). The trajectory of the Anthropocene: the great acceleration. *The Anthropocene Review 2* (1): 81–98.

33. Crutzen, P.J. (2006). The "Anthropocene". In: *Earth System Science in the Anthropocene* (eds. E. Ehlers and T. Krafft), 13–18. Berlin, Heidelberg: Springer.

34. Engelman, R. (2013). Beyond sustainababble. In: *State of the World 2013: Is Sustainability Still Possible?* (ed. L. Starke), 3–16. Washington, DC: Island Press/ Center for Resource Economics.
35. Geissdoerfer, M., Savaget, P., Bocken, N.M.P., and Hultink, E.J. (2017). The circular economy: a new sustainability paradigm? *Journal of Cleaner Production 143*: 757–768.
36. Rome, A. (2015). The launch of spaceship earth. *Nature 527*: 443–445.
37. Pearce, D. and Turner, K. (1990). *Economics of Natural Resources and the Environment*. Baltimore: Johns Hopkins University Press.
38. McDonough, W. (2002). *Cradle to Cradle: Remaking the Way We Make Things*, 1e. New York: North Point Press.
39. Benyus, J. and Biomimicry, M. (1997). *Innovation Inspired by Nature*. New York: Quill.
40. Schröder, P., Anantharaman, M., Anggraeni, K. et al. (2019). Introduction: sustainable lifestyles, livelihoods and the circular economy. In: *The Circular Economy and the Global South*, Pathways to Sustainability, 1e (eds. P. Schröder, M. Anantharaman and K. Anggraeni). London: Routledge.
41. Interreg Danube Transnational Programme MOVECO. The circular economy concept and its schools of thought. http://www.interreg-danube.eu/uploads/ media/approved_project_public/0001/20/34dbda672272975f4faa289f3041cb712a c0f925.jpeg (accessed 17 February 2020).
42. Ellen MacArthur Foundation (2013). *Towards the Circular Economy. Economic and Business Rationale for an Accelerated Transition*. Ellen MacArthur Foundation.
43. European Commission (2015). *Closing the Loop: An EU Action Plan for the Circular Economy; Communication from the Commission to the European Parliament, the Council, the European Economic and Social Committee and the Committee of the Regions COM(2015) 614 Final*. Brussels: European Commission.
44. Rizos, V., Tuokko, K., and Behrens, A. (2017). *A Review of Definitions, Processes and Impacts*, Circular Impacts; Research 2017/8, 44. Brussels: CEPS.
45. ISWA (2015). *Task Force on Resource Management. Circular Economy: Resources and Opportunities*. Vienna, Austria: International Solid Waste Association ISWA.
46. Homrich, A.S., Galvão, G., Abadia, L.G., and Carvalho, M.M. (2018). The circular economy umbrella: trends and gaps on integrating pathways. *Journal of Cleaner Production 175*: 525–543.
47. Kirchherr, J. and van Santen, R. (2019). ISWA: 114 + 1 definitions for a circular economy – finding a common language. *ISWA President's blog*, 2019.
48. Kirchherr, J., Reike, D., and Hekkert, M. (2017). Conceptualizing the circular economy: an analysis of 114 definitions. *Resources, Conservation and Recycling 127*: 221–232.
49. Korhonen, J., Honkasalo, A., and Seppälä, J. (2018). Circular economy: the concept and its limitations. *Ecological Economics 143*: 37–46.
50. McCarthy, A., Dellink, R., and Bibas, R. (2018). *The Macroeconomics of the Circular Economy Transition: A Critical Review of Modelling Approaches*, OECD Environment Working Papers 130. OECD: Paris.

51. Scoones, I. (2015). *Sustainable Livelihoods and Rural Development*, Agrarian Change and Peasant Studies Series. Rugby: Practical Action Publishing.
52. Ellen MacArthur Foundation. Circular economy system diagram. Ellen MacArthur Foundation, SUN, and McKinsey Center for Business and Environment: Drawing from Braungart & McDonough, Cradle to Cradle (C2C). https://www.ellenmacarthurfoundation.org/circular-economy/concept/infographic (accessed 17 February 2020).
53. Rockström, J., Steffen, W., Noone, K. et al. (2009). Planetary boundaries: exploring the safe operating space for humanity. *Ecology and Society 14* (2) https://doi.org/10.5751/ES-03180-140232.
54. Daly, H.E. (1990). Toward some operational principles of sustainable development. *Ecological Economics 2* (1): 1–6.
55. Goodland, R. and Herman, D. (1990). The missing tools (for sustainability). In: *Planet under Stress: The Challenge of Global Change* (eds. C. Mungall and D.J. McLaren), 269–282. Toronto: Oxford.
56. World Bank. GDP (constant 2010 US$) | Data. https://data.worldbank.org/indicator/NY.GDP.MKTP.KD (accessed 8 February 2020).
57. Ball, J. (2018). The emerging seven countries will hold increasing levels of global economic power by 2050. Global Security Review, 2018.
58. Wikipedia (2020). Deepwater Horizon oil spill.
59. Desjardins, J. (2017). Visualizing the World's deepest oil well. https://www.visualcapitalist.com/visualizing-worlds-deepest-oil-well (accessed 8 February 2020).
60. Turner, G.M. (2008). A comparison of the limits to growth with 30 years of reality. *Global Environmental Change 18* (3): 397–411.
61. Meadows, D. (2004). *Limits to Growth*. Chelsea Green Publishing.
62. West, G.B. (2018). *Scale: The Universal Laws of Growth, Innovation, Sustainability, and the Pace of Life in Organisms, Cities, Economies, and Companies*. Penguin Press.
63. Rammelt, C. and Crisp, P. (2014). A systems and thermodynamics perspective on technology in the circular economy. *Real-World Economics Review 68*: 25–40.
64. Nielsen, S.N. and Müller, F. (2009). Understanding the functional principles of nature: proposing another type of ecosystem services. *Ecological Modelling 220* (16): 1913–1925.
65. Paul, B. (2013). Cycles, spirals and linear flows. *Waste Management and Research 31* (1) https://doi.org/10.1177/0734242X13501152.
66. Turner, D.A., Williams, I.D., and Kemp, S. (2015). Greenhouse gas emission factors for recycling of source-segregated waste materials. *Resources, Conservation and Recycling 105*: 186–197.
67. Bartl, A. (2015). *Circular Economy: Cycles, Loops and Cascades*; ISWA Task Force on Resources. Vienna, Austria: ISWA.
68. Velis, C. (2015). *Circular Economy: Closing the Loops*; ISWA Task Force on Resources. Vienna, Austria: ISWA.

69. Iacovidou, E., Millward-Hopkins, J., Busch, J. et al. (2017). A pathway to circular economy: developing a conceptual framework for complex value assessment of resources recovered from waste. *Journal of Cleaner Production 168*: 1279–1288.
70. Krausmann, F., Lauk, C., Haas, W., and Wiedenhofer, D. (2018). From resource extraction to outflows of wastes and emissions: the socioeconomic metabolism of the global economy, 1900–2015. *Global Environmental Change 52*: 131–140.
71. Schiller, G., Müller, F., and Ortlepp, R. (2017). Mapping the anthropogenic stock in Germany: metabolic evidence for a circular economy. *Resources, Conservation and Recycling 123*: 93–107.
72. Baccini, P. and Brunner, P. (2012). *Metabolism of the Anthroposphere*, 2e. Cambridge MA: MIT Press.
73. Fellner, J., Lederer, J., Scharff, C., and Laner, D. (2017). Present potentials and limitations of a circular economy with respect to primary raw material demand: present potentials and limitations of a circular economy. *Journal of Industrial Ecology 21* (3): 494–496.
74. Haas, W., Krausmann, F., Wiedenhofer, D., and Heinz, M. (2016). How circular is the global economy? A sociometabolic analysis. In: *Social Ecology* (eds. H. Haberl, M. Fischer-Kowalski, F. Krausmann and V. Winiwarter), 259–275. Cham: Springer International Publishing.
75. Velte, C.J. and Steinhilper, R. (2016). Complexity in a circular economy: a need for rethinking complexity management strategies. *Proceedings of the World Congress on Engineering 2016 Vol II WCE 2016*, London (29 June to 1 July 2016).
76. EC (2017). Waste statistics: statistics explained. https://ec.europa.eu/eurostat/statistics-explained/index.php/Waste_statistics#Total_waste_generation (accessed 30 December 2019).
77. Circle Economy – Ecofys (2016). Implementing circular economy globally makes Paris targets achievable. p. 18.
78. Velenturf, A.P.M., Archer, S.A., Gomes, H.I. et al. (2019). Circular economy and the matter of integrated resources. *Science of the Total Environment 689*: 963–969.
79. Task Force on Globalization and Waste Management (2014). *Globalisation and Waste Management*, 36. Vienna, Austria: Task Force on Globalization and Waste Management; International Solid Waste Association ISWA.
80. Costas, V. (2015). Circular economy and global secondary material supply chains. *Waste Management and Research 33* (5): 389–391.
81. Ellen MacArthur Foundation (2015). *Growth Within: A Circular Economy Vision for a Competitive Europe*. Ellen MacArthur Foundation.
82. Pino, S.P.D., Metzger, E., Drew, D., and Moss, K. (2017). *The Elephant in the Boardroom: Why Unchecked Consumption Is Not an Option In Tomorrow's Markets*, Working Paper, 36. Washington, DC: World Resource Institute.
83. Ward, J.D., Sutton, P.C., Werner, A.D. et al. (2016). Is decoupling GDP growth from environmental impact possible? *PLoS One 11* (10): e0164733.

84. Parrique, T., Barth, J., Briens, F. et al. (2019). *Decoupling Debunked: Evidence and Arguments Against Green Growth as a Sole Strategy for Sustainability*, 80. European Environmental Bureau.

85. Murray, A., Skene, K., and Haynes, K. (2017). The circular economy: an interdisciplinary exploration of the concept and application in a global context. *Journal of Business Ethics 140* (3): 369–380.

86. Tim, J. (2009). *Prosperity Without Growth Economics for a Finite Planet*. London; Sterling, VA: Earthscan.

87. van den Bergh, J.C.J.M. and Kallis, G. (2012). Growth, a-growth or degrowth to stay within planetary boundaries? *Journal of Economic Issues 46* (4): 909–920.

88. Jänicke, M. (2012). "Green Growth": from a growing eco-industry to economic sustainability. *Energy Policy 48*: 13–21.

89. Daly, H.E. (1991). *Steady-State Economics: Second Edition with New Essays*. Island Press.

90. Mavropoulos, A. (2016). *Shaping the Social Footprint of Circular Economy*. Wasteless Future.

91. Valley, P. (2006). Forever in chains: the tragic history of Congo. *The Independent* (28 July 2006). https://www.independent.co.uk/news/world/africa/forever-in-chains-the-tragic-history-of-congo-6232383.html (accessed 2 January 2020).

92. Sanderson, H. (2019). Congo, child labour and your electric car. https://www.ft.com/content/c6909812-9ce4-11e9-9c06-a4640c9feebb (accessed 2 January 2020).

93. Kelly, A. (2019). Apple and Google named in US lawsuit over Congolese child cobalt mining deaths. *The Guardia* (16 December 2019).

94. Berg, A. (2015). The Nazi rag-pickers and their wine: the politics of waste and recycling in Nazi Germany. *Social History 40* (4): 446–472.

95. Weber, H. (2013). Towards 'Total' recycling: women, waste and food waste recovery in Germany, 1914–1939. *Contemporary European History 22* (3): 371–397.

96. Denton, C.B. and Weber, H. (2018). Bones of contention: the Nazi recycling project in Germany and France during World War II. In: *Coping with Hunger and Shortage Under German Occupation in World War II* (eds. T. Tönsmeyer, P. Haslinger and A. Laba), 119–139. Cham: Springer International Publishing.

97. Haider, H. (2017). Modern Slavery in the Democratic Republic of Congo; K4D Knowledge, Evidence and Learning for Development. Helpdesk Report; 2017; p. 20.

98. Cole, R. (2017). Recycling slavery gang jailed for 32 years. *Resource* (1 June 2017). https://resource.co/article/recycling-slavery-gang-jailed-32-years-11885 (accessed 2 January 2020).

99. SKY (2016). Health fears for thousands of toxic dump workers. *SKY News* (27 January 2016). https://news.sky.com/story/health-fears-for-thousands-of-toxic-dump-workers-10145266 (accessed 2 January 2020).

100. Lowman, S. (2017). *Modern Slavery: A Rational Approach to Reducing Human Trafficking in the United States*, POLS 4330, 22. University of North Georgia.

101. Williams C. (2019). A waste sector campaign against modern-day slavery (31 August 2019). https://www.mrw.co.uk/knowledge-centre/a-waste-sector-campaign-against-modern-day-slavery/10043943.article (accessed 2 January 2020).

102. Lemille, A. (2017). The circular economy 2.0. Medium. https://medium.com/@AlexLemille/the-circular-economy-2-0-6cb543497d4d (access 30 September 2017).

103. Houston, J., Casazza, E., Briguglio, M., and Spiteri J. (2018). Stakeholder views report enablers and barriers to a circular economy; R2π the route to circular economy.

104. Lane, E.L. (2016). Volkswagen and the high-tech greenwash. *European Journal of Risk Regulation 7* (1): 32–34.

105. Yamaguchi, S. (2018). *International Trade and the Transition to a More Resource Efficient and Circular Economy: A Concept Paper*, RE-CIRCLE. Paris: OECD.

106. Preston, F., Lehne, J., and Wellesley, L. (2019). *An Inclusive Circular Economy, Priorities for Developing Countries*, Research paper, 82. London: Chatham House.

107. Wilson, S., Benton, D., Brandmayr, C., and Hazell, J. (2017). *How Will Europe's Ecodesign Measures Affect the Circular Economy in Low-Income Countries?* 12. Tearfund.

108. Potting, J., Hekkert, M., Worrell, E., and Hanemaaijer, A. (2017). *Circular Economy: Measuring Innovation in the Product Chain*, Policy Report 2544, 46. The Hague: PBL Netherlands Environmental Assessment Agency.

109. Schroeder, P., Dewick, P., Kusi-Sarpong, S., and Hofstetter, J.S. (2018). Circular economy and power relations in global value chains: tensions and trade-offs for lower income countries. *Resources, Conservation and Recycling 136*: 77–78.

110. Stuebing, S. and Vries, C. (2018). *Governance for the Circular Economy: Leadership Observations*, 72. The Netherlands: Origame.

111. Cambridge Econometrics, Directorate-General for Environment (European Commission), ICF, Trinomics (2018). *Impacts of Circular Economy Policies on the Labour Market: Final Report*. Brussels: European Commission/Directorate-General for Environment.

112. Moreau, V., Sahakian, M., van Griethuysen, P., and Vuille, F. (2017). Coming full circle: why social and institutional dimensions matter for the circular economy: why social and institutional dimensions matter. *Journal of Industrial Ecology 21* (3): 497–506.

113. Benjamin, A. (2007). Stern: climate change a "Market Failure". *The Guardian* (29 November 2007).

114. Freedman, A. (2019). Alexandria Ocasio-Cortez and the problematic framing of the 12-year global warming deadline. https://www.axios.com/climate-change-scientists-comment-ocasio-cortez-12-year-deadline-c4ba1f99-bc76-42ac-8b93-e4eaa926938d.html (accessed 2 January 2020).

Chapter 4
Redefining Resources and Waste

The invention of the ship was also the invention of the shipwreck.

— Paul Virilio

Recommended Listening
American Idiot, **Green Day**

Because it has the right vibes to tune you with the pace of change around us.

Recommended Viewing
Peaky Blinders

Because it shows the unrest, the challenges, and the opportunities involved as the second industrial revolution unfolds in the United Kingdom.

Industry 4.0 and Circular Economy: Towards a Wasteless Future or a Wasteful Planet?, First Edition.
Antonis Mavropoulos and Anders Waage Nilsen.
© 2020 John Wiley & Sons Ltd. Published 2020 by John Wiley & Sons Ltd.

4.1 IND4.0 REDEFINES RESOURCES

Each industrial revolution brings new products and new ways to manufacture, distribute, and use them. This means that new resources and raw materials are utilized and new types of waste materials are produced in each industrial revolution. But the role of energy is of special importance, as Jänicke and Jacob have stressed [1]: "'Industrial Revolution' should be perceived as a radical and abrupt but also long-lasting ('secular') change at all levels of society. Due to fundamental technical innovations in the energy field, especially in generation and utilization, a new balance between the economy and the institutional framework is developing."

Around 1750, we discovered that there is a new source of energy named coal, which became Great Britain's primary raw material. When we burn coal, it releases energy that had been stored for hundreds of millions of years. James Watt, in 1776, designed an engine in which burning coal produced steam, which drove a piston assisted by a partial vacuum. By 1900, steam engines were burning coal 10 times more efficiently comparing to 100 years before. The steam engine stimulated the development of railways and steamships revolutionizing travel. The steam engine also changed radically the way iron was produced, resulting in a serious increase of iron production and relevant products. Besides coal, Wrigley describes [2] other changes in raw materials: "the most important change in raw material provision which took place was the substitution of inorganic for organic sources of supply, of mineral for vegetable or animal raw materials....The iron industry...was able to expand without prejudice to its future prosperity only up to the point at which annual cut of timber equaled the yearly increment of new growth."

The second industrial revolution (1870–1914) brought a plethora of new manufacturing technologies and was the period that a lot of the existing industries were developed, creating new demands for raw materials. The advances in petroleum drilling and refining resulted in the creation of kerosene, which was much more efficient and less expensive compared with whale or vegetable oil. Even so, kerosene was also substituted by electricity for street lighting around 1890 and for households almost 30 years later [3]. Interestingly, gasoline was an unwanted by-product of oil refining until automobiles were mass-produced after 1914, and gasoline shortages appeared during World War I. The development of automobiles and bicycles resulted in a huge growth for the demand of rubber. The advances of chemistry created the chemical industries and industrialized the use of ammonia as fertilizer. The invention of machines that extract fibers from wood stimulated paper production and made paper much cheaper. The use of concrete became the rule and not the exception.

It is clear that electrification, motorization, chemicalization, and use of concrete became gradually "shaky since the end of the 20th century. The limitation of the fossil energy base, which carried the first two industrial revolutions, has become apparent. The fact that clear alternatives have already been heralded justifies the concept of another, ... industrial revolution" [1]. As now we are heading towards IND4.0, we have to discuss what are the important raw materials of the future.

Talking about resources, it is clear that today, information is a key resource of IND4.0. The digitalization of everything aims to create the basis for the supply of the information required. But information means energy: energy to collect, organize, transmit, store, and retrieve information pieces. Let's see how much energy is required for our digital world.

4.1.1 The Energy Footprint of Digitalization

There is a usual misunderstanding about digitalization and dematerialization. Dematerialization is delivering the same product or service using a percentage or none of the mass or material types. A simple example of product dematerialization is the transition from vinyl discs to CDs and then to digital MP3 and then to live streaming of music. Although there is a clear reduction in material consumption, digitalization creates new energy demands. And the digitalization of everything, which is emblematic for IND4.0, brings both new opportunities to optimize and reduce energy consumption and an important and growing energy footprint.

According to the International Energy Agency [4], more than 1 billion households and 11 billion smart appliances could participate in interconnected electricity systems by 2040, thanks to smart meters and connected devices. This would allow demand-side responses – in building, industry, and transport – to provide 185 GW of flexibility and avoid US $270 billions of investment in new electricity infrastructure. Some scientists believe [5] that the digital technology sector is probably the world's most powerful influencer to accelerate action to stabilize global temperatures well below 2 °C and can take a strong lead in accelerating demand for 100% renewable energy.

On the other side, the benefits from digital–smart innovations in energy management are undermined by the increasing use of other digital services and the continued expansion of digital infrastructures [6]. The energy footprint of digitalization becomes more and more important. Cloud services drive a continuous growing energy demand to deliver the services we all enjoyed. Antivirus company McAfee reported [7] that "the electricity needed just to transmit the trillions of spam e-mails sent every year is equivalent to powering two million homes in the United States and generates the same amount of greenhouse gas emissions as that produced by three million cars." The Shift Project [8], a recent study on energy consumption by digital technologies, concluded that:

- The direct energy footprint of ICT (which includes the energy for the production and the use of equipment) is increasing rapidly by 9% per year and increases the demand on electric production, which already struggles to decarbonize.
- The share of digital technologies in global greenhouse gas (GHG) emissions has increased by half between 2013 and 2018, from 2.5 to 3.7% of global emissions.
- The CO_2 emissions of digital technologies increased by about 450 million tons since 2013 in OECD countries.

Anders Andrae, a Swedish researcher, has assessed that global computing power demand from Internet-connected devices (high-resolution video streaming, emails, surveillance cameras, and a new generation of smart TVs) is increasing by 20% a year, consuming roughly 3–5% of the world's electricity in 2015 [9]. It is expected that, as *The Guardian* put it, "a tsunami of data" could consume 20% of the global electricity by 2025. James Bridle, author of the book *New Dark Age* [10], refers to a study in Japan that forecasted that by 2030, the power requirements of digital services will outstrip the nation's entire current generation capacity. *The Guardian* [11] mentions an American report from 2013 that found out that using either a tablet or smartphone to wirelessly watch an hour of video a week used roughly the same amount of electricity (largely consumed at the data-center end of the process) as two new domestic fridges.

Kris De Decker describes [12] that "If we were to try to power the (2012) internet with pedal-powered generators, each producing 70 watt of electric power, we would need 8.2 billion people pedaling in three shifts of eight hours for 365 days per year." He concludes that we need a speed limit to control the continuous growing energy consumption related to the web. Interestingly, as many reports have shown [13], wireless access networks are clearly the biggest and most inefficient consumer of energy in the cloud environment, while wired connections are the most energy-efficient way to communicate digitally. Can we, at least, cover all this growing energy demand by renewables? De Decker answers, for 2012, that the Internet energy consumption was three times the electricity supplied by all wind and solar energy facilities worldwide.

The future of IND4.0 depends on the future of electricity [6]. Security in electricity supply becomes a key concern for securing the smooth, seamless, and continuous delivery of digital services. Electricity storage and innovations in batteries are consequently a central challenge. In addition, the electricity sector experiences three serious disruptions:

1. Electrification of large sectors of the economy such as transport and heating;
2. Decentralization, spurred by the sharp decrease in costs of distributed energy resources;

3. Digitalization of the grid (with smart metering, smart sensors, automation, and other digital network technologies) and, due to the advent of the Internet of Things (IoT), a surge of power-consuming connected devices.

All the disruptions will deliver more energy efficiency and decarbonization – but all of them require much more metals than we currently use.

4.1.2 Metals Are Key for IND4.0

Besides energy, what are the other resources required for IND4.0? The technical components of IND4.0 include sensors, batteries, RFID tags, flat and touch screens, industrial robots, microchips, and all kinds of network technologies. To have an idea of the exponential growth of connected devices, consider that according to Statista website [14], the connected devices to the IoT will rise from 15 billion in 2015 to 275 billion in 2025. So, it is not a surprise that a study by the German Mineral Resources Agency [15] examined 42 future technologies and found out that their development and application will require:

*   The quadruple of the present lithium production;
*   A threefold increase in heavy rare earths;
*   One and a half times increase in light rare earths and tantalum.

A study [16] on copper demand, supply, and associated energy use to 2050 estimated that, due to the increased use of electronics, the global demand for copper will grow between 231 and 341% until 2050. The study concluded that the cumulative demand for copper is expected to exceed its reserves and reserve base in most scenarios by 2050 and that the energy required to produce copper is expected to constitute between 1.0 and 2.4% of the total energy demand by 2050.

It is well known and pretty acceptable that the world requires a serious shift to renewable energy. What is less known is that this shift has a dirty side: the serious increase in the demand for the metals. For an idea of how much metals are required for the shift to renewables, a recent study [17] of the University of Technology of Sydney assessed the cumulative demand of metals (exclusively for renewable and storage technologies, without any other sector included) by 2050 for a theoretical scenario of shifting 100% to renewables. It came out that cobalt demand will be 430% of the available reserves, lithium demand about 270% of the available reserves, and nickel demand about 130%. Table 4.1 summarizes the key metals required for advancing renewable energy.

Electric vehicles (EVs) are a key element of IND4.0 and the new urban environment. It is expected that EVs will outsell fossil fuel-powered vehicles until 2035 [18]. Sales at this level would mean that EVs could make up 5–10% of total vehicle stock by 2030 [6] and up to 57% by 2040 [19]. But as Hannah Pilgrim mentioned [20], "…compared

Table 4.1 Key metals for renewable energy and storage technologies.

Technology	Key metals: aluminum and copper in all cases plus...
Lithium-ion batteries	Cobalt, lithium, nickel, manganese
Electric vehicles	Rare earths (neodymium and dysprosium)
Solar photovoltaics	Cadmium, indium, gallium, selenium, silver, tellurium
Wind power	Rare earths (neodymium and dysprosium)

Source: Based on data from Ref. [16, 17].

Table 4.2 Critical raw materials.

2017 Critical Raw Materials

Antimony	Gallium	Magnesium	Scandium
Baryte	Germanium	Natural graphite	Silicon metal
Beryllium	Hafnium	Natural rubber	Tantalum
Bismuth	Helium	Niobium	Tungsten
Borate	HREEs	PGMs	Vanadium
Cobalt	Indium	Phosphate rock	
Fluorspar	LREEs	Phosphorus	

Source: From Ref. [22], with kind permission from the European Commission.

to 'old' engine technologies, electric motor technology needs the fourfold of copper and also larger quantities of other metals like cobalt, lithium and heavy and light rare earths. Additionally, a large part of lithium-ion batteries consists of graphite. Thus, with the production of electric cars, the material consumption by the automobile industry will certainly not decrease. In fact, the opposite is the case. The replacement of diesel and gasoline engines with electric cars leads to a deadlock." A recent market forecast [21] assessed that by the year 2025, lithium demand is expected to increase to approximately 1.3 million metric tons of lithium carbonate equivalent (LCE) – over 5 times today's levels.

As a consequence, the competition for some raw materials will increase in the future as key countries such as China and the United States, together with the European Union (EU), are all highly reliant on imports for the same materials (e.g. niobium, chromite ore, platinum, palladium, rhodium, tantalum, etc.). A study [22] on behalf of the European Commission, in 2017, assessed 61 materials (58 individual and 3 grouped) and identified 26 materials (and groups) as critical, as shown in Table 4.2.

According to OECD [23], physical scarcity is not considered to be a source of supply risk until 2030. Potential disruptions are instead perceived to come from the nexus of production concentration and geopolitical risks. In several cases [24], the exhaustion of economically competitive mineral deposits in industrialized countries has made supplies increasingly dependent on the political stability of mineral-rich emerging economies. At the same time, increasing demand from these emerging markets, new technologies that require large amounts of rare minerals, low substitutability in applications, and low rates of recycling have made economies more vulnerable to potential supply disruptions.

4.1.3 More Food and More Water

The growing population and the rise of the new middle class, as we discussed them in Chapter 1, will also drive a serious increase in food requirements. According to the World Resources Institute [25], feeding 10 billion people sustainably by 2050 requires closing 3 gaps:

- A 56% food gap between crop calories produced in 2010 and those needed in 2050 under "business as usual" growth;

- A 593 million-hectare land gap (an area nearly twice the size of India) between global agricultural land area in 2010 and expected agricultural expansion by 2050;

- An 11-gigaton GHG mitigation gap between expected agricultural emissions in 2050 and the target level needed to hold global warming below 2 °C (3.6 °F), the level necessary for preventing the worst climate impacts.

To manage such a challenge, it is important to remember that agriculture is based on the use of nonrenewable resources in agricultural systems, particularly phosphate. Phosphorus is a major mineral nutrient required by crop plants for optimal growth and productivity. Phosphate is the only form of phosphorus that plants can absorb – it is often applied to crops as phosphate fertilizer. It is forecasted [26] that by 2035, the demand for phosphorus might outpace the supply and current global reserves may be depleted in 50–100 years causing problems of a global dimension, if replacement is not found.

The availability of water will dramatically shrink due to changes in land use, climate change, pollution, overuse of drinking water, and further increases in industrial and agricultural demand in many parts of the world [1]. The 2019 World Water Development Report [25] assessed that global water demand is expected to increase by 20–30% till 2050, mainly due to rising demand in the industrial and domestic sectors. Over two billion people live in countries experiencing high water stress, and about four billion people experience severe water scarcity during at least one month of the year.

It will be useful to close this section with a representative comment on mining, as mining, for fossil fuels and metals, is the key to acquire the resources required for IND4.0. In their 2014 book *Resource Revolution: How to Capture the Biggest Business Opportunity in a Century*, [27] Stefan Heck and his colleagues Matt Rogers and Paul Carroll describe the landscape concerning the emerging difficulties in mining resources: "as oil went from use in kerosene lamps to providing the fuel for most of the modern transportation system, drillers have had to go deeper and deeper and spread into less accessible territories as the easy pickings were used up. We've gone from scooping ready-to-burn oil off the ground to running giant offshore rigs that pull oil to the surface from 25,000 feet below the Gulf of Mexico…a remarkable engineering feat, but a lot more difficult and expensive than the original wells… The cost of bringing each new oil well online has more than tripled over the past decade." Few lines later they continue: "What is true for oil is true for just about every other traditional natural resource: The amounts being discovered each year have leveled off, while the money being spent on exploration have significantly increased. When copper was first used, big nuggets could be found lying on the ground. Those nuggets were 95% pure. Today, copper typically accounts for 0.4 to 1% of the rock…and the purity is in steady decline." We started 5% waste from copper nuggets and end up with 99% waste from the current copper mines. So, it is crucial to talk about the waste streams of IND4.0.

4.2 REDEFINING THE TERM "WASTE"

As we already mentioned each industrial revolution redefines the meaning of "raw materials," revolutionizes at least a big part of manufacturing (redefining input and outputs – waste), and creates new products that will finally become new types of waste. Before we go further, it is important to consider that what we consider waste is socially and historically defined, as it is obvious if you try to answer the following questions [28], for any given society or historical period:

- What is waste?
- When does a thing become waste?
- Which things shall be considered/treated as waste?
- How can waste be turned into non-waste?
- What is non-waste?
- When shall we stop considering a thing waste?
- Can every waste be turned into non-waste?
- Is there such thing as ultimate waste?
- What is the relationship between the concepts of artefact, non-artefact, waste, non-waste, and a natural thing?
- What is waste management?
- How does waste prevention/minimization fit into the description of waste management?

4.2.1 A Brief Historical Overview

Michael Thompson, in his classical book *Rubbish Theory: The Creation and Destruction of Value*, [29] explains in details the changing nature and the meaning of the term "waste" in different social contexts. He analyzes that rubbish is a socially defined category and describes how objects have different values, or non-values, for different groups in society. He also suggests that to understand rubbish we need to understand and explain the flowing relationships between status, the possession of objects, and the ability to disregard them. Besides that, recognizing the economic value of what is considered "scrap," we are able to follow the transformation of what is rubbish (for a specific entity or owner) to a useful asset (for another entity or owner). Chappells and Shove built further on Thompson's concept [30] and described that "A long history of informal scavenging activity in the UK bears witness to the differential values in both aesthetic or financial terms which different people have. In any event, the valuing of things sets the scene in which rubbish is defined and in which the bin acquires its specific contents. The valuing of novelty and the valuing of durability are also relevant for they influence the rate at which items defined as rubbish flow into the bin. …What goes into the household bin and what stays out also depends on the range of disposal options available and the time and effort people invest in finding new homes or uses for things they no longer need or want. The local second-hand market, the availability of repair shops, the ability to pass things on to relatives, all influence the rate and range of rubbish passing through the bin."

Thompson's effort to develop a "Rubbish Theory" is very useful to understand waste as a social category for three reasons [31]. "First it helps us to explore more fully the material dimensions of markets thus contributing to a 'thingly turn' in the study of consumption. Second it highlights the importance of thinking in terms of movement, flow and circulation and moves us away from means-end, supply-demand, production-consumption linearities in thinking through the consumption process. Third it suggests that value, rather than being an inbuilt property of an object, emerges through our ways of seeing and placing objects."

Eva Pongrácz and her colleagues, trying to develop further a waste management theory, put the notion of waste in a remarkable way [28]: "the label 'waste' simply indicates that something will be treated as waste. Problems therefore arise when authorities are adamant on labelling a substance as waste, even when there is a potential for re-use or recovery. The unfortunate consequence of such labelling, at the insistence of regulative bodies, is that something may end up just being treated as waste, that is, being discarded. The definition of waste not only fails to prevent waste, it actually amasses waste." After analyzing several legal definitions of the term waste, they concluded that the current legal waste definitions are ambiguous and do not really give an insight into the concept of waste. Moreover, despite the wishes about waste prevention, when according to current legislation a thing is assigned the label of a waste, it is going to be treated like waste; implicitly, legislation thus amasses waste.

With those remarks in mind, we can better understand the changing nature of waste as the world moves from one industrial revolution to another. Starting from the end of eighteenth century, urbanization increases quantities of urban waste for the first time. Urban waste includes human and animal excrement, rubble from demolitions, various mineral and wood debris, and ashes. Burning waste inside the houses is a common practice. Wilson mentions [32] that "...in London a reasonably effective, formal waste management system appears to date back at least to the end of the eighteenth century, when the resource value of the waste began to provide a much more systematic driver. Due to domestic heating and cooking with coal, the ash content of household waste was high. The industrial revolution and rapid urban expansion led to an excess in demand for bricks and 'breeze' for building, for which municipal waste became an important raw material. So, the London parishes began to let contracts, effectively granting an exclusive franchise to a private contractor to collect the waste in their area.... The waste was taken to a network of dust-yards across London, where a small army of workers were employed to sift through the waste, separating coal, breeze, 'soil' for use as a fertilizer or in brick making, and a whole range of saleable materials."

For the same period, Barles remarks [33] that "Throughout Europe, scientists and intellectuals stressed the need for cities to return their food as fertilizer to the countryside. It was the only way to ensure both salubrity (through an efficient collection of organic materials scattered throughout cities) and food production... Furthermore, an important part of emerging industry was reliant on using raw materials that could be supplied only by cities. This is the case, for example, with vegetable rags used for papermaking for several centuries, but became much more needed once the papermaking machine was developed. In the 19th century, rags became a strategic industrial issue (1.5 kg of rags were needed to produce 1 kg of paper), such that France banned their export from 1771, followed by Belgium, Holland, Spain, Portugal and a few other countries during the first half of the 19th century."

The rise of the second industrial revolution creates several social changes. Women go to work and leave their households. Consequently, household reuse and repair practices are gradually abandoned. The first organized collection systems appear worldwide. Food waste is increasing; paper, glass, and metals appear in the waste stream. Recycling of rags, bone horse manure, and tin cans is the norm in urban centers. Ash continues to be a major component of waste. Ravetz mentioned [34] that "Throughout the 19th century and during the early 1900's ash made up almost all household waste with unburned refuse representing only a small proportion of household debris and deposited in co-existing dustbins, which were often improvised receptacles provided by the household." Lewis Herbert stresses [35] that "It is little appreciated that by 1800 London had both an informal recycling collection system, and an organized 'residual' waste management system. This was driven by the resource value of household waste rather than any legislation or public health concerns. Since medieval times, an active network of waste-buyers and 'street finders' had removed saleable items from the capital's waste mounds." Velis, Wilson, and Cheeseman analyzed the phenomenon of dust-yards in the nineteenth century in London and concluded [36] that "The overall system developed in response to the market value of constituents of municipal waste, and particularly the high coal ash content of household 'dust'...The dust-yard system had been working successfully for more than 50 years before the Public Health Acts of 1848 and 1875, and was thus important in facilitating a relatively smooth transition to an institutionalized, municipally-run solid waste management system in England. The dust-yards can be seen as early precursors of modern materials recycling facilities (MRFs) and mechanical-biological treatment (MBT) plants."

Things were gradually changing, slower till World War II and faster after it. Chappells and Shove explain [30] that "Although the first moveable containers appeared by the end of the 18th century, it took until the 1960s for the standardized dustbin to become a reality. During this time the form of the dustbin changed significantly, most markedly in terms of size. Early experiments with small improvised containers such as biscuit tins in the early 1900s gave way to medium sized galvanized metal bins in the 1950s, and onto larger plastic bins in the 1960s. During this time the dustbin has seen the demise of both ash and the dust elements after which it was named, with bins containing increasing proportions of higher volume, lower density wastes such as packaging, associated with a new household waste generation." Heather Rogers describes the atmosphere in the post-World War II United States as the golden age of waste [37]: "Central to this society-as-machine aesthetic was a plethora of new commodities made for easy use and quick disposal...It was the age of the paper plate, polyester, fast food, disposable diapers, TV dinners, new refrigerators, washing machines and rapidly changing automobile styles. Most of all it was the epoch of packaging – lots of bright, clean, sterile packaging in the form of boxes, bags, cellophane wrappers and throwaway beer cans. The golden era of consumption had arrived, bringing the full materialization of modern garbage as we know it: soft, toxic, ubiquitous." By the arrival of the third industrial revolution, around the end of 1960s, fast food was already mainstream in the United States and expanding in Europe. Plastic is becoming a dominant material. Single-use products are increasing. Packaging waste is increasing; plastics and papers go up to 30% of the solid waste in cities. Hazardous waste becomes a serious problem. Recycling, treatment, and disposal are industrialized. Global secondary material markets are developed, together with waste trafficking practices.

By 1995–2000 the impacts of automation and digitalization are becoming obvious to businesses, labor productivity, and factory's design. This is when Andrew McAfee and Erik Brynjolfsson put the beginning of the Second Machine Age [38] (a more descriptive but still equivalent with IND4.0 term, from a historical perspective): "Now comes the second machine age. Computers and other digital advances are doing for mental power – the ability to use our brains to understand and shape our environments – what the steam engine and its descendants did for muscle power."

So, now it's time to see the future waste streams relevant to IND4.0. But a final note is worthy. For waste management professionals, students, and researchers that want to have a deeper understanding of the historical changes in waste management, it is strongly suggested to read CIWM's "Centenary History of Waste and Waste managers in London and South East England," [35] David Wilson's paper "Development drivers for waste management," [32] and of course Susan Strasser's emblematic book *Waste and Want: A Social History of Trash* [39].

4.2.2 Future Waste Streams

The waste streams of IND4.0 will be a mix of the existing waste streams and the future ones, the ones that are expected to grow as IND4.0 is advancing. Of course, there will be different waste streams from country to country, depending on the level of industrialization and progress of IND4.0 and also depending on the role of each country in the global supply and value chains. In this section we will focus on the streams that will have a special global interest and require more attention. In brief we will focus on anthropogenic stocks, food waste, plastics, e-waste, and other important rising waste streams.

4.2.3 Anthropogenic Stocks

If we want to think forward and get prepared for the future waste streams, then clearly anthropogenic stocks are the most important element. We already discussed about stocks in Chapter 3; now it's time to understand them better.

What do we mean with the term "anthropogenic stocks"? We prefer the description given by Schiller et al. [40]: according to it anthropogenic stocks are "The accumulated wealth of assets in the form of buildings, infrastructure and other durable goods." They include roads, buildings, infrastructural constructions, cars, building inventory, etc. Consequently, stocks should be considered as a valuable reservoir of secondary raw materials that are deposited in the items and can be reused. Anthropogenic stocks are a key concept of the *urban metabolism* model, which is "the sum total of the technical and socio-economic process that occur in cities, resulting in growth, production of energy and elimination of waste" [41].

Why is it important to think about stocks? Because a large part of all primary materials extracted globally, almost half of them (31 billion tonnes/year in 2015) according to a recent study [42], accumulates in stocks of manufactured capital, including buildings, infrastructure, machinery, and equipment. These in-use stocks of materials provide important services for society and the economy and drive long-term demand for materials and energy. Consequently, configuration and quantity of stocks determine future waste flows and recycling potential and are key to closing material loops and reducing waste and emissions in a circular economy. Prof. Krausmann and his colleagues found out that the rate we accumulate anthropogenic stocks per year has grown 69 times between 1900 and 2015, from 0.5 to 31 billion tonnes/year [43]. The result is that by 2015 anthropogenic stocks were assessed at 961 billion tonnes, an equivalent of 11 years of materials extraction (global material extraction was close to 90 billion tonnes in 2018) – in these stocks 1.11 billion tonnes of metals are included.

What other materials are included in these stocks? Aggregate, bricks, stones, asphalt, and concrete are the vast majority in terms of weight. As mentioned, stocks also include important metals like cooper (from electrical installations and equipment), aluminum, iron, and lead and of course a lot of plastics and papers. Table 4.3 shows an assessment of the metal stocks provided by a report of the International Resource Panel [44].

Although it is obvious that the stocks of metals are a primary target, the importance of aggregate stocks should not be underestimated on a global scale. It might seem a paradox, but our world is actually running out of a material that seems to be abundant: sand!

Sand and gravel are the largest portion of the primary material extracted. In 2017 we extracted and use close to 50 billion tonnes of sand and gravel, or 6.5 tonnes/capita [45]. Sand and gravel are the most extracted group of materials worldwide, exceeding fossil fuels and biomass. In most regions, sand is a common-pool resource, i.e. a resource that is open to all because access can be limited only at high cost. When it comes to concrete production, roughly 3 metric tons of sand are needed to make 1 metric ton of cement, so in 2014, the UNEP estimated that between 26 and 30 billion metric tons of sand are poured into cement mixers every year worldwide. The demand for sand in expanding metropolises such as Singapore, Shanghai, or Dubai is always growing. Mega construction projects in these cities are already devouring

Table 4.3 In-use metal stock estimations for the major engineering metals.

Metal	Number of estimates	Percent of all estimates	Global per capita stock	MDC per capita stock	LDC per capita stock
Aluminum	9	7.4	80	350–500	35
Copper	34	27.0	35–55	140–300	30–40
Iron	13	10.7	2200	7000–14000	2000
Lead	20	16.4	8	20–150	1–4
Steel	1	0.8		7085	
Stainless steel	5	4.1		80–180	15
Zinc	14	11.5		80–200	20–40

MDC, most developed countries; LDC, less developed countries; all figures are in kilogram/capita.
Source: From Ref. [44], with kind permission from Global E-Waste.

vast quantities of sand, as we see with the numerous major projects in Dubai. The raw material is not only needed to erect huge skyscrapers like the Burj Khalifa. The foundation for Dubai's Palm Jumeirah artificial island in the Persian Gulf is also formed from a massive deposit of these fine grains. According to media reports, more than 150 million tons of sand were delivered from Australia [46]. Besides the rapid urbanization, further strains on sand deposits arise from escalating transformations in the land–sea interface as a result of burgeoning coastal populations, land scarcity, and rising threats from climate change and coastal erosion. Even hydraulic fracturing is among the plethora of activities that demand the use of increasing amounts of sand [47]. In addition, quartz sand is more or less 70% of the glass we use, and it's an important raw material for the construction of microchips due to its high silicon content.

Someone can easily say "so what? Our planet is full of sand and gravel." Well, this is completely misleading because even countries full of deserts are importing sand! This is because the sand we need for concrete is the more angular stuff found in the beds, banks, and floodplains of rivers, as well as in lakes and on the seashore. In contrast, desert sand grains are small, smooth, and uniform, thus completely unsuitable for concrete [46]. Finally, the result is that the type of sand we need to make concrete is available only in riverbeds, beaches, and quarries, and its extraction is creating huge environmental and biodiversity impacts. In practice, easily accessible quartz sand deposits are gradually depleted and become scarce. A review of construction materials and their flows and stocks concluded that stock accumulation is still ongoing and that nonmetallic mineral secondary resources would be insufficient to totally meet future demand [48]. Consequently, concrete recycling will become mainstream in the future, and the use of anthropogenic stocks from buildings and outdated infrastructure is already tested in several countries, like Netherlands and Switzerland, as an alternative [46].

To get a more granulated idea of the importance and the extent of anthropogenic stocks, it's useful to see some more information. As an example, the total weight of all consumer goods in Kawasaki City was calculated as 3 667 000 tonnes. Materials stocks in Japanese households were between 5.7 and 5.84 tonnes/household [49]. A study on paper stocks in Germany found that in 2010 they were equivalent to 410 kg/person [50]. It was found [40] that in Germany the annual per capita growth in Germany's anthropogenic material stock is almost 10 tonnes. In the last 50 years, an estimated 42 billion tonnes of material has been added to the anthropogenic stock, and around 28 million tonnes of material has been consumed by buildings, infrastructure, building services, and durable consumer goods. According to a city metabolism study for Vienna in 1997, the total stock of lead in Vienna is 340 000 tonnes, mainly accumulated in buildings/infrastructure, and it is still growing by 1% per year [51]. Landfill and urban mining, depending on the market conditions and trends, might become a key trend of the future. According to a report [44], the waste disposal sites worldwide contain up to 225 million tonnes of copper, and in the United States it is estimated that landfills contain approximately 850 million tonnes of iron. According to the European Enhanced Landfill

Mining Consortium (EURELCO), Europe comprises most likely more than 500 000 landfills, and 90% of these landfills are to be considered as "non-sanitary" landfills. Many of these landfills will at some point require expensive remediation measures that should be combined with resource recovery measures as they contain important material resources [52].

4.2.4 Food Waste

There is a growing need for more food, and consequently more food waste is expected, unless a serious intervention will take place on a global scale, especially in the developing world. Studies commissioned by FAO [53] estimated yearly global food loss and waste by quantity at roughly 30% of cereals; 40–50% of root crops, fruits, and vegetables; 20% of oilseeds, meat, and dairy products; and 35% of fish. This amounts to 1.3 billion tonnes/year. The value of food lost or wasted annually at the global level is estimated at US $1 trillion [53]. Food losses and waste represent wastage of resources, including the land, water, labor, and energy used to produce food. It strongly contributes to climate change because GHGs are emitted during food production and distribution activities and methane is released during the decay of wasted food [54]. It also affects food supply chains by lowering income for food producers, increasing costs for food. The World Resources Institute [25] estimates that reducing food loss and waste by 25% by 2050 would close the food gap by 12%, the land gap by 27%, and the GHG mitigation gap by 15%. In addition, if bioenergy competes with food production by using food or energy crops or dedicated land, it widens the food, land, and GHG mitigation gaps. According to FAO, the issue of food waste is high on the political agenda in industrialized countries, but it is expected to constitute a growing problem in developing countries given the changes that food systems in these countries are undergoing because of such factors as rapid urbanization, expansion of supermarket chains, and changes in diets and lifestyles. In many developing countries the lack of infrastructure and poor harvesting/growing techniques will continue to contribute and drive higher food waste generation. Despite the fact that the reduction of losses to post-harvest systems is considered a very serious problem, only 5% of the funding for agricultural research is allocated to this problem [55].

Going to waste, on a global scale, the largest waste category is food and green waste, making up to 44% of global waste generated in 2016 (almost 880 million tonnes/year), according to the World Bank [54]. According to ISWA's report "Global assessment of municipal organic waste production and recycling" [56], available data show that currently 2.6 million tonnes/day of organic waste is generated as municipal solid waste (MSW) every day, and, due to increasing urbanization and improving living conditions, this amount is expected to reach 4.5 million tonnes/day by 2050. The report demonstrates the huge impact of organic waste on the management of MSW in cities: 37 of the world's megacities currently generate 115 million tonnes/annum of organic waste, an amount comparable with the biowaste fraction that arises across the EU. ISWA suggests that to address the problem, there is a need for a density of composting and anaerobic digestion facilities that range from 1 plant per 200 000 inhabitants up to 1 per 22 000 inhabitants in areas with complex logistics. Another ISWA report titled "Circular Economy: Carbon, Nutrients and Soil" [57] shows that, within OECD countries, an estimated 177 million tonnes of organic municipal waste is generated annually, of which only 66 million tonnes is recycled through composting and anaerobic digestion. Assuming an overall maximum capture rate of 70%, this means that potentially 124 million tonnes a year of municipal organic waste could be collected for biological treatment, which equates to an additional 58 million tonnes a year over and above what is currently collected. The potential for organic commercial and industrial wastes, crop residues, and manures is largely unknown but is likely to be considerable and well in excess of the municipal waste fraction. The resources contained within this 124 million tonnes of municipal organic waste are significant, holding anywhere between 0.1–3 million tonnes of nitrogen and 4–41 million tonnes of carbon.

Closing the loop for the organic fraction involved in municipal and food waste is the main and, in many cases, the only measure that enables circular economy on a local level, no matter how rich, populated, urbanized, or globalized it is. It is also the easiest way to involve almost everyone in an effort towards an improved resource management.

4.2.5 Plastics

Plastics are an increasingly important waste stream. According to the World Bank [54], in 2016, the world generated 242 million tonnes of plastic waste – 12% of all MSW. Plastic waste is choking our oceans, yet our consumption of plastics is only increasing. Cities and countries are rapidly developing without adequate systems in place to manage the changing waste composition. The famous article "Production, use, and fate of all plastics ever made" [58] assessed that by 2017 8300 million metric tons of virgin plastics have been produced. As of 2015, approximately 6300 million metric tons of plastic waste had been generated, around 9% of which had been recycled, 12% was incinerated, and 79% was accumulated in landfills or the natural environment. If current production and waste management trends continue, roughly 12 000 million metric tons of plastic waste will be in landfills or in the natural environment by 2050. ISWA's Task Force on Marine Litter considers [59] that plastic marine litter is a challenge of planetary scale and implications. It is associated with four key systemic failures, which relate directly to the waste and resources management sector:

- Poor or absent solid waste management (SWM) services and infrastructure (mainly in low-income countries) and insufficient monitoring and law enforcement (mainly in high-income countries);
- Problematic and vulnerable markets for secondary plastics, resulting in poor and very fragile incentives for material recovery;

- Lack of a systemic and in-depth understanding of:
  - The technical challenges and the restrictions of material properties and the flows of plastics;
  - The effects of social consumption patterns and littering behaviors on solid waste generation.

Plastic waste has attracted a lot of attention recently, especially the single-use plastic streams, mainly due to the massive research findings regarding the impacts of plastic pollutions to oceans [60]. Leading the way for richer countries but also facing heavily polluted beaches especially in the Mediterranean Sea, the EU launched its plastic strategy in 2018 [61]. This aims to ensure all plastic packaging is reusable or recyclable by 2030. It also calls for 90% of all plastic bottles to be recycled by 2025. It is expected that the strategy will transform the way plastic products are designed, used, produced, and recycled in the EU. Better design of plastic products, higher plastic waste recycling rates, and more and better quality recyclates will help in boosting the market for recycled plastics.

The 2016 Ellen MacArthur Foundation report titled "The New Plastics Economy – Rethinking the Future of Plastics" [62] found out that (i) most plastic packaging is used only once (95% of the value of plastic packaging material, worth US $80–120 billion annually, is lost to the economy) and (ii) plastic packaging generates negative externalities, valued conservatively by UNEP at US $40 billion: given projected growth in consumption, in a business as usual scenario, oceans are expected to contain, in terms of weight, more plastics than fish by 2050, and the entire plastics industry will consume 20% of total oil production and 15% of the annual carbon budget. This report created a huge impact, and actually it pushed the packaging industry for a response.

Several multinational giants voluntarily committed to reduce the amount of virgin plastic they used by 2025 and to increase the amount of recycled plastic. The global commitment on plastic, introduced in late 2018 to get corporations to pledge to use less and recycle more, grew to more than 400 of the world's biggest companies. Together, these companies are responsible for more than 20% of all the plastic packaging produced. According to the first annual "New Plastics Economy Global Commitment" progress report [63], companies set out actions to eliminate problematic plastic packaging and increase the use of recycled plastic in packaging by more than fivefold by 2025, equivalent to keeping 25 million barrels of oil in the ground every year. However, there are serious doubts regarding the real impact of those actions for three reasons:

First, because a recent report [64] from S&P Global Platts revealed that recycled plastic now costs an extra US $72/tonne, compared with newly made plastic. According to the analysts this trend is driven in part by the growing demand to include recycled plastics in new products. Meanwhile, new plastic is becoming cheaper to make due to a flood of petrochemicals production from the United States driven by the shale gas boom.

Second, because, since 2010 the petrochemical industry has invested about $200 billion and with $100 billion more planned to be spent, plastic production is expected to grow 40% by 2030, as *The Guardian* [60] revealed. Dozens of new ethane-cracking plants with capacities at the order of million tonnes of plastics per year planned. Latest figures, as published in *The Guardian*, suggest that 359 million tonnes were produced in 2018. Nearly one-third went to single-use packaging, and less than 10% was recycled. In addition, it seems that the industry is heavily lobbying to make sure that it will be allowed to increase substantially the production of plastics.

Third, because recycling can't manage the growing quantities of plastic, in the best case it can accommodate less growth in waste than in plastic production, and in the worst case it could be just a nice alibi for the continuously growing plastic production. Especially because the companies' commitments are really reflected in their budgets where you can find billions of dollars for advertising to stimulate more consumption and millions of dollars, usually under corporate social responsibility or marketing budgets, to promote recycling in their countries. Take the example of the largest spender on global advertising and marketing of any other soft drink producer. In 2018, the company spent a whopping $5.8 billion on global advertising [65], almost 0.8 billion in the United States alone. At the same time, despite the announcements and the global commitments to collect and recycle the equivalent of every bottle or can it sells by 2030, as CNN revealed [66], it will start by investing $5.4 million in the United States for recycling efforts. The readers are advised to compare the figures and make their own conclusions. In addition, several reports indicate that despite the commitments, the same company rejects [67] any proposal for extended producer responsibility or deposit schemes.

According to a recent UNEP report [68], as of July 2018:

- One hundred and twenty-seven out of 192 countries have adopted some form of legislation to regulate plastic bags.

- Twenty-seven countries have enacted legislation banning either specific products (e.g. plates, cups, straws, packaging), materials (e.g. polystyrene), or production levels.

- Twenty-seven countries have instituted taxes on the manufacture and production of plastic, bags while thirty charge consumers fees for plastic bags at the national level.

- Forty-three countries have included elements or characteristics of extended producer responsibility for plastic bags within legislation.

- Sixty-three countries have mandates for extended producer responsibility for single-use plastics, including deposit refunds, product take-back, and recycling targets.

- Eight out of 192 countries worldwide (4%) have established bans of microbeads through national laws or regulations.

In an ISWA Guest Blog, Professor Linda Godfrey argues [69] that for developing countries "there is no one single solution to addressing the leakage of plastic into the environment, but that the solution is likely to be a combination of three approaches. These three approaches include (i) banning single-use plastics, as many countries and cities are doing; (ii) replacing petroleum-based single-use plastics with alternative bio-benign materials, such as paper, glass or biodegradable plastics; and (iii) improving waste collection and sending the collected waste to engineered landfills, or recycling or recovery centres, thereby ensuring that plastic has little opportunity to 'leak' into the environment. Countries, cities, and businesses must find their 'position' within these three options, depending on their local conditions and on what they have control over." In a relevant paper [70], she adds that "...the intention of the author is to highlight that the solution to waste plastic is not a binary one, but a combination of the three responses, tailored to fit the reality of a city or country depending on what is realistically achievable" as shown in Figure 4.1.

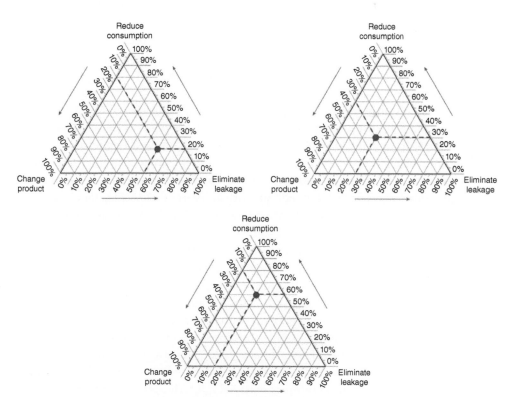

Figure 4.1 Indicative plots of different responses to reduce plastic leakages. *Source*: From Ref. [70], with kind permission from MDPI.

4.2.6 E-Waste

E-waste is the most emblematic waste stream of IND4.0, and it is expected to continue to grow. The report "Global E-waste Monitor 2017," jointly authored by ITU, the United Nations University (UNU), and ISWA [71], shows a staggering 44.7 million metric tonnes (Mt) of e-waste generated in 2016 – up to 3.3 Mt or 8% from 2014. In 2016 the world generated e-waste – everything from end-of-life refrigerators and television sets to solar panels, mobile phones, and computers – equal in weight to almost nine Great Pyramids of Giza, 4500 Eiffel Towers, or 1.23 million fully loaded 18-wheel 40-ton trucks, enough to form a line 28 160 km long, the distance from New York to Bangkok and back. Experts foresee a further 17% increase – to 52.2 million MT of e-waste by 2021 – the fastest-growing part of the world's domestic waste stream. Only 20% of 2016's e-waste is documented to have been collected and recycled despite rich deposits of gold, silver, copper, platinum, palladium, and other high-value recoverable materials. The conservatively estimated value of recoverable materials in last year's e-waste was US $55 billion, which is more than the 2016 gross domestic product (GDP) of most countries in the world. About 4% of 2016's e-waste is known to have been thrown into dumpsites; 76% or 34.1 Mt likely ended up incinerated, in landfills, recycled in informal (backyard) operations, or remained stored in our households. On a per capita basis, the report shows a rising trend as well [72]. Falling prices now make electronic and electrical devices affordable for most people worldwide while encouraging early equipment replacement or new acquisitions in wealthier countries. As a result, the average worldwide per capita e-waste generated was 6.1 kg, up to 5% from 5.8 kg in 2014. The highest per capita e-waste generators (at 17.3 kg/inhabitant) were Australia, New Zealand, and the other nations of Oceania, with only 6% formally collected and recycled. Europe (including Russia) is the second largest generator of e-waste per inhabitant with an average of 16.6 kg/inhabitant. However, Europe has the highest collection rate (35%). The Americas generate 11.6 kg/inhabitant and collect only 17%, comparable with the collection rate in Asia (15%). However, at 4.2 kg/inhabitant, Asia generates only about one-third of America's e-waste per capita. Africa, meanwhile, generates 1.9 kg/inhabitant, with little information available on its collection rate. Figure 4.2 shows the countries with, without, and with proposed legislation for e-waste [73], demonstrating that there are many countries that have not even started to face the problem.

Another recent report [74] assessed that the global economy generates approximately 50 million tonnes of e-waste every year. This is a huge amount, representing the mass of all the commercial aircraft ever produced. Less than 20% of this is waste formally recycled. In terms of material value, this presents an opportunity worth over $62.5 billion/year, more than the GDP of most countries and three times the

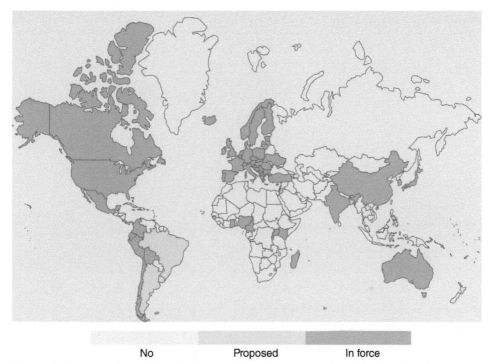

No   Proposed   In force

Figure 4.2 Countries with existing (in force), proposed, and no legislation relevant to e-waste.
Source: From Ref. [73], with kind permission from Global E-Waste.

output of the world's silver mines. There is 100 times more gold in a tonne of e-waste than a tonne of gold ore.

In this report, the UNU predicts e-waste could nearly triple to 120 million tonnes by 2050 if nothing changes.

A report titled "Future E-Waste Scenarios" [75] stated that "As a combined result of population growth, higher purchasing power, and availability, the demand of e-products will increase. However, with more efficient technologies, the total number of devices used on average may not rise in the same proportion. Smartphones are an example of how one product can replace several others." Still, the report assessed that, under a baseline scenario, the amount of e-waste will more than double by 2050 to approximately 111 million tonnes/year. However, as the report stresses, the growing quantities of e-waste alone do not tell the entire story. "The size and severity of the future e-waste problem will ultimately depend on our production and consumption models...The framework for optimizing the 'technical cycle,' laid out by the concept of circular economy, is of high relevance in tackling the e-waste problem. This includes

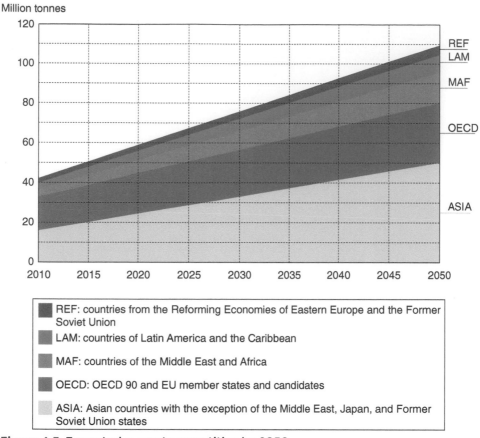

Figure 4.3 Expected e-waste quantities by 2050.
Source: From Ref. [75], with kind permission from Global E-Waste.

better product design in order to facilitate a) lifetime extension through repair and reuse of products and components and b) better recovery of valuable resources by making products easily recyclable." As Figure 4.3 shows the biggest quantities are expected in Asia and OECD countries.

4.2.7 Other Important Waste Streams

Table 4.4 presents some of the most important new waste streams that are expected in increasing quantities, and, unless a real revolution happens regarding their recyclability and reuse, they will be also found in waste disposal sites. For each waste stream, facts and figures, key issues, and impacts of waste disposal and health and safety are presented. Figure 4.4 presents a snapshot of the evolution of solid waste during industrial revolutions.

	Raw materials	Energy	Solid waste evolution
1765 The first industrial revolution	Coal, iron, cotton, wood	Coal steam	Urbanization increases quantities of urban waste, for the first time. Urban waste includes human and animal excrement, rubble from demolitions, various mineral and wood debris, and ashes. Burning waste inside the houses is a common practice
1870 The second industrial revolution	Oil, gas, iron, aluminum, copper, phosphorus, rubber, chemicals	Oil and gas, electricity	Women go to work; household reuse and repair practices gradually are abandoned. The first organized collection systems appear worldwide. Food waste is increasing; paper, glass, and metals appear in the waste stream. Recycling of rags, bones, horse manure, and tin cans is the norm in urban centers. Ash continues to be a major component of waste
1969 The third industrial revolution	Oil, gas, iron, aluminum, copper, phosphorus, rubber, chemicals	Oil, gas, fracking, nuclear, renewable, electricity	Fast food is becoming mainstream. Plastic becomes a dominant material. Single-use products are increasing. Packaging waste is increasing; plastics and papers go up to 30% of the solid waste in cities. Hazardous waste becomes a serious problem. Recycling, treatment, and disposal are industrialized. Global secondary material markets are developed, together with waste trafficking practices
Nowadays Industry 4.0	Oil, gas, iron, aluminum, copper, phosphorus, rare earths, lithium, chemicals	Oil, gas, fracking, nuclear, renewable, electricity	Food waste and plastics continue to rise. E-waste becomes the emblematic waste stream. E-commerce contributes to the further increase of packaging waste. Medicine and chemicals are present in urban waste. Batteries, spent photovoltaics, wind generators, wearables, and clothes require special emphasis. Hazardous and C&D waste are increasingly important. Circular Economy redefines the solid waste industry. Global markets are re-arranged

Figure 4.4 A historical snapshot of the solid waste evolution during industrial revolutions.

Source: With kind permission from Nick Rigas, D-Waste.

4.3 WASTE HIERARCHY: UPGRADED OR OBSOLETE?

The waste management hierarchy indicates an order of preference for action to reduce and manage waste. It is usually presented diagrammatically in the form of a ladder or a pyramid, as it is shown in Figure 4.5. The hierarchy, in different forms and phrasing in different countries, is the cornerstone for waste management policies during the last 30 years, inspiring and influencing legislation and delivering decision-making guidance on a global scale. However, as the concept of circular economy becomes a mainstream policy narrative that determines waste management, the suitability of waste hierarchy is rediscussed. Ad Lansink, the founder of the waste hierarchy, which is also known as Lansink's Ladder, in his great book *Challenging Changes – Connecting Waste Hierarchy and Circular Economy*

Table 4.4 Important future waste streams.

Waste stream	Facts and figures	Key issues	Issues related to waste disposal, health, and safety
Lithium-ion batteries	300–550 000 tonnes of Li-ion batteries from electric cars are expected to reach their end of life annually in 2040, compared with around 100 000 tonnes in 2025 [76]. The total amount of lithium-ion batteries available for recycling in 2030 will be above 1.2 million tonnes; China alone produces already more than half a million tonnes per year [77]	By 2030, the amounts of metals that are recoverable from end-of-life batteries are 125 000 tonnes of lithium, 35 000 tonnes of cobalt, and 86 000 tonnes of nickel	These batteries have a high potential to become hazardous pollutants. Cobalt, copper, nickel, and lead, depending on the specific waste disposal site conditions, can easily leach in high concentrations posing serious health and environmental risks [78]
Wind turbine blade waste	It has been forecasted [79] that by 2050 almost 43 million tonnes of wind turbine blade waste will be available (40% in China, 25% in Europe, 19% in the United States)	The blades are considered unrecyclable because (most of them) they are constructed from polymer composite materials (high-grade epoxy and polyester) that is strengthened with glass and carbon fiber or hybrid combinations of both [80]	The disposal of wind turbine blade waste in landfills constitutes a major challenge because the blades are heat and sunlight resistant; thus their degradation will take hundreds of years [81]. In addition, they are so big items that special preparation for size reduction would be required. In case this preparation takes place on-site, additional health and safety risks will be created

Photovoltaics (PV)	The waste PV, worldwide, was assessed at 43–250000 tonnes in 2016, but it is expected to grow to 5.5–6 million tonnes in 2050 [82], with huge amounts expected around 2030, when the first PV panels are expected to complete their life cycle. The cumulated amount of PV waste is expected to be close to 80 million tonnes in 2050	The cumulative value of the recoverable raw materials will be 0.45 billion dollars by 2030, which is similar to the materials required to manufacture roughly 60 million new panels [82]	Waste PV includes toxic materials, although in small quantities. However, it is not yet clear what is the leachability of these substances in waste PV. Several studies have shown conflicting results and it seems that in the United States the major problem is the lead leachability [83]
Wearables (connected clothes, earwear, watches, wristbands, and others)	Worldwide shipments of wearables are expected to reach at 279 million in 2023, a 60% increase compared to 2018. The expected market compound annual growth rate (CAGR) for the next 5 years is almost 9% [84]	Wearables are composite materials that combine clothing, electronic devices, elastomers, smart fabrics, and even solar cells [85]. They hold batteries, and their design is pretty compact and ultralight to provide easiness to the users. This design makes almost impossible their disassembly and to remove the batteries [86]. With the current design trends, their recycling is actually impossible	E-wastes involved in wearables have a potential to release heavy metals and thus create human exposure through different pathways. This is linked with several adverse health impacts. In addition, if informal recyclers are involved in managing wearables to recover the small amounts of metals involved, the fire is the only way to do it, which means increased emissions of dioxins, furans, and PMs [87]

(Continued)

Table 4.4 (Continued)

Waste stream	Facts and figures	Key issues	Issues related to waste disposal, health, and safety
Apparel and Footwear	In 2015, the total clothing sales were 50 million tonnes, and they are expected to grow to 150 million tonnes in 2050. Worldwide, each second, the equivalent of a garbage truck is landfilled or burned. 73% of the textile waste is incinerated or landfilled. Recycling rate (textile waste to new clothes) is less than 1%, resulting in $100 billion of annual material losses [88]	In 2015, the total GHG emissions from the fashion industry was 1.2 billion tonnes of CO_2 equivalents, more than the flights and shipping combined. The extended use of chemicals and plastic fibers releases microplastics and chemicals into waterways during washing. The global supply chain creates huge amounts of packaging waste	Synthetic textile waste is not going to be decomposed in landfills; however woolen and cotton clothes will decompose, producing GHGs. In terms of health and safety, clothing in dumpsites will increase the risks for informal workers especially when it can be easily used as a secondhand [89]
E-commerce packaging	E-commerce sales are expected to reach US $6.5 trillions by 2023 [90], a 275% increase compared to 2017. The e-commerce packaging market was valued at US $27.04 billion in 2019 and is expected to reach US $61.55 billion by 2025 [91]. In 2017, in China only, 40 billion sets of parcels were delivered corresponding to 8 million tonnes of packaging waste [92]. 165 billion parcels are shipped annually in the United States using cardboard that corresponds to 1 billion trees [93]. The freezer pack waste used to pack 96 million meals annually (from one company only) is about 192 000 tons/year, or the weight of nearly 100 000 cars or 2 million adult men [94]	No organization has quantified just how much additional plastic packaging is being used due to e-commerce, or the net impact on the environment. The average box is dropped 17 times during the logistics [94]. Online purchases must be packaged one extra time (one B2B and one B2C) to protect them during shipping. In addition, return rates for items bought online are as high as 30%, which implies multiple purchases of the same item [95]	E-commerce packaging creates an additional burden to city waste management systems and drives higher the relevant costs. A lot of the used materials are not easily recyclable, especially the plastics ones or the freezer packs. It is urgent to redesign e-commerce packaging to avoid the fact that 70–80% of the delivered parcels still have unnecessary packaging [96]

Sources of data are referenced.

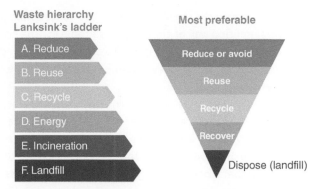

Figure 4.5 Lansink's ladder (left) and the pyramid form of waste hierarchy.
Source: With kind permission from Nick Rigas, D-Waste.

[97], explains that "The waste hierarchy seems to be a linear concept, but in fact (it is) a sequence of circular processes. This would be all the more evident if end-of-life solutions were screened out...the upper rungs of waste hierarchy (prevention, reuse of products and materials after recycling) are essential for achieving circular economy. The waste hierarchy stimulates innovation of and between each of the rungs and forms the basis for ecodesign. The waste hierarchy also points the way to development of statutory and other instruments, such as educational and informative projects. This last point is important regarding the challenging rung of prevention...The same is true for system leakages."

Lansink explains that waste prevention is by far the most challenging rung of the waste hierarchy. The prevailing growth-focused political and social mindset isn't easily compatible with waste prevention's goals of decreasing consumption by cutting waste. Awareness of the looming threat of shortages of raw materials has increased, but consumer spending also increased and continuous economic growth is the dominant economic model. Lansink makes an interesting comment regarding economic growth [97]. "Some argue that so long as we design products well, consumption can continue indefinitely. Michael Braungart considers those who urge waste prevention to have 'an air of guilty people with a pessimistic view on the challenge of a full cycle of materials.' In his view, there need be no limit to consumption, if only products are properly designed to be non-toxic, and at the end of their lives can act as technical or biological nutrients, to be reincorporated into new products or to biodegrade. Even if we accept that Cradle to Cradle design is widely achievable, Braungart seems to ignore the energy involved in the creation and distribution of products, which will always impede the realization of total circular economy. At the least, we need to consider how we can prevent the waste associated with products that don't meet the Cradle to Cradle standard."

4.3.1 Critiques About Waste Hierarchy

Several questions and critiques have been raised about the role of waste hierarchy. The fact that the waste hierarchy is a simple and single preference order is definitely what makes it so powerful and easily understood and communicated. However, the same fact is also the reason for which the concept has been criticized as a very linear or oversimplified way of thinking, unsuitable for the rising complexity of the waste management world. We are going to present the main discussion about waste hierarchy because this discussion is eye-opening for some of the most controversial issues that we already face.

A first set of arguments concerns the preference order itself. In practice, this means that the preference order is not necessarily the right one for every case or for every type of waste. As an example, a paper [98] that revisited the waste hierarchy for the case of wood waste, in the EU context, based on life cycle assessment (LCA), concluded that wood recycling and waste to energy are essentially equal in terms of environmental assessment; thus prioritizing recycling seems to be a political intervention and requires a closer examination. DEFRA has found that for food waste, garden waste, and lower-grade wood, evidence suggests that the best waste management options are not in accordance with the waste hierarchy [99]. Price and Joseph, based on their analysis for demand management, concluded [100] that "Whilst the philosophy of the hierarchy is based on an integrated approach to waste management, the reality is a prescriptive approach that does little to alleviate the reliance on end of pipe solutions. Little regard is given to demand management and the development of efficient processes, which reduce energy and resource usage and have a direct impact on waste generated." This last argument, although valid in many cases, has nothing to do with the real value and the offer of the waste hierarchy as a guiding principle. The point here is that the waste hierarchy is every time implemented in a political and social framework that determines the outcomes much more than the principle itself. As a general principle, it is usually squeezed, interpreted, and reshaped through the availability of funds, the institutional framework, the political conflicts, and the market mechanisms that finally impose the bridge that links the principle with the tangible outcome.

As an example, an EU brochure on the importance of life cycle thinking as a supportive tool for waste management decisions [101] states that "the waste hierarchy will generally lead to the most resource efficient and environmentally sound choice. However, in some cases refining decisions within the hierarchy or departing from it can lead to better environmental outcomes. The 'best' choice is often influenced by specific local conditions and care needs to be taken not to simply shift environmental problems from one area to another." In the same document, comparing recycling against incineration of PET bottles, it is concluded that recycling usually results in lower energy consumption than incinerating of bottles and producing new ones from raw material; however this is based on the assumption that the plastic is not heavily soiled and is not degraded in the recycling process. In a project that examined the suitability of waste hierarchy for the case of waste paper in Denmark [102] using LCA, it was concluded that although the waste hierarchy is a sound principle regarding the handling of waste paper, it does not necessarily apply for other types of waste. In similar studies, it was shown that the principle does not work for inert materials such as glass, because incineration is a far worse alternative than landfills, while it appears that some types of recycling, such as composting or even making biogas, are often worse than incineration.

However, it has to be considered that the applicability of LCA for waste management planning and policy making is restricted by certain limitations, some of which are characteristics inherent to LCA methodology as such and some of which are relevant specifically in the context of waste management [103, 104]. A comparative analysis [105] of selected 20 LCAs in waste management found that LCAs differ substantially in their system boundaries. Half or more of the LCAs either do not mention or are unclear in whether or not life cycle emissions from energy inputs or capital equipment are included in the calculation of results. Only four impact categories are common to more than half of the reviewed LCAs. The human and ecological toxicity impact categories are much less common than global warming potential, acidification, and eutrophication. A financial life cycle costing is present in eight of the reviewed LCAs, while an economic valuation of the environmental impacts is observed in five. The conclusion was that "A comparison of the relative environmental performances of MSW treatment scenario types within each study did not provide a clear confirmation or repudiation of the waste hierarchy. This paper concludes that many recently published LCAs do not ensure that the methodological assumptions are made clear to the reader. Lack of transparency makes the results difficult to interpret and hampers meaningful comparisons between the LCA results." The selection and relative weighting of environmental impacts in LCA is also another controversial issue that demonstrates the inconsistencies and the uncertainties of LCA methodologies [106]. Finally, it is well known that LCA requires huge amounts of data regarding material and waste flows as well as environmental impacts. When these data are not available, the validity and the subjectivity of the assumptions made determine the reliability of the outcomes [107].

Another critique to waste hierarchy comes from a rather techno-political view. In an interesting debate that took place in 2018 in the Wasteless Future blog (wastelessfuture.com), Simon Penney argued that we need a new narrative for waste management because there is ample evidence that the waste hierarchy does not work conceptually or practically. He explained [108] that the reason is that the hierarchy itself is a product of capitalist neoliberal governance, which has an agenda that is diametrically opposed to the concept of reducing consumption, and the only certain way in which we can achieve anything like a sustainable form of consumption is to redefine what consumption means and radically alter its distribution. With his words, "My argument is that our institutional, economic, social structures are set up to stop us from doing this. Whilst we pay attention to recycling rates and other aspects of flawed structural thinking, our collective imagination is not engaged to come up with alternatives to this because we are mostly unaware that it is not working." Ad Lansink replied immediately, arguing that [109] "sustainable waste policy is not obstructed by the waste hierarchy but by the political struggle between economy and ecology. The latter often tasted defeat against economic and social trends." Lansink stressed that he shares the view of Simon Penney that institutional, economic, and even social structures hinder in solving global waste problems, and he concluded that "We must build a bridge between ecology and economy, by telling the people what should be done in a sustainable future. But a new narrative for waste management is not necessary, while we already have the waste hierarchy, with two challenging sides: the top with prevention and reuse, and the bottom line, abolishing or diminishing landfilling, and reducing or changing incineration as much as possible."

4.3.2 Looking for More Systemic and Complex Tools

It is clear that the shift to resource management requires a more holistic approach. The readers can have an idea about this shift looking at Figure 4.6 in which the size of the worldwide revenues of different industrial sectors is represented by the area they cover, based on a relevant report by Statista [110]. The first three sectors in terms of worldwide revenues were manufacturing with $35.5 trillion in 2016, wholesale and retail trade with $20.3 trillion, and the energy sector with almost $8 trillion. As it shown in Figure 4.6, if we consider the whole economic system as a sequence of gears, municipal waste management is the smallest gear at the end of the system, while the big gears are manufacturing, wholesale and retail trade, and extraction of energy resources.

This is true in terms of waste quantities as well. As we discussed in Chapter 3, as an example in EU, MSW is a tiny 8–10% of the total waste generated; still it is the waste stream that is more heavily targeted by regulations and affected by the shift to circular economy. Actually, if someone wants to stimulate change, the main issue that should be considered is to develop circular loops in manufacturing and wholesale and retail trade sectors, because the possibility of circular loops in the energy resources is rather restricted. Thus, optimizing the performance of the last small gear will not change the systemic performance of the economy; there is a need to see waste hierarchy and circular economy in a broader context – but waste hierarchy itself does not become less valid just because the system's economic and social limits prevent its full application.

Moving deeper to the resource management aspects, the limitations of the waste hierarchy for achieving absolute reductions in material throughput have been discussed in details by van Ewijk and Stegemann [111]. Their point of view is that the reduction of material throughput is the key element for achieving a sustainable future and a real circular economy that will respect the biophysical planetary limits. They concluded that there are three major problems. First, the inclusion of an option in a priority order legitimizes its existence. For instance, instead of categorically rejecting landfill, the waste hierarchy states that other options are better than landfill. Second, the hierarchy informs the user on direction of change rather than on the end goal that needs to

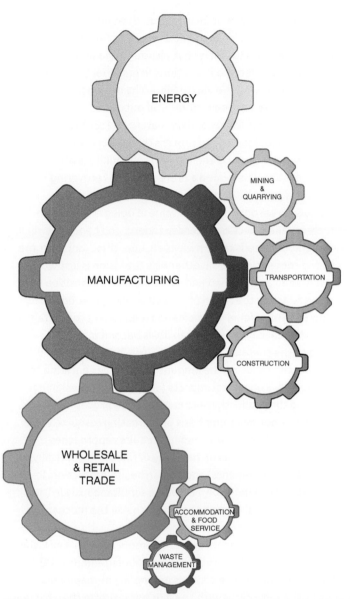

ENERGY

MINING & QUARRYING

MANUFACTURING

TRANSPORTATION

CONSTRUCTION

WHOLESALE & RETAIL TRADE

ACCOMMODATION & FOOD SERVICE

WASTE MANAGEMENT

Figure 4.6 The gears of linear economy and the role of waste management – the size of the gears corresponds to the revenues of each sector worldwide. *Source:* From Ref. [110], with kind permission from Nick Rigas, D-Waste.

be reached. As a result, the hierarchy provides a helpful framework for stimulating incremental progress rather than radical change. Third, the hierarchy merely communicates the relative desirability of waste management options but does not give any pointers regarding trade-offs with activities outside waste management. For instance, the relative desirability of recycling over incineration does not communicate whether limited investment funds need to be directed at a recycling facility rather than a public transport project. Their overall conclusion is that although the waste hierarchy and its current policy implementation can lead to reductions in material throughput by diverting materials from landfill, following the waste hierarchy does not necessarily save natural resources, nor guarantee the best environmental outcome.

Costas Velis, discussing the global markets for secondary materials [112], has stressed the need for a more multidimensional decision-making process that will serve the shift to circular economy writing that "The scientific and engineering community is often preoccupied with technological innovation: it is indeed a prerequisite to recovering resources – closing of the loop. But, can we please dare ask ourselves, what is the fundamental need for this? And how do we obtain optimal value in doing so? Value (in secondary resources) is multifaceted, and facets are interdependent, and hence complex; some value to the society is inherent, but needs to be measured, quantified, and become tangible through a transparent and robust evaluation, otherwise the vision is just words or even a Chimera. We desperately need an informed and evidence-based transition to optimally efficient resource recovery within the realm of circular economy, along with other related concepts." A review [113] of guiding principles and approaches for sustainable products suggested that although the concept of sustainability is becoming more coherent, its application is limited to the contexts in which they have been developed. Some approaches put more emphasis on reducing the amount of resources entering the waste stream and of their associated environmental impacts (e.g. through redesigning components and products or altering the production and consumption processes across all stages of the life cycle), while others put more emphasis on increasing resource value retention. It was concluded that the shift to circular economy "requires approaches that move beyond end-of-pipe solutions that focus on narrowly defined and politically, geographically and/or time-restricted environmental and human health protection. It requires approaches that can assess and evaluate complex value simultaneously in the environmental, economic, social and technical domains and can provide the mechanisms that enable radical and systemic interventions to become mainstream."

Dealing with the transition from waste to resource management, ISWA's Key Issue Paper on Waste Prevention, Waste Minimization and Resource Management [114] highlights that resource management means the process and policy of managing materials and energy throughout their life cycle with the aim to maximize the efficiency of material and energy utilization and minimize loss of material as waste for disposal. In this view resource management is the broadest term and covers the complete chain from "cradle to cradle." It can be seen as evolution of waste management. In contrast to waste management, which exclusively deals with the end-of-life phase, resource management is a holistic approach and covers a variety of actions. Waste is no longer in the foreground but an action or policy that results in material cycle process, as it is shown in Figure 4.7.

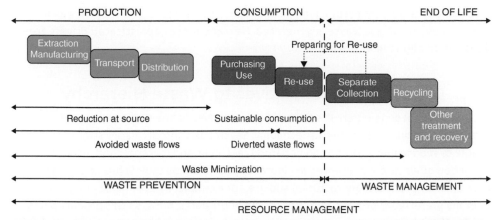

Figure 4.7 Waste management and waste prevention in a resource management context.
Source: From Ref. [114], with kind permission from ISWA.

According to ISWA, a large set of instruments and policies is available to trigger this development [114]. The question is not which instruments we should apply, but how we can apply and combine the available instruments in order to achieve the best results. The tools available are:

- Introduce or enhance recycling schemes;
- Introduce financial stimuli;
- Extended producer responsibility;
- Research and development policy;
- Integration of waste prevention in environmental permitting of SME;
- Integration of environmental criteria in product regulation;
- Product–service systems;
- Sustainable design.

Andreas Bartl, vice chair of ISWA's Recycling and Waste Minimization Group, has analyzed the shift from recycling to waste prevention [115]. He has found out that there are certain barriers for waste prevention, including conflicts of interests, the lack of evidence for decoupling between economic growth and waste generation, the lack of measurable indicators that will make waste prevention accountable, and planned obsolescence and waste exports – trafficking. He stressed that the current economic system is not feasible to realize a high rate of reuse. The overall conclusion was that waste prevention is difficult to measure and counteracts the economic interests of various stakeholders. Producers and retailers are interested to increase sales and turnaround, which are both drivers for waste generation. Even the waste management sector, collectors, landfill operators, incinerators, and recyclers will have less revenue if less waste is generated. New and more sophisticated drivers are necessary to align the interests.

4.3.3 Looking for Alternatives to Waste Hierarchy

There are several ongoing efforts to provide alternatives to the traditional waste hierarchy, more suitable for circular economy. Several existing circularity strategies capable to reduce the consumption of natural resources and materials and minimize the production of waste are presented in Figure 4.8 prioritized by their levels of circularity [116]. Smarter product manufacturing and use, for example, by-product sharing, are generally preferred overextending the lifetime of products, because this product being used for the same product function or more users being served by one product (strategy with high circularity). Lifetime extension is the next option and is followed by recycling of materials through recovery. Incineration from which energy is recovered has the lowest priority in a circular economy, because it means the materials are no longer available to be applied in other products (low-circularity strategy). As a rule of thumb, more circularity equals more environmental benefits. However, there are of course exceptions to this rule. For example, making a product chain more circular may lead to increased natural resource consumption, usually in the form of (fossil) fuels. This occurs in chemical recycling of contaminated plastics that usually requires relatively large amounts of energy to decompose the material to its initial building blocks and then synthesize these building blocks back into material (back-to-monomer recycling).

Zero Waste Europe has also proposed an alternative "zero waste hierarchy" for circular economy, as it is shown in Figure 4.9. The focus is on keeping resources in use for as long as possible. Therefore, the top two levels focus on "products," not waste, expanding the "prevention" level of the traditional hierarchy out into "Refuse, rethink, redesign." This part incorporates any barriers to waste generation including behavioral aspects, bans on single use, and rethinking the business models. The second level, "Reduce and reuse," focuses on keeping used products in use and preventing them from becoming waste, while the third level, "Preparation for reuse," refers to taking waste products and refurbishing them so that they can become usable again. Level four mirrors the traditional hierarchy, with recycling, composting, and anaerobic digestion the ideal "last option" for keeping materials in use. Interestingly, level five puts emphasis on "Material and chemical recovery" over simple energy recovery – this puts more of a focus on retaining materials and resources, rather than converting valuable material into energy through, say, incineration. Finally, at the bottom of the proposed hierarchy are disposal options, with residuals management the last option for the waste that is left over after all valuable materials have been reclaimed.

ISWA has rightly commented [118] that the waste management sector is predominantly occupied with materials that have already entered the waste stream;

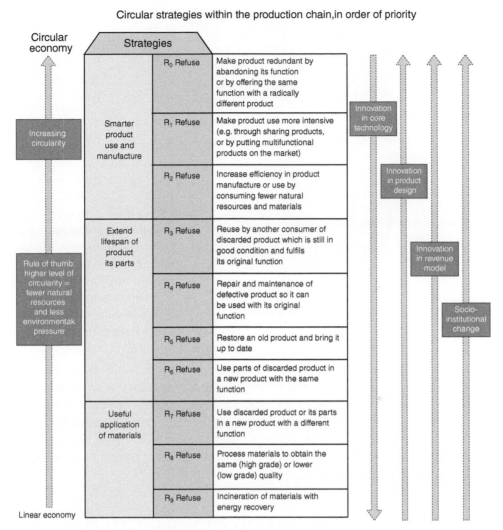

Circular strategies within the production chain, in order of priority

	R_0 Refuse	Make product redundant by abandoning its function or by offering the same function with a radically different product
Smarter product use and manufacture	R_1 Refuse	Make product use more intensive (e.g. through sharing products, or by putting multifunctional products on the market)
	R_2 Refuse	Increase efficiency in product manufacture or use by consuming fewer natural resources and materials
Extend lifespan of product its parts	R_3 Refuse	Reuse by another consumer of discarded product which is still in good condition and fulfils its original function
	R_4 Refuse	Repair and maintenance of defective product so it can be used with its original function
	R_5 Refuse	Restore an old product and bring it up to date
	R_6 Refuse	Use parts of discarded product in a new product with the same function
Useful application of materials	R_7 Refuse	Use discarded product or its parts in a new product with a different function
	R_8 Refuse	Process materials to obtain the same (high grade) or lower (low grade) quality
	R_9 Refuse	Incineration of materials with energy recovery

Circular economy — Strategies

Increasing circularity

Rule of thumb: higher level of circularity = fewer natural resources and less environmentak pressure

Linear economy

Innovation in core technology

Innovation in product design

Innovation in revenue model

Socio-institutional change

Figure 4.8 Circularity strategies within the production chain in order of priority. *Source:* **From Ref. [116], with kind permission from the Netherlands Environmental Assessment Agency.**

thus waste prevention has not yet been fully integrated into the activities and business models of the sector. However, the waste management sector is already contributing considerable services and expertise towards sustainable material and energy management. Most recognized and of particular significance are the provision of secondary raw materials that can substitute virgin materials, replacements for fossil fuels, and the return of carbon matter and nutrients to soils. As a result, resource management is not new to the SWM sector. Historically, the main driver for keeping

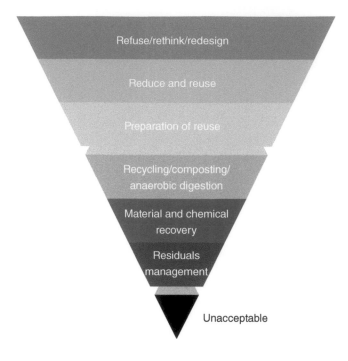

Figure 4.9 The zero waste hierarchy for circular economy. *Source*: **From Ref. [117], with kind permission of Resource Media Limited, adapted by Nick Rigas, D-Waste.**

a selection of resources, such as metals, in circulation has been a simple matter of economics. More recently an increased environmental awareness, focus on global warming mitigation, and concerns on energy and material scarcity have added further drivers. Resource management is also becoming a stand-alone driver for capturing material value from waste. In all these cases economic sustainability depends on secure markets for the recovered raw materials. Now, it's time to discuss the controversial role of recycling in a circular economy.

4.4 SORRY, RECYCLING IS NOT CIRCULAR ECONOMY

We have already mentioned some barriers to recycling in Section 3.5. In this section we will try to dive deeper in the role of recycling in a circular economy. This is an important discussion because as we already showed in Chapter 3 a lot of the circular economy definitions portray circular economy as a more advanced recycling, excluding (or only wishing) in practice, waste prevention, reuse, and repair practices from it. In addition, politicians and the public audience usually perceive circular economy in the same way, like an upgrade to the already existing recycling efforts. The truth is that by identifying circular economy with recycling, you lose the substance (resource consumption and depletion, biophysical limits), but you gain convenience (you do not need to explain that the too much focus on recycling, for so many years, was kind of misleading if not an alibi for increasing consumption). This is, as Al Gore has put it, an inconvenient truth: if we talk about circular economy in order to rebalance the human footprint on ecosystems, then recycling is one of the least preferred options – if we talk about circular economy as a cover that will allow us to continue the business as usual activities, then identifying circular economy with recycling is a great and convenient way to go on. I believe that for several years, politicians, NGOs, citizens, and of course a big part of the waste management sector were facing recycling more as a religion and less as a real day-to-day activity with costs and benefits. The end of recycling as we know it became official with the first efforts of China to ban certain types of dirty recyclables from its imports. In January 2018, ISWA's president blog post [119] was calling to utilize the China ban in order to rethink recycling and its role:

> If we want to be bold and ambitious, we have to grab the opportunity of the China ban to promote another adaptation plan. A plan that will prioritize waste prevention and reuse as the most urgent priorities of any system. A plan that will recognize the current technical and economic limitations of recycling. A plan that will boost eco and modular design, utilizing the unbelievable technological advances of the fourth industrial revolution. A plan that will demand not only the consumers to develop 'greener behaviors,' but also, and mainly, the industries to develop new business models and manufacturing patterns. A plan that will stimulate Circular Economy as a realistic opportunity for specific materials and industrial sectors, rather than as an obligation of the waste sector.

> China's ban is a great opportunity to rethink Circular Economy and to prioritize the development of local closed loops, as a basic condition for the long-term viability of our systems... China's ban is a great opportunity to move away from the fallacy that everything can be and should be recycled. It's an opportunity to face materials' recycling as just one intermediate, imperfect and sometimes costly solution that does not always contribute to Circular Economy. There are scientific works that prove that the more we push people to recycle, the more we cultivate the wrong idea that recycling (and not waste prevention, reuse, ecodesign and the necessary industrial shift of the Circular Economy) is the solution.

In the next sections we will deliver some more arguments that aim to inspire a serious rethink about recycling and its role in circular economy.

4.4.1 Weight-Based Targets Are Misleading

For the last 20 years, EU is continuously increasing the recycling targets setting higher and higher weight-based goals. This has resulted to some good achievements, and undoubtfully it has contributed to better waste management systems, in terms of environmental impacts and resource efficiency. However, after 20 years, the whole approach needs to be reconsidered. A key element regards the weight-based targets. We know now that they are misleading and in many cases they result in losses of financial resources for recycling meaningless materials. Why is that?

First, because the weight of materials that is recycled says absolutely nothing for its purity or for the contaminants involved in it. You may recycle 1 tonne of "dirty" plastics, as the ones that were sent to China to be used as cheap fuel [120]. And you may recycle 1 tonne of pure PET bottles with a high commercial value close to US $300–400/tonne. That is a completely different achievement, both financially and environmentally. The focus on quantity can mean that quality is compromised, with low quality/contaminated materials sent for recycling. Even worst, if 1 tonne of mixed plastics is collected for recycling, we do not even know if all the materials are recyclable. Take the example of food packaging: it is completely unreasonable to require a durable (protective of the food quality), fancy, and flexible food packaging and at the same time to expect that this should be easily recycled or magically turned to something else without posing environmental impacts.

Second, because the benefits of recycling a tonne of each material are not the same, so using weight-based tonnes the environmental outcomes are levelized. As an example, ISWA has assessed [121] that the savings due to recycling (in CO_2 tonnes equivalents) are 10 for aluminum, 2 for steel, 0.5 for glass, and around 1 for plastics. Instead of looking at these benefits, based on LCA analysis, weight-based targets push to recycle heavier materials as garden waste, while the right solution might be, as an example, home composting. As it was clearly recognized in the report "Our Waste, our Resources: a Strategy for England" [122], some lightweight materials have large environmental footprints, like plastics, while some heavy materials have small footprints, like aggregates. The report concludes that "This can encourage behaviours that do nothing to help meet our goals. We will therefore develop new indicators and metrics that help us understand and act more in line with our strategic aims, focusing particularly on greenhouse gas emissions and natural capital."

Third, because with weight-based recycling targets, the pollution aspects and their economic, environmental, and health impacts are ignored. Arne Ragossnig and Daniel Schneider describe [123] exactly how difficult, complicated, and costly are some forms of recycling: "For example, lead, cadmium, and other additives used as stabilisers for polyvinyl chloride (PVC) cannot be removed from this type of plastic. In order to avoid negative health effects associated with these types of contaminants one needs to (1) eliminate the use of deleterious additives to plastics as a means to improve the quality of subsequent plastic wastes, (2) design the waste collection schemes accordingly in order to exclude problematic wastes from the recycling stream to improve the quality of current plastic waste, (3) optimise technical processes for recycling to remove contaminants at least indirectly, for example by removing the polymer PVC potentially containing lead and cadmium as contaminants, and (4) match the type of recycled plastics to specific uses (and reuses) that can tolerate feedstocks with known chemical constituents." The last point is in many cases the most difficult.

Christophe Scharff from the Institute for Water Quality, Resource and Waste Management, Vienna University of Technology, explained that [124] mandatory recycling targets lead to an artificial increase in supply (i.e. not driven by demand). The hope is that this supply will eventually create a market and demand as the legal environment fosters innovation and investment that serve the objectives of growth and environmental protection. However, this line of argument ignores two aspects: even though it is quite safe to assume that in 10–15 years there will still be a market for consumer recyclables like packaging made of glass, paper, metal, PE, or PET, it remains to be seen whether the materials obtained due to the massive increase in recycling targets will actually meet the required quality standards, and if they do, at what competitive cost. Scharff adds that "The second aspect that, surprisingly, has been completely ignored so far is what raw materials the EU industry will actually need over the (specific) planning horizon. It would be presumptuous to assume that an artificially high supply level of secondary raw materials of unspecified quality is enough of an incentive for corporations in a global competitive environment to make radical shifts in their raw materials procurement policies."

4.4.2 Searching for New Metrics

The substitution of weight-based targets with LCA and GHG metrics has been proposed [125] by several studies. A relevant project [126] has suggested that the first step is to widen the examination to encompass the "whole product life" approach to product management because "Weight-based targets have encouraged a simplistic approach based on collecting materials from householders once they have become waste, with little consideration of the relative impact of each element of the waste stream, other than how much it weighs or what value it has as a secondary material." The study interestingly points out that many environmental impacts can be avoided only by redesigning products in the right way and it's important that producers take responsibility for the products that they create and sell (producer responsibility).

Paul Brunner and Costas Velis suggested that [127] there is an urgent need to measure recycling on a more sustainable basis. A recycling metric must be based on clear definitions of inputs and outputs, considering the time axis and material stocks along this time axis. Furthermore, the different qualities and constituents of materials must be taken into account. Plastics and metals are composed of numerous additives and alloying metals. When recycling targets are defined, these mixtures require individual attention in order to yield an optimum and balanced solution regarding economic, environmental, and resource aspects. They concluded that "...the need for safe energy recovery from non-recyclables (or final storage) has to be taken into account... What is needed is an agreement on the types of added value and the contributions towards the higher-level goals aspired to. The potential problematic side effects have to be taken into account, and contributions and side effects can be combined in an evaluation matrix."

It is beyond the scope of this book to get into more details, but we think that now it is clear that the quantity of materials recycled says very few or nothing regarding the environmental benefits achieved. There is a big discussion on what could be a suitable substitute of the weight-based targets, but at least, we should be certain that these weight-based targets are too restricted to reflect the complexity of materials, the relevant environmental benefits, and the pollution involved. They might be simple and convenient for policy makers, but they are misleading, especially when the transition to circular economy is considered.

A very interesting discussion about what are the proper metrics of recycling has already started. We will use the example of paper recycling to demonstrate the importance and the perspectives of this discussion. First of all, there is an issue of consistency. There are at least twelve different definitions for paper recycling [128], with significant differences regarding the term "recycling." In addition, the United States and Japan do not use the term "recycling" but the term "recovery," but the mathematical formulas they use are not comparable. A comparison between these definitions resulted in the following outcome:

> (European recycling) = (European collection) = (Japanese recovery) = (American recovery, excluding uses of recovered paper outside the paper and board industry) = (European utilization, including both domestic utilization and net trade of recovered paper)

Interestingly, the collected recyclable paper plays an important role in all definitions. Recycling is commonly calculated by dividing the collected paper for recycling by total production of paper and cardboard. For the global paper system, this results in a collection rate of 54% [128]. This indicator provides a really distorting picture of recycling in the framework of the global paper material flows. First, it compares a quantity from the pulping stage (paper collected for recycling) with a quantity of the papermaking phase. As it was highlighted, "The metric omits the losses that occur in between the two stages and ignores that not all paper is discarded and therefore not available for recycling. The metric also lacks meaning because its value does not reflect the purpose of recycling. The main goal of recycling is the reduction of impacts by displacing virgin production. A recycling metric can only reflect the avoidance of virgin inputs by focusing directly on the harvest stage of the life cycle. A recycling metric that is both consistent and meaningful should compare waste paper inputs (paper for recycling) with total inputs (paper for recycling plus virgin fibrous harvest)."

Graedel developed [129] a relevant indicator for the case of several metals and named it recycled input rate (RIR). Applying the concept on the global material flows of paper, it comes out that the value of the RIR is 38% while the usually referred recycling rate is 54% [130]. The difference reflects the relatively high yield ratio of recycled pulping compared with chemical pulping. In other words, an increase in paper for recycling inputs does not imply a proportional decrease in virgin input requirements. Due to the differences in pulping efficiencies, 1 mass unit of paper for recycling may either displace 0.9 units of wood for mechanical pulping or 1.7 units of wood for chemical pulping. When paper for recycling substitutes virgin inputs without affecting the ratio between mechanical and chemical pulp inputs, the average global substitution rate is around 1.5. In practice, it depends on the desired properties of the final product whether recycled pulp will substitute mostly mechanical or chemical pulp.

Another important effort is the "reuse potential" that specifies how "resource-like" versus how "waste-like" specific materials are on a continuum [131]. The reuse potential indicator is developed to aid management decision making about waste based not on perception but more objectively on the technical ability of the materials to be reused in commerce. This new indicator is based on the extent of technological innovation and commercial application of actual reuse approaches identified and cataloged. The reuse potential represents the usefulness of a waste with a score between 0 (complete waste) and 1 (complete resource). For example, a score of 0.45 indicates that 45% of the waste can be reused. The reuse potential shows what is technically feasible before other factors such as market demand and government regulation are considered. A similar concept is the "recovery potential." As an example, for the case of paper, the term "recovery" includes recycling (substituting the original material), non-energy recovery (substituting other materials), and energy recovery (substituting fuels); these three activities represent the most widely observed uses of waste in the paper life cycle. For the case of paper, it has been calculated [132] that only 67–73% of fibrous inputs can be supplied by waste paper and the rest needs to be virgin fibers. The current performance for the metric is 38%, which is just over half of the technical potential.

4.4.3 Losing My (Recycling) Religion

Now it's time to put recycling in the right framework. As we have discussed in Chapter 3, even if we recycle 100% of the waste stream, despite the second law of thermodynamics and the problems relevant to quality of materials, the emissions per capita will be reduced by a tiny 1.6%, and energy savings would at maximum contribute to a reduction of 1.8% [133]. This is because of two main reasons: first because of the growing importance of building and managing the continuously rising anthropogenic stocks and second because, as we also discussed in Chapter 3, almost 50% of the GHG emissions are related to primary materials [134]; thus recycling the municipal waste makes really a tiny impact because they represent only 2–2.5% of the global material extraction, something that can't enhance circular loops that will result in domino effects such less extraction, less transport, less energy consumption for manufacturing, lifetime extension, etc.

Redesigning products targeting waste prevention, lifetime extension, reuse, and advanced recyclability is in fact what we urgently need, not to abandon recycling but to put it in the right framework. But when we talk about "products," in this chapter, we mean elements of our daily lives far beyond the horizon of daily consumption and municipal waste. As an example, we urgently need to redesign buildings. The carbon emissions embodied in building materials make up around 40–50% of the carbon footprint of an office building, primarily due to the production of cement and steel required [135]. Nearly 80% of the energy used during construction comes from the preparation of construction materials, primarily the cement and concrete for the foundations and structures of our infrastructure and buildings. The remaining 20% is used to transport materials, remove waste, and use energy on-site [136].

The Circular Construction Challenge that was organized by Realdania and the Danish Design Center, with ISWA's participation, called innovators to help solve the problem of reducing waste in the built environment [137]. As an example of the first steps and innovations required, the three winners provided (i) a new sustainable building material – like baffle plates and insulation – produced by growing fungus/mycelium spores in waste flows, which can replace existing building materials with negative environmental impacts; (ii) reused timber in large-scale retail sale that is a system where timber from temporary constructions at construction sites is collected and sold for reuse via the building market; and (iii) a building system for sheds (e.g. for bikes, waste bins, or office supplies) made from recycled timber and tiles from roof material waste. A view of the six more interesting ideas will also be very helpful for the reader [138].

In another interesting project that promoted the necessary dialogue between designers and the waste sector, focusing on the challenge of circular economy regarding jeans, ISWA made several recommendations [139] to improve the design of jeans. The most important one was regarding the materials used. Mixed-material textiles were found difficult to recycle and often end in downcycled applications. Inexpensive, garments of low quality encourage disposal and replacement rather than repair and reuse. Designers are in a position to improve the sustainability and recyclability of textiles by using mono-material textiles in garments where possible, making it easier to produce high-quality recycled textiles. In addition, using high-quality textiles and supply repair accessories will result in longer-lasting, better clothes. The relevant report concluded that designers should partner with the waste management sector that has great experience with end-of-life products, including textiles.

The fashion industry has to play an important role in the path towards sustainability and the circular economy – a relevant discussion shows both the limits and the opportunities of circular economy and recycling. Indeed, the fashion industry is a sector with a high environmental impact; it involves a very long and complicated supply chain, which is associated with large consumption of water and energy, use of chemical substances, water and air pollution, waste production, and finally microplastic generation. As an example, according to the report "The New Textile Economy," more than US $500 billion of value is lost every year due to clothing underutilization and the lack of recycling. Furthermore, as we already mentioned, total GHG emissions from textiles production are assessed at 1.2 billion tonnes annually, more than those of all international flights and maritime shipping combined [88]. However, despite some positive trends (e.g. H&M's website now shows that the company currently makes 57% of its products from recycled or "sustainably sourced" materials; Zara has developed its own recycling program), only 12.5% of the global market players have set goals relevant to circular economy [140]. In addition, the industry is expected to grow by 4–5% every year, resulting in 170% growth between 2019 and 2030. As Gulnaz Khusainova, a fashion start-up founder, wrote in Forbes [141], "with these growth rates, it's impossible and irresponsible to expect circularity to fix the problem. This becomes even bleaker considering that the above-mentioned report admits that the industry's sustainability efforts are already stalling…In fact, the report finds that 40% of all fashion companies have not even begun to take sustainability seriously by setting targets and rethinking their supply chain. Among the rest of the 60%, a lot of the improvement is happening with small companies (or those with less than $100 million in revenue per year, which includes many startups) and mid-sized companies (which make less than $1 billion in revenue per year). Among the biggest players in the market, which make billions in revenue every year, the pace of improvement has basically stalled out. The only way the industry can address the climate catastrophe is to slow the growth. No amount of reusing or recycling will offset the continuous growth of the industry as it stands, so the solution is to simply produce less."

We believe that now it's clear, through the examples mentioned above, that recycling is actually a part of the linear economy and the business as usual model. Recycling, especially in the developed world, starts when materials and products are already thrown away and the effort is to divert them from waste disposal operations. Thus, by definition, this is a practice that is not designed to increase or preserve the value of materials and products but rather to minimize the environmental impacts of waste management. As successfully has been mentioned by Alexandre Lemille [142], "We need to understand that recycling is not an effective strategy for dealing with unused resource volumes in a growth model. We will find ourselves in a never-ending pursuit of continuously generated waste, rather than seeing the avoidance of waste as a path to beneficial innovations on many levels. Of course, it is easier to think about recycling. This avoids changing the whole of our volume-based production model. But in a world where we have to shift our consumption patterns and use less energy, recycling no longer has all the answers." But recycling is not the same in developing and developed countries. Now it's time to answer again the question: why do we recycle?

4.4.4 Why Do We Recycle? Service vs. Value Chain

In general terms, it seems that recycling activities are based on different drivers in developed and emerging countries and this has serious impacts regarding the required interventions and policies for improvements. In brief, recycling in rich and middle-income countries is a service driven by the cost of treatment and disposal, as well as the need to reduce the environmental impacts of waste management, and it is usually organized by municipalities or private companies working for them. On the other hand, in poor and emerging economies, recycling is a profitable activity driven by the value of resources and organized by private sector companies, cooperatives, or individuals. This difference explains why the efforts to promote municipal recycling in poor countries result almost always in serious failures.

The following text is based on Anne Scheinberg's work for UNEP-ISWA Global Waste Management Outlook 2 that is expected to be published by the end of 2020. Anne is the chair of ISWA's Recycling and Waste Minimization Group and a great expert in recycling and informal sector aspects [143, 144].

In high income countries, the world of solid waste is where the public services have a goal and an earning model based on the idea of *removal of waste*, or to say it another way, the *removal of disvalue*. The product of the service chain is the clean empty space left behind when waste is picked up or streets are swept or litter bins are emptied. The service chain in a specific country or city may include both public and private stakeholders, such as collection companies or landfill operators, but the core is always the public responsibility for maintaining clean streets and a safe environment, which makes solid waste a governance function. In an increasing number of countries and small- and medium-sized cities, public governing bodies such as a Ministry or a Public Works Department may choose to contract basic solid waste and street sweeping services to private companies or to require the individual service users to hire a contractor and pay them directly. There are many different flavors of private sector participation in the service chain, but if the streets are dirty or the waste is building up on empty lots, it is the government that has to solve the problem. The public companies or private contractors are paid *by the hour or the service unit* (such as a house or a km of street) for the work of creating the clean, empty space that is left behind after they have done their work.

The other landscape is the private recycling value chain as we can see it in both rich and poor countries. It starts from individual or private entrepreneurs – the so-called guy with a truck, who goes around and collect used furniture or bricks or glass bottles or textiles. They are not interested in that clean empty space left behind – they are happy to leave that to the service chains. Their interest is in materials that have value, which they can recover and trade. All of these private companies – even informal or micro recyclers – are in the business of trading. They are paid by the kg for the *conserved value added* in the products and materials that they extract from the waste or capture directly from the generators.

It is quite difficult for someone in the government, or even a person looking out their window and seeing a truck stopping to pick up materials, to understand the difference between the *removal function* of collecting waste and the *extraction* or *urban mining* function of recycling collectors, and from the outside the distinction is not always clear.

What pays for *municipal recycling as a service* as it has evolved in high-income countries is not the value of the materials, but the *avoided cost of disposal*. In a typical municipal recycling system as it was operated in the period 1995–2010 in North America, Australia, or Northwestern Europe, the separate collection and processing of recyclables and biowaste is a cost reduction measure, motivated by the need for environmental protection. Municipal recycling is not motivated by any scarcity of resources, nor by the value of the materials or their potential revenues. Recycling (and composting or other forms of organics recovery) represent a less expensive destination for a significant part of the materials that must be managed. Municipal recycling is an activity of the service chain that is directly related to a waste management system that needs to provide affordable disposal via a controlled or sanitary landfill that delivers adequate levels of environmental protection, in Europe frequently in combination with a waste-to-energy (WtE) incinerator.

Resource management is a nice idea and a way to conserve natural resources, but it does not motivate government institutions in high-income countries to invest in recycling. Robust demand for recyclables is a necessary condition for the public sector to organize recycling, although it is an important factor in keeping the costs of municipal recycling low enough to contribute to overall service chain functioning. Demand, then, is a necessary but not a sufficient condition: the real driver is the price for disposal. In general, the price for disposal has to be higher than the supplemental cost to add recycling to a functioning service chain. While the specifics differ from place to place, a good estimate of the *tipping fee* price that is sufficient to motivate a commitment to municipal recycling is around US $40–50/tonne. When the price for disposing a tonne of mixed waste is more than this, municipal recycling creates a virtuous circle of system change, with investment in separate collection, increasing diversion from disposal producing more and more recycling (and composting), which reduces the waste to landfill so that the price of landfilling (or incineration) of residual waste gets higher and higher and it is easier to see the benefits.

In summary, in high-income countries solid waste systems municipal recycling is adopted when the price of landfilling exceeds the ancillary cost of adding recycling operations to a waste management system and is motivated by *the driver to protect the environment*.

The value chains represent the "other face" of SWM and are collectively referred to as *valorization*. Valorization or material recovery involves diverting used products and materials from the waste stream, or extracting them from legal or illegal accumulations of waste, in order to trade them. The trade is based on the *retained value added*, which was produced by making the materials or items. Retained value added is thus the residual value that is still available at the time the user discards them. Valorization begins with knowing what is valuable, finding and claiming the materials, removing contaminants, upgrading them, and trading them to the value chains. In emerging economies, most recycling is driven by real economic demand from the value chains.

Entrepreneurs in recycling or reuse extract materials for which they know that there is real demand and established prices and trade them to the industrial value chain.

Value chain recycling is an industrial activity that is globally organized and completely private. Value chain recycling occurs when the commercial value of the recycled materials exceeds the cost of recovering them from the waste stream, upgrading them to meet industrial specifications, and transporting them to a company with real demand for precisely that material at that grade of purity. Prices and standards are set by the global industrial leaders – usually large producers located in East Asia – and "cascade" down to locally paid prices that depend on quality and quantity of materials supplied, adjusted for transport costs and distance to the large buyers. This means that there is a kind of hierarchy of types of recyclables according to their value and marketability.

4.4.5 Usual Misconceptions About Recycling

All the potential benefits of recycling can be categorized in two major categories: one related to the diversion of materials from landfills and incinerators and one related to displacement of primary materials. However, there is a serious question: is the displacement of primary materials really happening? Zink and Geyer have demonstrated that displacement of primary production from increased recycling is driven by market forces and is not guaranteed. Very often part of the recyclables becomes a kind of stock due to the market prices, and if some time passes without finding outlets, this stock might become a new type of waste, or its quality might be deteriorated. In this case, this stock will end up in a landfill or in an incinerator, as it happened with many plastic recyclables in the United States, Australia, and EU after the China ban. Improperly assuming all recycled material avoids disposal underestimates the environmental impacts of the product system. This is a serious problem for the relevant LCA models that need to be updated so as not to overstate the benefits of recycling. But the most important thing is that policy and decision makers "should focus on finding and implementing ways to increase the displacement potential of recyclable materials rather than focusing on disposal diversion targets" [145].

A simplified example, developed for this book by the authors, for the impact of recyclables that do not displace primary materials on disposal ratios is shown in Box 4.1.

The graph shows that what really matters for the disposal ratios is the sub-stream S, the materials that actually substitute primary materials and not the diversion rate. When market conditions create stocks of recovered materials that can't be sold and become obsolete, the increase of the diversion rate creates limited impact to the disposal ratios. For stocks above 20%, the system is stabilized after some years, and the increase of the diversion rate becomes practically irrelevant. Both the assumptions that the waste generation remains stable and the 3% annual growth of the diversion rate are rather optimistic.

Box 4.1 Disposal Ratios Affected By Recyclables That Do Not Displace Primary Materials

Assumptions:

a. Waste generated remains stable at 100 units/year.

b. The diversion rate (source separated waste) starts from 20% and increases by 3% each year reaching 26.1% after 10 years.

In Scenario 1, the diverted waste is divided in two sub-streams (Re+S = 100%) where:

- Re = 20% residual (impurities, materials that are not recyclable)
- S = 80% recovered materials used to substitute primary materials

In Scenarios 2–5, we have three sub-streams (Re+St+S = 100%, St+S = 80%):

- Re defined as before.
- St (from 10–40%) refers to recovered materials that are stocked due to market conditions and become obsolete after one year, joining the Re stream.
- S (70–40%) refers to recovered materials that are used to substitute primary materials.

Figure 4.10 shows the time evolution of disposal ratios for different scenarios.

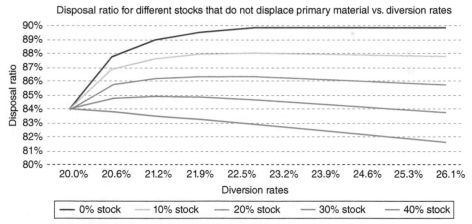

Disposal ratio for different stocks that do not displace primary material vs. diversion rates

Figure 4.10 Diversion rates might be completely misleading when there are important quantities of recovered materials that are not used to displace primary materials and become obsolete stocks due to market conditions.
Source: With kind permission from Nick Rigas, D-Waste.

Figure 4.10 shows that, in accordance with other findings, the increase of the weight of the waste quantities that are diverted to recycling programs does not necessarily means that they will be diverted from disposal. If, due to market conditions, the materials are not used to substitute primary materials (Scenarios 2–5), then finally a high disposal ratio is stabilized independently of the increase of the diversion rate. As it was mentioned in this case, "…any material that is produced must eventually be disposed; recycling cannot alter that eventual fate. Rather, recycling only reduces disposal if and to the extent that it displaces primary production. If recycling fails to prevent primary production, for instance, by displacing other secondary materials or by expanding markets, it also will not prevent disposal – it will merely delay that fate" [145].

Table 4.5 presents usual misconceptions about recycling that ignore the science and the market functions behind it, based on the article "Common misconceptions about recycling" [149] and other relevant research papers.

Table 4.5 Usual misconceptions on recycling.

Misconceptions about recycling	
Recycling means displacement of primary materials	As it was shown in Figure 4.11, market conditions will determine the fate of the recovered materials. Thus, recycling does not automatically bring the benefits of the displacement of primary materials, which are the main benefits [145]
Recycling materials multiple times is better than once	Cascading effects and chemical pollution undermine the recycling benefits, and at a certain point the inclusion of virgin materials is definitely required, even for materials like aluminum [146] and steel [147] that are rather easily recycled and without too much cascading effects
Closed loops are better than open ones	This is also a confusing argument because the logistics of the closed loop and the exact type of substituted materials finally determine the preference order
Recycling is a decentralized activity with small financial cost and huge financial benefits	This is completely misleading for municipal recycling in middle- and high-income countries. The cost of source separation programs and the cost pf reprocessing materials in material recovery facilities (MRFs) are usually 4–5 higher than the revenues from the sales of recyclables. MRF operations demonstrate important scale economies with processing costs reaching the range of €70–100/tonne [148]

Source: Created with ideas from Ref. [149].

Geyer and his colleagues concluded [149] that the environmental benefits of recycling are finally described by the following equation, which presents the whole concept in a realistic and tangible way:

$$\text{Environmental benefits} = D \times (E_{prim} + E_{landfill}) - E_{repro}$$

where

D: the quantity of primary materials substituted by recyclables
E_{prim}: the environmental benefits from avoiding primary materials extraction
$E_{landfill}$: the environmental benefits from avoiding landfilling
E_{repro}: the environmental burden for collecting and reprocessing recyclables

To conclude this section about recycling in a circular economy framework, we think it is very useful to put the right questions that will allow us to answer how much recycling we need and for which materials, as they were phrased in the article "Common misconceptions about recycling" [149]:

- What and how much primary resource does a recycling activity displace?
- Which uses of a recycled material have the highest impact reduction potential?
- What is the best level of reprocessing? Does further increasing the value of the secondary resource increase its impact reduction potential?
- How can the triad of collection, recycling, and displacement rates be increased in a way that increases impact reduction potential?

4.5 WASTE MANAGEMENT GOES BEYOND WASTE

Waste management is reshaped by the interaction of five global trends. These five trends made their influence obvious during the period 2015–2020, and their combined effect stimulates a systemic change in waste management theory, practices, business, and legal instruments.

- Global warming increases the importance of waste management as a way to achieve immediate reductions in CO_2 emissions while reducing pollution and improving health and environmental protection. The incorporation of waste management practices in resilience and adaptation plans is an urgency for existing and historical infrastructure.

- Circular economy became a mainstream narrative for the last five years. Although there is no concrete definition and theory behind it, circular economy pushes waste management towards a better integration with resource management and integrates waste management practices in each and every supply chain. Circular economy concerns a much broader change than waste management practices. The concept should be discussed and further adapted to the needs and practices in low- and middle-income countries in which informal practices are already shaping circular practices.

- Marine litter and plastic pollution have also become top priorities on a global scale due to the massive evidence regarding their leakages and impacts in marine ecosystems. Plastic pollution in oceans and water is considered as a failure of the plastic industry and our modern waste management systems. As plastic has a central role in human societies, marine litter and plastic pollution have triggered a global wave of new regulations and measures to reduce plastics (starting from single use ones), advance their recycling, and redesign them for recyclability.

- The rise of the fourth industrial revolution (IND4.0) redefines the terms "resources" and "waste" through the new technological advances that reshape manufacturing, raw materials, and products. A new massive e-waste stream is expected as a result of the growing impact of IND4.0, while IND4.0 is already transforming the waste management industry delivering new solutions and data driven business models (see Chapters 5 and 6).

- Starting with China, many countries that were receiving recyclables from the United States, EU, and Australia raised quality standards to avoid pollution related to "dirty recyclables" imports. These moves, supported by a recent Basel Convention resolution, create a new global landscape especially in plastic recycling and oblige the exporting countries to find new domestic solutions. As a result, governments, local authorities, and the recycling industry in many countries have to face the end of recycling as we know it and deliver innovative business models for sustainable recycling programs.

Those five trends (and their combination) are stimulating the end of business as usual in waste management, something that the waste management industry, the local authorities, and the governments are already experiencing in their daily business. All those five trends have many interlinkages between them, but finally, from the view of waste management, they all converge to reshape what we call waste management and its daily practices. In addition, they put new serious questions that require scientific answers pushing practitioners, researchers, academics, and public officers towards the formation of a more cohesive and systematic theory of waste management. Table 4.6 shows the impacts of those trends on the "software" and "hardware" of waste management, while Figure 4.11 presents a mind map with the most important correlations between the five trends and waste management, from a systemic point of view.

The obvious impact of the five trends is that they push waste management closer to resource management, integrating more the waste management systems in the

Table 4.6 Five trends that reshape the "software" and "hardware" of waste management.

Main trends	Why is it important?	Impacts on "hardware"	Impacts on "software"
Global warming	Waste management can substantially contribute to carbon dioxide emission reduction. Emphasis in material recovery can trigger domino effects and further savings in the whole supply chain	Uncontrolled dumpsites must close. Landfill gas recovery becomes necessary. Resource recovery is upgraded; organic fraction should be diverted from disposal sites. Landfills and compost in soil as final sinks are becoming important topics. Energy recovery becomes more beneficial, and carbon capture and storage is considered for WtE plants	Carbon credit and funds related to GHG reduction are becoming key tools for finance. Life cycle analysis of emissions becomes the key tool for decision making. Waste infrastructure becomes a central component for resilience and adaptation plans. Emphasis to recycling becomes necessary
Circular economy	Circular economy is a global trend that redefines the relationship between economic activities and their environmental footprint. Waste management is a key element for the development of circular material loops	Waste prevention, reuse, and resource recovery are prioritized. Recycling is downgraded. Landfills and WtE are becoming suitable only for residual streams, with no valuable recyclables. Technological and business model innovations on circular loops are stimulated	The role of informal recyclers in developing countries becomes more important. Closed material loops on a global scale require international coordination and better markets for secondary materials. The control–avoidance of waste trafficking in the name of circularities is a key for a successful transition. EPR and deposit return systems are reconsidered. Recycling targets are getting higher
Marine litter	Plastic pollution is becoming a planetary challenge, like global warming. Proper waste management practices are key for the elimination of plastic leakages in oceans and rivers. Circular economy models for plastics is necessary	The advances in plastic recycling, innovation on plastic design, and the development of closed loops is becoming a priority. Waste collection and proper disposal for the nonrecyclable part, especially in the developing world, are an urgency in order to reduce leakages	New regulations and legal frameworks are developed worldwide. Proper standards for eco-design should be developed. Funding for marine litter prevention and management is becoming more available. EPR and deposit systems are becoming very important for plastic management

(Continued)

Table 4.6 (Continued)

Main trends	Why is it important?	Impacts on "hardware"	Impacts on "software"
IND4.0	IND4.0 redefines raw materials, manufacturing, and waste. In addition, it transforms business models and technologies in the waste management sector. IND4.0 and circular economy will determine the future of the waste industry	E-waste becomes the emblematic waste stream of IND4.0 creating serious challenges in existing waste management systems. Automation and robotics gradually appear in all waste infrastructure. The evolution of data-driven collection and recycling systems transforms the traditional business models and drives the rise of on-demand tailor-made services. Driverless collection is still far, but visible	The rise of software platforms and mobile apps provides new governance tools. Data-driven systems allow better financial recovery practices. Monitoring and performance assessment are becoming much easier through sensors and the Internet of Things. Users' and providers' inclusivity are stimulated by digital interfaces and tools. Institutions should be built from the very beginning with proper information systems and new governance patterns are required
The end of "dirty" recycling	The end of "dirty" recycling brings new emphasis in the quality of recyclables. It redefines the viability and the business models of recycling programs in the western world and the rich countries	Global markets and supply chains, especially for plastics, are reconsidered and rearranged. The viability of recycling programs should be better assessed and based on less global and more domestic solutions. Cost and benefits for each potential recycling stream should be carefully considered	Quality- versus quantity-based recycling targets are becoming necessary. In addition, with the rise of circular economy, recycling should be reconsidered as a component of every waste management system. Legal, administrative, and regulatory frameworks for the development of local and domestic closed loops are becoming very important

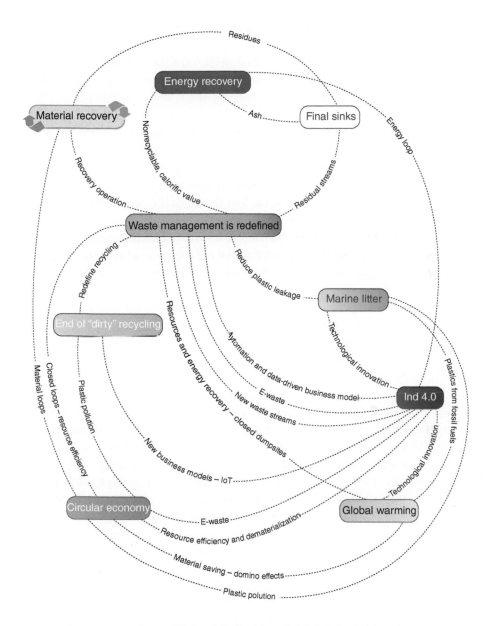

Figure 4.11 A mind map with the interconnections of the five trends.
Source: With kind permission from Nick Rigas, D-Waste.

broader resource management. A closer examination of Figure 4.11 shows that
the triangle IND4.0, circular economy, and waste management are very strongly
interconnected. Clearly, the future of waste management depends on the dynamics
between IND4.0 and circular economy, as those three nodes have the most connections

with other elements and between them. However, there are more specific issues that should be discussed.

The final sinks (on top of the figure and left – middle) are the final receptors of all the flows; after them there is no other receptor. As we will discuss in the next section, the role of final sinks in the context of a circular economy is much more important than in the linear economy.

The system boundaries are much broader than the traditional waste management systems. Actually, someone can say that the combined effect of the five trends is that they integrate waste management as an organic component of each and every supply chain and they upgrade it to a global component that becomes crucial for the viability of any industrial activity. The business case will be similar to the software expansion. Software development and applications are required in each and every modern industry, and they are becoming a core element of it – this is a blue ocean for the software industry, despite the fact that most of the big companies have internalized their software applications and they have their own IT departments. A similar case should be expected for waste management too [150].

As it is demonstrated in Table 4.6, both the "software" and the "hardware" of waste management are going through substantial changes. The changes are pushing traditional waste management systems towards their limits and clearly require a systemic change rather than partial improvements. Such a systemic change requires to include, in the description of waste management systems, upstream subsystems that link the transformation of raw materials into waste. This approach is also helpful to the analysis of waste prevention activities and the quantification of the biogenic carbon present in waste [151]. However, such an approach requires a redefinition of the functional units of the waste management system and different metrics than we are used too. As an example, a circular waste management system that focuses on the management of food waste might serve the aim to provide food for the population of a given region and to valorize the generated organic waste into a fertilizer that is looped back into the food production subsystem. One parameter that could quantify the primary system function (waste valorization into a fertilizer) better than the incoming mass of raw materials into the system would be the area of land that is fertilized. In addition, the circular loops for different materials are not always compatible between them, not only in terms of cost or in terms of markets available for the recyclables, but technically speaking.

4.5.1 Complexity and Uncertainties Become the New Normal

Let's see a realistic circular waste management system for textiles, as the one shown in Figure 4.12.

The smooth operation of such a system depends on the assumption that textiles sent to recycling are free of environmental burden and that reused products and

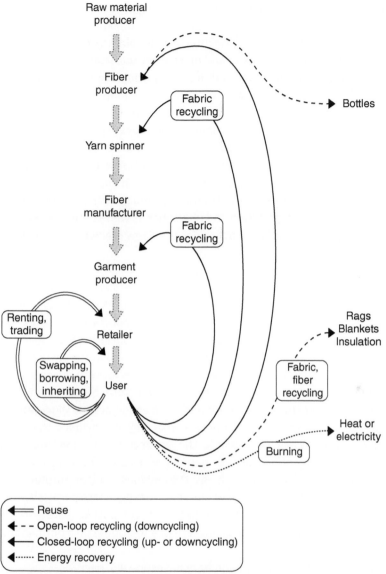

Figure 4.12 A circular system waste management system for textiles.
Source: From Ref. [152], © 2018 The Authors, published by Elsevier Ltd. under the CC BY license, adapted by Nick Rigas, D-Waste.

products made from recycled materials replace products made from virgin fibers. Such an assumption is by definition leading to idealized but not implementable systems. In addition, the smooth functioning of the system depends on the smooth functioning of five closed loops (two reuse loops on the left side and three closed loops on the right side), two recycling open loops (on the right side), and an energy recovery linear end. In total, the system's function depends on the smooth function of eight subsystems, eight markets and prices, and eight technological linkages. It is obvious that such a system has an exponentially higher complexity and much more uncertainties than the traditional waste management system. In addition, the pricing, logistics, and the specifications required for the closed loops and recycling open – loops will be very difficult to be simultaneously optimised and fit each other, especially in the case of global supply chains. However, in this case, the impact of IND4.0 is expected to be a very substantial one. The full digitalization of the supply chains will allow better control of the emerging complexities and early identification of the potential interactions that might disrupt or even destruct the whole system (more about it in Chapter 7). It will also stimulate better alignment between markets and by-products within the supply chain. Still, the problems of chemical pollution and impurities as well as the problems due to the complexity of the materials involved are not problems that will be managed by the digitalization process.

Two other practical examples, for the uncertainties involved when waste management expands into supply chains, are provided by Genovese et al., one for food waste that turned to biodiesel and one for a chemical supply chain [153]. Both supply chains are proven vulnerable to external economic factors, like imports and exports, manufacturing costs, the carbon credit prices, etc. The authors concluded that "while environmental benefits may be obvious, the implementation of circular supply chains may be challenging from an economic point of view. Thus, bottom-up initiatives at a supply chain level might need to be incentivized through some form of top-down governmental support." Another study [154] examined the management of digestate from anaerobic digestion in the framework of a circular economy. It concluded that "Although waste can be reused as much as possible, the aim of closing the material loop cannot be achieved without confronting the management of residues from production and recycling process. As for digestate management, the residues could be surplus digestate, ashes from digestate incineration, and bio-char from digestate pyrolysis. …the destinations of these residues may include utilization of compost/digestate in agriculture, production of landscape restoration materials, and disposal in a permanent waste deposit site."

As we have already discussed in Chapter 3 and here in Section 4.3, decision-making tools have to be reconsidered in the context of a circular economy. According to a study, what is required to assess the impacts of circular economy [155] is (i) a macroeconomic model, (ii) a systems' engineering model for waste management, and (iii) an LCA model for waste management. This set of tools is primarily suited to assessing market-based instruments and environmental regulations. The authors concluded that "Considerable resources were needed for developing and using the set, and there are clear limits as to what can be addressed. However, if only one of the models had been used, neither the range of instruments nor the scope of impacts would have been possible to cover." The most impressive part of this approach is that a new macroeconomic model needs to be embedded in decision making in order to evaluate the policy options relevant to circular economy. What we need more to understand that a systemic change is required in waste management theory?

Professor Massarutto has described that the economic analysis of circular systems should consider the law of diminishing returns [156] in the capacity of the system to improve the separation level in terms of materials flows: this means that when the source separation levels increase as a percentage of the total waste generated, the purity of materials collected is becoming lower and the marginal unit costs for separate collection are increasing too. He has also proved [157] that the critical threshold of optimal recycling seems to be around 50%: trying to push source separation level beyond this level implies higher financial costs (in the range of 30–60%), and these are not compensated by positive externalities. Moreover, considering externalities reinforces this statement, at least when energy recovery is optimized and once the benefit arising from reduced emissions of the displaced energy sources is accounted for. This outcome contradicts the one of many LCA studies, where recycling most often results as a preferable option. In the same paper it was concluded that intensive recycling makes sense in economic terms only if the actual recycling rates are very high: which implies not only source separation levels above 75% but also that the quality of materials collected is high as well. As it was written, "This target can sometimes be achieved, as some experiences in Northern Italy already show, at least in small centres; yet it is questionable whether they can be extended at higher territorial scales. In any case, they require a high degree of public participation, active involvement and education, since households are required not only to separate many different waste flows, but also to prevent waste generation by making smart shopping choices, sorting and washing of materials at home, domestic composting, etc."

It is clear that the increasing complexity of the emerging waste management systems and the integration of all the subsystems' uncertainties are serious threats for any broad shift to circular economy. For that reason, the viability of the transition to a circular economy, as well as the viability of any circular economy closed loop, depends on the existence and capacity of final sinks. In addition, as we will discuss right now, high-quality recycling requires more capacity in final sinks than low-quality recycling.

4.6 FINAL SINKS DURING THE ANTHROPOCENE

We have already discussed and outlined the role of final sinks in Chapter 3. Here, we will focus more on them, as final sinks are a key element for the shift to circular economy. In Section 3.2, we presented that the primary flows that are coming out of the technosphere (like emissions and waste) enter the biosphere and, after their transformational use, finally they are disposed of in final sinks, like land, atmosphere, rivers, oceans, etc. We also mentioned that for millions of years, the available primary flows from biosphere and the available final sinks were much larger than the requirements and the outputs of the technosphere. In these conditions, the biosphere was able to offer (primary resources and energy) and receive (emissions and waste) without undermining the continuous flow. In these conditions, the biosphere takes easily care of recycling, and it is capable to absorb and metabolize the outcomes of the technosphere.

Circularity occurs as the natural rule for the flows between biosphere and technosphere as a result of the high capacity of the final sinks in comparison with the outflows of the human activities. This changed with the industrial revolutions and the systematic exploitation of fossil fuels, which is directly linked with the rise of human population and the urbanization process. Due to its increasing size and rate of growth, technosphere now requires a much larger and faster supply of primary flows, while the emissions and waste going into the biosphere require much larger capacity of final sinks. In Section 3.4 we explained that in this case, the biosphere is not able to metabolize–circulate the waste and emissions coming out of the economy. The case of the high carbon dioxide emissions in atmosphere is a characteristic example of a final sink (atmosphere) that is beyond its limits to receive and metabolize the outflows of the global economy. The case of plastic marine litter in oceans is another case that illustrates the importance of final sinks for the global ecosystem. Now, let's try to see the role of final sinks and waste management in a more systematic way. Our guide for this discussion will be Professor Paul Brunner that has developed this concept in many works and demonstrated its high importance for waste management.

Generally speaking, the exponential per capita growth in material consumption and the introduction of millions of new materials and chemicals are the basic characteristics of our era, sometimes also called the Anthropocene. Paul Crutzen wrote [158] that "Human activities are exerting increasing impacts on the environment on all scales, in many ways outcompeting natural processes. This includes the manufacturing of hazardous chemical compounds which are not produced by nature, such as for instance the chlorofluorocarbon gases which are responsible for the 'ozone hole.' Because human activities have also grown to become significant geological forces, for instance through land use changes, deforestation and fossil fuel burning, it is justified to assign the term 'anthropocene' to the current geological epoch. This epoch may be defined to have started about two centuries ago."

If we combine the huge material consumption and the myriads of new substances with the growth in population, the result is a continuously increasing and complex system of material flows and stocks, particularly in urban areas. The view of urban metabolism is very helpful here to outline the conceptual relationships. According to Christopher Kennedy [41], urban metabolism is "the sum total of the technical and socio-economic process that occur in cities, resulting in growth, production of energy and elimination of waste." In this view, recycling should be considered as a tool with different impacts in growing and shrinking economies. On the one hand, recycling can help growing (in size) economies to control their environmental impacts. On the other hand, shrinking economies can use recycling to ensure more self-sufficiency and smooth transitions towards a less intensive resource management.

4.6.1 Chemical Pollution Is a Serious Threat

US EPA has more than 85 000 chemicals [159] listed on its inventory of substances that fall under the Toxic Substances Control Act (TSCA). But the agency is struggling to get a handle on which of those chemicals are in the marketplace today and how they are actually being used. To get an idea of the material flows, Brunner and Kral noted [160] that "The mass flows range between 0.75 kg/cap/yr for zinc and 1,100 kg/cap/yr for carbon. The per capita substance stocks amount from 1 kg of P to 420 kg of iron... one has to take into account that about 30,000 substances are produced and consumed in large amounts, and that close to one million new (mainly organic) substances are discovered and described every year." The European Environment Agency has visualized [161] the chemical risks in the form of an iceberg, as shown in Figure 4.13. Out of the 100 000 individual chemicals produced and used worldwide, only 500 have been extensively characterized for their hazards and exposure risks, even though it is almost recognized that 70% are hazardous to health and/or environment with 30% out of these 70% creating long-term effects.

To understand the importance of final sinks, someone has to consider that for specific substances the amounts emitted to water, air, or soil are huge and, in case they are accumulated in an uncontrolled way in the environment, they will create serious health and environmental impacts. Let's start with carbon: it's a main feature of all human beings and animals, of combustion engines, and of coal-fired power plants to oxidize organic carbon to carbon dioxide to utilize energy. Men as well as machines depend upon the atmosphere to dilute the resulting CO_2. In the case of carbon emissions, atmosphere acts as a final sink. Or let's take phosphorus: the amount of phosphorus and nitrogen directed to soils and water (both of them are final sinks) is large because of the need for nutrients in agriculture. For agricultural purposes also we use too much pesticides: large amounts are applied to natural systems such as soils or biocenosis in order to fight pests. Each of these substances is directed to a substance-specific sink. The important thing is that due to the high material turnover, some of these sinks are overloaded and lose their important regulating functions [160].

The outflows of the technosphere are either recycled or accommodated in sinks. Heading towards a circular economy means that the amounts of materials for recycling and disposal will be much larger due to the fact that the large amount of materials in stocks reach the end of their lifetime and must be managed together with the output flows. So, it seems reasonable to consider that the objective of future materials management is to ensure long-term supply of resources on one hand and safe disposal

The unknown territory of chemical risks

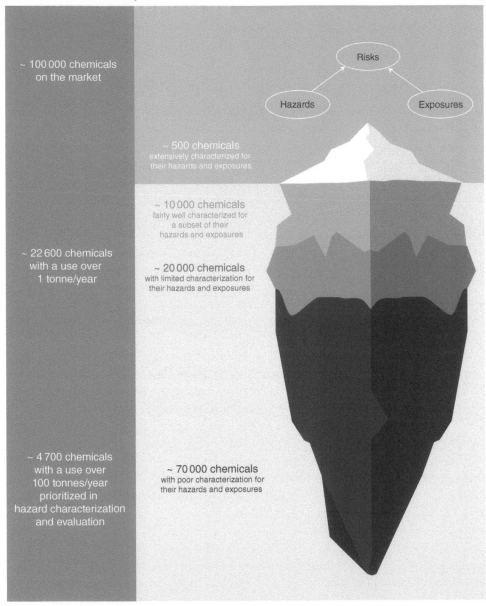

Figure 4.13 The unknown territory of chemical risks.
Source: **From Ref. [161], with kind permission from the European Environment Agency.**

of emissions and wastes on the other hand. If we agree on that, the questions that emerge are the following:

- Where all these substances are going?
- How we can make sure that we will recycle the materials and substances we want and we will avoid contaminants that might have very serious health and environmental impacts?
- In other terms, how we can make sure that we will have clean and not contaminated cycles in the framework of a circular economy?

Someone might doubt for the importance of contamination in recycled materials but allow us to provide some examples. Carcinogenic substances contaminate the asphalt cycle [162] as proven by studies in St. Gallen, Thurgau, Zurich, and Liechtenstein [163]. Brominated flame retardants have been found in children toys [164], and phthalates that were not removed in recycling operations were found in households and industrial applications [165]. The presence of chemicals in recycled paper has been linked to increased toxicity, particularly in food-contact applications [166]. So, regarding the

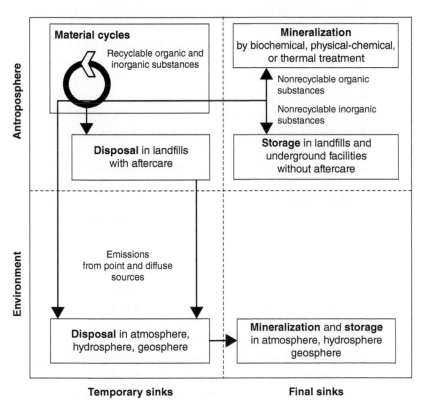

Figure 4.14 The concept of final sinks.
Source: From Ref. [168], © 2018 The Authors, published by Springer.

hazardous substances, we have two available strategies [167] in order to ensure that our material cycles will be kept clean and without serious health and environmental risks. Either we will remove the contaminants and transform them into harmless substances, or we will dispose them of in safe storages for a very long period of time. Final sinks can also be defined as these transformation and storage processes, and they are presented in Figure 4.14.

4.6.2 Clean Material Cycles in Circular Economy Require More Final Finks

Final sinks can be located in both the anthroposphere and the environment and are substance specific. Figure 4.15 demonstrates the direct relationship between final sinks and waste management. Landfills and WtE plants are the final sinks for the solid waste we produce, especially for the streams that can't be recycled due to either technical or financial limitations. Organic substances are subject of mineralization in transformation processes in WtE plants. Inorganic substances are, for instance, subject of safe storage in landfills depending on the composition of the disposed waste and on technical as well as geological barriers [168]. Underground storage in deep mines is the final sink of nuclear waste. There is a simple truth that is demonstrated here: as technologies and industrial revolutions transformed our societies resulting in new and more complex materials and waste types, we were obliged to create—invent artificial final sinks that can contain the pollution involved in the more and more complex solid waste we produce, because the natural final sinks are not able to metabolize them. As Brunner and Kral put it, "...waste management plays a key role for society: on one hand, it transforms waste and energy in quality proven resources for the next cycle by separating non-recyclable hazardous materials from the waste stream. On the other hand, it takes care of the non-recyclables and is instrumental for safe disposal of hazardous residues of recycling" [160].

Here we have a political problem: while the value of circular loops and material recycling can be, more or less, easily monetized, the value that final sinks offer to our societies is much more intangible because we realize the added value of good waste management only when we miss it. Thus, it is important to have a look in two examples that will make the importance and the social value of final sinks more visible. These case studies are from the paper "Cycles and sinks: two key elements of a circular economy" [168].

The first case study regards the use of polycyclic aromatic hydrocarbons (PAHs) in asphalt pavements. PAHs pose severe risks for human health, including skin and lung cancer, as it is very well documented [169]. Although in the past PAHs in tar for asphalt road pavement were banned and replaced by bitumen, they are still present in road pavements. When, at the end of their life cycle, road pavements are recycled, PAHs become part of future construction materials. This practice keeps PAHs in the loop

IND4.0 will deliver a plethora of new products, materials, and chemical substances that will become, sooner or later, part of the waste streams. The heterogeneity of MSW is expected to increase

Final sinks are becoming a key element for a more circular economy

Circular economy requires the development of closed loops to provide clean materials to displace material extraction. The cleaner the cycle, the higher the demand for safe final sinks for the contaminants removed

Waste management has at its core the need to protect human health and environment by ensuring safe removal and disposal of hazardous and nonrecyclable materials and substances

Figure 4.15 The role of final sinks in a circular economy.
Source: With kind permission from Nick Rigas, D-Waste.

and prolongs PAH exposure risks. Thus, we need to explore ways to remove PAHs from cycles and to manage the remaining PAHs, or we need to provide a safe final sink for them. The second case study is about copper residues from WtE plants. Copper has a high economic importance, and it is widely used in many applications as it is the metal with the highest electric conductivity. It is also well suited for recycling, resulting in less environmental impact than primary production. However, copper also represents an impurity for specific materials such as steel. Collection and waste treatment must be directed towards "clean" scrap to avoid copper traces in steel. Leachates from copper-containing wastes in landfills pose a risk for human and environmental health. Measures to reduce Cu in landfills and to control its mobility are needed to lower these risks. Hence, copper waste management needs to be optimized towards maximum recycling of high-quality materials and least environmental risks from landfilling.

Analyzing those two case studies, for the Canton of Zurich in Switzerland, the authors of the paper concluded that there is a clear link between clean material cycles, thus also for circular economy, and the need for final sinks. Clean material cycles require the removal of contaminants, so the cleaner the materials cycle requirements, the highest capacity of final sinks is required.

Especially for PAHs it was found that a reduction of the recycling rate (from 88 to 75%) will remove from the material cycle the fractions with higher PAHs concentrations and, consequently, decrease the exposure risks for construction works. As a result, less recycling and higher need for final sinks increase the quality of the final product. In addition, it was found that in certain cases we need to prioritize thermal treatment over recycling because this is the only way to eliminate PAHs and thus exposure risks.

For the case of copper, it was found that the increase in the copper recovery rate in WtE plants from 53 to 95% reduces the demand for sanitary landfills as final sink but does not entirely remove their necessity. This is because certain amounts of particularly nonmetallic copper quantities remain in bottom ash because of technological and economic constraints of the recovery process. However, the copper concentration in bottom ash changes from about 6000 to 3000 mg Cu/ton, which reduces the leachate potential and exposure risks below legal standards. In addition, the authors comment that the grade of chemical homogeneity of the waste plays a key role for material recovery. Highly heterogeneous waste flows, such as MSW, contain in part materials that cannot yet be mechanically sorted and recovered efficiently. Since a WtE plant acts as a huge homogenizer, the combination of incineration and bottom ash treatment allows effective recovery of copper (and other metals) from heterogeneous MSW flows.

4.6.3 Circular Economy Needs More and Better Waste Management

In this point it is important to remember the discussion held in Section 1.4. There are two important key points in this discussion that (hopefully) would become much clearer after reading this book. First, what we know as modern SWM is the response of human societies to the pollution that the industrial revolutions brought and the vast health and environmental impacts they created. To put it in another way, the primary scope of waste management, both historically and actually, is to protect human health and environment. Second, both circular economy and IND4.0 will bring new forms of pollution, and they will generate new types of waste, as we explained in detail in the beginning of this chapter. As we already wrote in Section 1.4, "Actually, the circular economy as well as the zero waste approaches require much more and not less waste management. Their implementation requires advanced management of multiple streams of materials, before they become waste but after they have been discarded from the main production. Those streams must be as clean as possible in order to have high added value, so advanced treatment for removal of residuals will be required, in one or another way. And of course, there will be always residuals looking for appropriate final sinks, including energy recovery where this is possible. For that reason, the road to circular economy passes through substantial improvements of the waste management sector and IND4.0 advances can deliver the required shifts."

After these remarks, we think that now it is easy to understand the role of final sinks in a circular economy. Figure 4.15 presents that the final sinks are becoming a central element of the shift to circular economy as a result of the combined interaction of circular economy, IND4.0, and the core mission of waste management. Both landfills and WtE plants have a crucial role for the transition to a circular economy, as it will be explained below. But before that, it is important to consider another dimension that makes final sinks very important: the time required for the transition to circular economy.

How much time do we need for a shift to circular economy? Based on historical experiences a 30–50 years' time for the shift of the majority of the world population to more circular practices would be a huge achievement, although there is no evidence that this is possible. But let's assume that it can be done by 2050. In these 30 years, between 2020 and 2050, we need to advance the technologies, the governance, and the business models required to deliver closed loops at least for the most scarce and hazardous materials. In these 30 years, based on the assessment made by Krausmann and his colleagues [42], there will be at least 650 billion tonnes of total waste produced worldwide. How we will manage them without proper financial sinks? What is clear is that the transition to a circular economy will not happen automatically and that for a long period, for all the different waste streams and substances, there will be a growing need to ensure proper final sinks, with minimum health and environmental impacts. This will allow the world to develop the technologies, the governance, and the business models required for advanced circularities. And this is the good case – the bad case, as we have already discussed, is that, no matter how good we will become in circularities, there are global limits, the laws of thermodynamics, and the problem that energy is not recycled that restrict the impact of circular loops that is so often overestimated.

What we need is to develop integrated sustainable waste management systems that will allow us to develop circular loops and ensure safe final disposal for the most harmful waste and the nonrecyclable (for technical or financial reasons) part. In these systems, WtE plants and simple incinerators for hazardous waste as well as sanitary landfills have a crucial role to play. The safe disposal of waste is of equal (if not higher) importance with recycling activities. Modern incinerators are the only long-term solution that wipes out hazardous pollutants and chemical pollution, and sanitary landfills can provide the proper final (although not in the long term) storage for materials that can't be recycled or incinerated for technical or economic reasons.

The discussion in the next section aims to present the role of WtE and landfills as final sinks of pollutants. From this point of view, this is not a discussion about any type of preference to WtE or landfills in comparison with other technologies. It is well known and established that the suitablility of any technology can only be assessed in a systematic framework that considers the whole waste management system as well as the social, economic and cultural context of a given society.

4.6.4 Waste to Energy in Circular Economy

It is already clear that incinerators for hazardous materials like healthcare waste [170] or industrial hazardous waste are safe final sinks that protect human health and environment from their pollution. No one doubts about it; even the most fanatic enemies of WtE for MSW agree that incineration of hazardous waste streams in proper well-equipped facilities with state-of-the-art antipollution systems is the best solution available.

Moving to MSW, it is important to notice that energy recovery serves the same high-level objective as many material recycling activities. For instance, one objective of recycling plastic is saving oil or natural gas, which is normally used for energy purposes. In a similar manner, oil, gas, or other primary energy resources are saved through energy recovery of plastics in WtE facilities producing electricity. The selection of the right systemic approach depends on the outputs, processing efficiencies, and local circumstances. ISWA's report "Circular Economy: Energy and Fuels" [171] highlights that energy recovery supplements recycling by increasing the total achievable recovery and both serve the same purpose of saving natural resources. In this view, ISWA suggests that the following points should be considered:

- Residue from recycling processes can be utilized for energy recovery.
- When the quality of recycled products deteriorates in the course of several recycling circles and recycling no longer is feasible, they can still be used for energy recovery.
- Metals that are not captured in the collection system, e.g. because they are trapped in combined products, can be recovered from the inert residues after combustion.
- Waste contains contaminated materials and substances with, for instance, sanitary and health hazards and should therefore be taken out of circulation. Such materials can be safely destroyed by combustion while recovering energy.
- Countries with distinct and ambitious environmental targets for their waste management all have a combination of material and energy recovery. The countries with the highest degree of material recovery are mostly also those with highest degree of energy recovery.

Another important element regards the reliability of WtE. WtE facilities are continuously pushing towards two targets, namely, higher efficiency and better antipollution. Consequently, the new facilities are becoming more efficient when it comes to replacement of fossil fuels and more protective of human health and environment. Professor Arena highlighted that the performances of new air pollution control (APC) devices demonstrate that emissions are safe and the recovery of energy from nonrecyclable wastes is in compliance with severe limits [172].

Brunner and Rechberger have stressed [173] that incinerators are crucial and unique for the complete destruction of hazardous organic materials, to reduce risks due to pathogenic microorganisms and viruses, and for concentrating valuable as well as toxic metals in certain fractions. Bottom ash and APC residues have become new sources of secondary metals; hence incineration has become a materials recycling facility, too. In addition, as they highlighted, "WTE plants are supporting decisions about waste and environmental management: They can routinely and cost effectively supply information about chemical waste composition as well as about the ratio of biogenic to fossil carbon in MSW and off-gas."

4.6.5 Sanitary Landfills in Circular Economy

The role of sanitary landfills in sustainable waste management and circular economy is probably one of the most important ongoing debates. Some people might not agree that this is a real debate since there is a growing trend to talk about "zero landfills" as a policy target. After all, advanced waste management systems tend to move away from landfills towards more recycling and WtE technologies, as it has been demonstrated for EU (considered as a single geographical entity). Analyzing and visualizing the historical evolution in waste management in EU, it was concluded that one of the interesting coherences shown by the method was the linked development of recycling and composting (60–65%) with incineration (40–35%) performance over the last 20 years in the EU-28 [174].

However, the truth is that no waste management system can exist without sanitary landfills. Even countries that have very advanced systems require to have landfills, at least as a backup. Let's see the news in July 2019 [175]: "60,000 cubic meters of waste were temporarily discharged in the last two weeks at a dumping site in Amersfoort; at least until Renewi, the company that deals with garbage collection, decides what to do with the huge surplus of garbage in Amsterdam. In normal times this waste would have been incinerated at the Afval Energie Bedrijf Amsterdam (AEB). But since the AEB, the largest waste incinerator in the Netherlands, has largely closed its furnaces due to serious technical problems, the garbage remains." By the way, AEB is one of the most modern and bigger WtE facilities in the world. It seems that the problem was related to huge imports of secondary fuels from abroad and the consequence was that Netherlands decided to ban, temporarily, RDF imports [176]. Then, the impact was transferred to the other side of the supply chain, the exporters. Few days later, an article [177] was setting the tone for the UK waste management industry, the main RDF exporter: "The waste industry has warned it may have no option but to landfill hundreds of thousands of tonnes of waste after the Netherlands announced a temporary ban on refuse-derived fuel (RDF) imports last week." So, two rich countries, with decades of billions of investments in waste management and some of the most modern facilities in the world, turn to landfilling because of an emergency!

If this is the case, then it is clear that sanitary landfills will always be the last resort for any waste management system that faces a collapse of its service/supply chain or a technical, financial, or administrative problem with treatment and recycling facilities. Actually, this is also one of the most important lessons learned from the big crisis in Napoli [178] and the adventures of waste management in Rome. It is reasonably argued that "landfilling of waste and environmentally safe 'sustainable' landfills, as part of an efficient waste management, is crucial for overall effective, flexible and appropriate waste management sector" [179]. This is clear for three main reasons:

- First, landfill is by far the cheapest and the easiest (in terms of construction) option for containment of pollution, although this containment [180] will not stay intact for more than two to three decades [181]. This is why landfilling is suitable for emergencies.
- Second, there will always be hazardous and/or inorganic waste that can't be incinerated nor recycled. In addition, WtE capacities are not always available, and many countries do not have WtE facilities at all.
- Third, proper landfilling of waste could make future landfill mining [182] activities more efficient, if and when the markets or the technologies allow it. This potential has driven some experts to say that landfill mining might be the missing link of circular economy [52].

There is no doubt that landfills pose environmental impacts, as all the waste management facilities and all human activities. There is also no doubt that waste treatment reduces those environmental impacts [183] and provides better recovery of energy and materials; thus pretreatment prior to landfilling is an important way to control landfills' impacts. In addition, as Rainer Stegmann has noticed, "It should be always kept in mind that landfills are a long-term problem. That means also after closure and capping the landfill still has to be operated and/or controlled. When and under which construction a landfill can be released from aftercare cannot be answered today...The aftercare phase for 'normal' landfills without any enhancement of the biological processes also having low water infiltration rates will be significantly longer. Hazardous waste landfills and some types of inorganic waste landfills cannot be released form aftercare in the opinion of the author. The functionality of the liner and associated parts of the landfill has to be guaranteed forever...On the long-term landfills are not the appropriate method for waste management." Still, sanitary landfills have an important role to play, especially in the developing world, because moving from uncontrolled disposal to sanitary landfills is a huge step towards better health and environmental protection. Table 4.7 summarizes the role of sanitary landfills and WtE in a circular economy.

Table 4.7 Sanitary landfills and WtE in a circular economy.

Circular economy and final sinks	Sanitary landfills	Waste-to-energy plants
Main role	• Containment of contaminants for a certain period • Final sink for nonrecyclables, for low calorific value streams, and for non-incinerated hazardous waste and ashes from incinerators • The only way available for poor countries and emergencies	• Destruction of hazardous substances and chemical pollutants • Homogenization of heterogeneous MSW that allows recovery of important metals • Substantial reduction of volume
Secondary features	• Biogas-to-energy production • Storage of materials for future utilization based on technology and market developments	• Contribution to energy production • Analytical tool for the biogenic fraction of waste
Main problem	• Integrity and long-term behavior of containment systems • Leachate collection and treatment are necessary to avoid water and soil pollution	• Advanced air pollution systems required to ensure public health protection • Ash treatment is an important element for resource recovery

4.7 CIRCULARITIES FOR MATERIALS-LINEARITIES FOR PEOPLE

We are discussing how we will stimulate closed loops for the most important materials and how we will overcome geographical barriers and coordinate global supply chains, and at the same time, we are building walls to avoid war or climate refugees and put new restrictions on immigration.

So, we face the following strange situation. In EU, as an example, studies made for the European Commission [184] propose to develop a common Schengen area for waste in order to boost recycling and recovery and abolish notification requirements for transboundary movement within the EU (in accordance with a proposal that was made first by Gerben-Jan Gerbrandy [185], a member of the European Parliament from the Alliance of Liberals and Democrats for Europe). At the same time, strong barriers are built in order to make sure that all refugees (coming from the war in Syria and the intentional destruction of Libya) and poor economic immigrants coming from Africa and Asia will stay forever in the countries that first receive them, mainly Greece and Italy. Obviously, refugees and immigrants in EU are valued less than waste and recyclables, so no one is proposing a common Schengen area in which they can easily flow within EU. Actually, things are even worst, and some politicians suggest that it is better to let them get drowned. According to the UNHCR [186], an estimated 362 000 refugees and migrants risked their lives crossing the Mediterranean Sea in 2016, with 181 400 people arriving in Italy and 173 450 in Greece. This movement towards Europe continues to take a devastating toll on human life. Since the beginning of 2017, over 2700 people are believed to have died or gone missing while crossing the Mediterranean Sea to reach Europe, with reports of many others perishing en route.

Let's see the analogy. The vast majority of refugees and immigrants are coming to EU because their lives are wasted either by war or by extreme poverty, because they can't find any descent work and survival options in their countries and they are looking for an opportunity to recover their lost lives. They can easily be considered as (human) resources, which should not be disposed of in dumpsites (like most of the refugee camps) but find a way to be recirculated in the (declining and aging European) economy. We say that trying to expand the life cycle of useful material resources and identify second and third circular opportunities to reuse them is a key to sustainability – why can't we say the same for people? Should their lives be only linear and get wasted without any second opportunity? Or sustainability does not include human lives and poor refugees? Actually, it seems that the concepts of circular economy are much more familiar to poor refugees and immigrants.

Take the example of Sladjana Mijatovic, an expert in circular innovation [187]. Her relationship with the circular economy began at the age of five, when she experienced life in refugee centers in the Netherlands, where she lived for a while with her parents after fleeing the war in the former Yugoslavia. Mijatovic described how they lost everything – they had no material goods, and the refugees were not allowed to work. This meant that refugees had to develop certain skills such as learning crafts, building objects and subsisting through the exchange of services, and recycling and reuse of products, that is, what we know today as a circular economy. Life as a refugee was her first experience of a collaborative economy.

A relevant argument [188] was made by Hans Stegeman, head of Research and Investment Strategy at Triodos. As he explained, the principles of the circular economy can also be applied directly to human capital. "People flows" can be made circular instead of linear, based on exactly the same argument that applies to material flows: to prevent waste and shortages. Given the tightness of labor markets in many countries, this is more relevant than ever before. With his words, "Actually, the same applies here as for the careful use of materials: we want to prevent that there are not enough people to be able to work productively (scarcity) and we do not want to create (social) waste: people who drop out for a long period of time. In other words: we want to employ every person in a job that makes the most of what they have to offer and for as long as possible. For a number of reasons that is more important now than ever before." The fact that this obvious remark is completely ignored in the mainstream circular economy narrative is another sign of the corporate footprint on the content of circular economy, as we discussed in Chapter 3.

However, as this chapter refers to resources, the point we want to make is this one: circulating material resources and ignoring human resources is meaningless, misleading, and catastrophic for the long-term viability of any circular economy narrative. On the other hand, when you bring the "human resources" at the epicenter of circular economy discussions, then you can't avoid the discussion on the distribution of the social benefits of circular economy.

But there is something more. Refugees and economic immigrants are not the only ones that are ignored. Informal recyclers are also usually excluded from the circular economy narrative, although it is more than obvious that any effort to apply circular concepts in the developing world should consider and adapt to the high informality of the global south economies. To get an idea about the size of informal economies, in South Asia, it is estimated that it covers 82% of the labor, in East and Southeast Asia almost 65%, in Pakistan 78%, in India 85%, in Thailand 43%, and in Indonesia 73% [189]. In other words, circular economy, as waste management systems and democracy, can't be copied and pasted from one country to another; it has to be adapted to the local context and adopted with a domestic narrative (see also Section 3.8).

Informal recyclers are the social category in which human rights (the right to a decent life and the right to work) meet with resource efficiency and waste management too. The problem with informal recyclers is that in many cases they are ignored not only by the rich countries and the big companies but also by their own municipal authorities and frequently by a big part of their own societies. According to an ISWA report [190], almost 20 million people around the world depend for their livelihood on informal recycling from MSW. The main driver is poverty, working conditions are often unsafe and unhygienic, and child labor is commonplace. This is primarily a social issue, but the informal recycling sector is often achieving notable recycling rates as well as providing a livelihood to significant numbers of the urban poor.

Preston considers [191] that the large informal sectors in the developing countries already practice "circular" activities – in areas such as electronic waste (e-waste) and phone repairs, for example – and they have a high potential to engage in higher-value circular economy supply chains. He also suggested that a proactive engagement by multinational companies with suppliers in developing countries – including SMEs and those operating in the informal sector – will be necessary for circular activities to be scaled up in a manner that is inclusive and avoids the displacement of vulnerable workers.

A report by the World Business Council for Sustainable Development [192] describes the contribution of informal recyclers to plastic recycling explaining that the informal sector in developing countries have successfully built businesses on the collection, trade, and recycling of plastic waste. This informal sector consists of small businesses and self-employed persons with little or no legal recognition and low capital investments.

Its workers are often wrongly considered to be poorly skilled and to possess little technological know-how. Yet, the informal recycling sector as a whole contributes massively to a more circular plastic economy. Furthermore, this sector is one of the most dynamic and adaptive, catering to ever-changing demands in plastic products.

Take as an example Kabadiwalla Connect, a technology-based social enterprise based in Chennai, India, that has determined that leveraging the informal ecosystem of urban waste recyclers has the potential to decrease the amount of waste sent to landfills in by 70% [193]. Rather than approaching the informality as a problem and developing a new system for waste management, Kabadiwalla Connect uses its technology platform to leverage the already existing informal infrastructure towards a more efficient waste management system. The Kabadiwalla Connect platform makes the informal ecosystem more accessible to other players. Municipalities can utilize informal infrastructure to bring down operational costs; waste management firms can source from it; corporations can carry out their extended producer responsibility through it; apartments and small businesses can send their recyclable waste directly to informal stakeholders that are a part of the informal ecosystem.

Or consider the company Mr. Green Africa (MGA), a for-profit company located in Nairobi, Kenya [194]. Established in 2014, MGA was since that time active in trading and processing postconsumer plastic wastes sourced from the local informal recycling sector. The basic idea is to abandon the traditional value chain of informal recycling based on multiple middlemen and replace it by a direct fair trade-like relation between the individual waste picker and the recycler, the latter one acting as the immediate link to the formal economy. To do so, the company set up proprietary trading points throughout Nairobi where waste pickers sell collected plastic wastes to MGA buying clerks. The trading price is fixed at a rate of 19 Kenyan shillings (KSh) per kilogram plastics and hence not subject to market price volatility. That price is openly communicated and chosen such to be very competitive to that offered by other local scrap traders.

China's efforts to put an order in e-waste recycling (and largely formalize it) came into a break-even point because the state ignored the strong business networks that were developed for many years by the informal recyclers. As it was mentioned, "... the informal sector with extensive linkages with the global production network is by and large invisible in the public policy negotiation. The visible hand for planning needs better understanding on the invisible hand of the market" [195].

Costas Velis, from the Institute for Public Health and Environmental Engineering (iPHEE) of the University of Leeds, recognizes that waste pickers are here to stay, but there are also some clear limits on what they can achieve without a proper integration with the formal sector and access to finance [196]. He highlights that despite 30 years of effort to build upon waste picker activities, there is no clear global consensus on the best way forward, at least on the organizational and empowerment side. Cooperatives and associations are an obvious internal operational mode, but these terms conceal huge variability regarding realities on the ground. The circular economy emphasis on business models not only adds to the complexity but also opens opportunities via use of information technologies that may not be within the reach of waste pickers. An early but very good case study on information tools for informal recyclers concerns an application to map their spatial organization of waste collection and to develop software tools for coordinating with clients and planning operations [197].

Anne Scheinberg, in a very interesting report [144], criticizes five assumptions that mislead the decision makers regarding informal recyclers and create barriers to a proper cooperation with them. The first concerns the belief that a modern and technologically advanced waste management system represents the future; consequently any part that does not fit this future will be gradually eliminated. The second is that if informal recyclers are eliminated, cities can make money on recycling and use it to finance their solid waste program. The third assumption is that informal recyclers should try to be integrated, ignoring that integration is a two-way cooperative coexistence. Fourth, the undocumented belief that informality is a transitional situation and it will gradually go away as countries and waste management systems will be more developed. The fifth assumption is that integration is more important for informal recyclers themselves, so organizing social and educational safety nets should be the main focus of integration.

In a very interesting piece titled "Informal Workers: The Front Lines of Enabling Circular Economies" [198] written by Samit Aich, CEO of S3IDF – India, the author explains that in developing countries, the poor already operate in a circular economy; their small business viability and familial existence depend on extreme efficiency and optimization of energy and materials. Recognizing and incorporating informal workers into circular economies is critically important to bring the required systemic shifts that build long-term resilience, generate business and economic opportunities, and provide environmental and societal benefits. Then he presents seven necessary ecosystem principles that were identified in a relevant panel discussion at SOCAP18 called "Informal Workers: The Front Lines of Enabling Circular Economies." The principles are the following:

1. Acknowledge and include informality – don't demonize or exploit it.
2. Address the unique role women play in society and economy. Any solution that is designed through the lens of circularity and inclusivity must be designed with consciousness of gender at the core of its configuration.

3. Build on activism and citizen advocacy – many governments act only when citizens rise, ask questions, and push the status quo.

4. Empower ecosystem players on the level of urban decision makers. Ecosystem players, like capacity building bodies, nonprofits, and training institutions, have to have strategy and stamina to make these bodies work.

5. Promote policy interventions based on real experiences – geographic "ideas inter-portability" can be an important tool to transfer domain knowledge and learnings to like-to-like parts of the world.

6. Advance upskilling and capacity building in all the relevant levels, including policy makers, businesses, and community leaders.

7. Scale up and integrate – create multi-stakeholder alignments to ensure support of the necessary reforms.

Instead of conclusions, we would like to finish referring to Robert Neuwirth's fascinating book *Stealth of Nations* [199]. The red thin line that connects all the chapters of the book is that informality takes place in the gap between highly organized business activities and real needs that can't or it's not profitable to be covered, e.g. distributing water and generators into areas where larger-scale service providers find it too costly to work. In this gap, informal economy delivers both employment and services, so informality is actually reinforcing the reproduction of the relevant business models. So, informality comes as a consequence of the inefficiency of the business models of big companies to deliver specific services and/or goods to specific populations. In addition, formal and informal economies are closely connected, they are working together and rely on each other. The informal economy relies on formal enterprises to invent and manufacture goods that others can copy and forge for cheap sale and distribution around the world at no profit to the original manufacturers. The formal economy, on the other hand, relies on the informal economy to largely handle the lower levels of their supply chain at an affordable enough rate to provide manufactured goods at reasonable prices to buyers in the "regulated world" even if this results in some loss through forged and stolen goods.

So, let's close with a practical question: if you want to learn the daily social practice of resource efficiency, if you want to know how you can reuse and recover useful materials and resources for real needs, and if you want to understand how to create more with less, would you ask a CEO that travels to Davos with his own jet and speaks about circular economy drinking champagne and eating caviar or a poor worker that sustains his family with $15/week, making meaningful decisions for every single cent he spends and expanding the life cycle of each and every material as much as possible? Circular economy is a way of living for many poor people, by definition and by practice they know it better. Informal recyclers are not simply a part of the circular economy in the developing world – they are its tracers and pioneers.

Key Take-Outs of Chapter 4

Key take-outs	Why is it important?
IND4.0 new technologies redefine resources. 1.2 billion tonnes of metals are required to implement the shift towards a less carbon-intensive economy. The digital matrix is expected to consume 25% of the electricity produced worldwide in 2025	As metal ores are becoming poorer, circular economy is becoming a condition to ensure the provision of the materials required for a low-carbon economy avoiding catastrophic environmental impacts. In addition, special emphasis should be given to closed loops of the so-called critical raw materials
IND4.0 will revolutionize a big part of manufacturing creating new products, which will finally become new types of waste. The waste streams of IND4.0 will be a mix of existing waste streams and e-waste, wearables, wind turbine blades, photovoltaics, etc.	The future waste streams will be more composite and complex, and the waste management systems are not ready to treat them recovering the valuable resources involved. The way anthropogenic stocks will be managed will determine the availability of crucial elements in a circular economy
Waste hierarchy has been the guiding principle of waste management systems and legislation for the last 20–30 years. Its suitability for the shift to circular economy is debated because waste hierarchy can't reflect the complexities of circular loops	New multidimensional tools for decision making are required, capable to reflect the social, political, technical, and economic challenges and opportunities of closed loops. However, waste hierarchy can inspire such tools and remain a policy pillar for waste management policies
Recycling is one of the least preferred options in circular economy. Weight-based targets should be abandoned in favor of more representative forms of the environmental benefits and the overall targets set	It is time to reconsider the answer to the question "why do we recycle?" and put recycling in the right framework for both developed and developing countries. The framework of service vs. value chains for recycling provides the way forward
Waste management is redefined by the interaction of five global trends, namely, IND4.0, circular economy, global warming, marine litter, and the end of "dirty" recyclables. Those five trends (and their combination) are stimulating the end of business as usual in waste management, as it is already experienced on a daily basis by local authorities and businesses	Those five trends stimulate a systemic change in waste management theory, practices, business, and legal instruments. At the same time, they push waste management closer to resource management, integrating more the waste management systems in the broader resource management and supply chains. Increased complexity and uncertainties are becoming the new normal

Key take-outs	Why is it important?
Final sinks are a central element for a circular economy. Their role is crucial to ensure clean cycles and proper storage of the contaminants. The cleaner the cycles, the higher the demand for final sinks	WtE and sanitary landfills are artificial final sinks that ensure either destruction or containment of chemicals and hazardous substances, ensuring health and environmental protection and providing the time required for the transition to circular strategies

REFERENCES

1. Jänicke, M. and Jacob, K. (2009). *A Third Industrial Revolution? Solutions to the Crisis of Resource-Intensive Growth*, SSRN Scholarly Paper ID 2023121. Rochester, NY: Social Science Research Network.
2. Wrigley, E.A. (1962). The supply of raw materials in the industrial revolution. *The Economic History Review 15* (1): 1–16.
3. Wikipedia. Second industrial revolution. https://en.wikipedia.org/wiki/Second_Industrial_Revolution#Paper_making (accessed 4 January 2020).
4. IEA (2017). Digitalization and energy: analysis. https://www.iea.org/reports/digitalisation-and-energy (accessed 4 January 2020).
5. Ekholm, B. and Rockstrom, J. (2019). Digital technology can cut global emissions by 15%. Here's how. https://www.weforum.org/agenda/2019/01/why-digitalization-is-the-key-to-exponential-climate-action (accessed 4 January 2020).
6. Samson, O. (2018). Electricity and the fourth industrial revolution. Ascertaining the role of electricity in the fourth industrial revolution; At Abuja, 2018.
7. Schmidt, S. (2010). A growing digital waste cloud: our world. https://ourworld.unu.edu/en/a-growing-digital-waste-cloud (accessed 4 January 2020).
8. en:former (2019). The downside of progress: digital energy consumption on the rise. https://www.en-former.com/en/digital-energy-consumption-on-the-rise (accessed 26 April 2020)
9. News, C.H. (2017). Network, part of the G. E. 'Tsunami of Data' could consume one fifth of global electricity by 2025. *The Guardian* (11 December 2017).
10. Bridle, J. (2018). *New Dark Age: Technology and the End of the Future.* Verso.
11. Harris, J. (2018). Our phones and gadgets are now endangering the planet. *The Guardian* (17 July 2018).
12. De Decker, K. (2015). Why we need a speed limit for the internet. *Resilience.* https://www.resilience.org/stories/2015-10-21/why-we-need-a-speed-limit-for-the-internet (accessed 27 April 2020).

13. CEET (2013). *The Power of Wireless Cloud*, White Paper, 22. Melbourne: CEET.
14. Statista (2016). IoT: number of connected devices worldwide 2012–2025. https://www.statista.com/statistics/471264/iot-number-of-connected-devices-worldwide (accessed 4 January 2020).
15. DERA (2016). *Rohstof Fe Für Zukunf Tstechnologien 2016*. Bonn, Germany: DERA German Mineral Resources Agency.
16. Elshkaki, A., Graedel, T.E., Ciacci, L., and Reck, B.K. (2016). Copper demand, supply, and associated energy use to 2050. *Global Environmental Change 39*: 305–315.
17. Dominish, E., Florin, N., and Teske, S. (2019). *Responsible Minerals Sourcing for Renewable Energy, Report Prepared for Earthworks by the Institute for Sustainable Futures*. Sydney: University of Technology Sydney.
18. Bloomber (2017). The electric car revolution is accelerating. Bloomberg.com (6 July 2017).
19. Valdes-Dapena, P., Business, CNN (2019). By 2040, more than half of new cars will be electric. https://www.cnn.com/2019/05/15/business/electric-car-outlook-bloomberg/index.html (accessed 4 January 2020).
20. Pilgrim, H. (2017). *The Dark Side of Digitalization: Will Industry 4.0 Create New Raw Materials Demands?* Fact Sheet. Berlin: PowerShift e.V.
21. Ellsmoor, J. (2019). Electric vehicles are driving demand for lithium: with environmental consequences. https://www.forbes.com/sites/jamesellsmoor/2019/06/10/electric-vehicles-are-driving-demand-for-lithium-with-environmental-consequences (accessed 5 January 2020).
22. British Geological Survey; Bureau de Recherches Géologiques et Minières; Deloitte Sustainability; European Commission; Directorate-General for Internal Market, I., Entrepreneurship and SMEs; Toegepast natuurwetenschappelijk onderzoek (2017). *Study on the Review of the List of Critical Raw Materials: Final Report*. Brussels: European Commission.
23. Coulomb, R., Dietz, S., Godunova, M., and Nielsen, T. (2015). *Critical Minerals Today and in 2030: An Analysis for OECD Countries*, OECD Environment Working Papers 91. OECD Publishing.
24. OECD (2016). *Costs of Inaction and Resource Scarcity: Consequences for Long-Term Economic Growth*, CIRCLE; Policy Perspectives. Paris: OECD.
25. Ranganathan, J., Waite, R., Searchinger, T., and Hanson, C. (2018). *How to Sustainably Feed 10 Billion People by 2050, in 21 Charts*. World Resources Institute.
26. Cordell, D., Drangert, J.-O., and White, S. (2009). The story of phosphorus: global food security and food for thought. *Global Environmental Change 19* (2): 292–305.
27. Heck, S., Rogers, M., and Carroll, P. (2014). *Resource Revolution: How to Capture the Biggest Business Opportunity in a Century*. Boston, MA: New Harvest.
28. Pongracz, E., Philips, P., and Keiski, R. (2004). Evolving the theory of waste management: defining key concepts. *International Journal of Environment and Waste Management*, Rhodes, Greece *78*: 10.

29. Thompson, M. (2017). *Rubbish Theory: The Creation and Destruction of Value*, 2e. London: Pluto Press.
30. Chappells, H. and Shove, E. (1999). *Bins and the History of Waste Relations*, Consumption Everyday Life and Sustainability Summer School 1999, Centre for Science Studies. Lancaster: Lancaster University.
31. Parsons, L. (2007). Thompsons' rubbish theory: exploring the practices of value creation. In: *E - European Advances in Consumer Research*, vol. *8* (eds. S. Borghini, M.A. McGrath and C. Otnes), 390–393. Association for Consumer Research.
32. Wilson, D.C. (2007). Development drivers for waste management. *Waste Management and Research 25*: 198–207.
33. Barles, S. (2014). History of waste management and the social and cultural representations of waste. In: *The Basic Environmental History*, vol. *4* (eds. M. Agnoletti and S. Neri Serneri), 199–226. Cham: Springer International Publishing.
34. Ravetz, A., Ravetz, P.A., Turkington, R. et al. (2013). *The Place of Home: English Domestic Environments, 1914–2000*. Routledge.
35. Herbert, L. (2007). *Centenary History of Waste and Waste Managers in London and South East England*, 52. London: The Chartered Institution of Wastes Management.
36. Velis, C.A., Wilson, D.C., and Cheeseman, C.R. (2009). 19th century London dust-yards: a case study in closed-loop resource efficiency. *Waste Management 29* (4): 1282–1290.
37. Rogers, H. (2005). *Gone Tomorrow: The Hidden Life of Garbage*. New York, London: W.W. Norton & Company.
38. Brynjolfsson, E. and McAfee, A. (2014). *The Second Machine Age: Work, Progress, and Prosperity in a Time of Brilliant Technologies*. New York, London: W.W. Norton & Company.
39. Susan, S. (1999). *Waste and Want: A Social History of Trash*. New York: Macmillan.
40. Schiller, G., Müller, F., and Ortlepp, R. (2017). Mapping the anthropogenic stock in Germany: metabolic evidence for a circular economy. *Resources, Conservation and Recycling 123*: 93–107.
41. Kennedy, C., Cuddihy, J., and Engel-Yan, J. (2007). The changing metabolism of cities. *Journal of Industrial Ecology 11* (2): 43–59.
42. Krausmann, F., Wiedenhofer, D., Lauk, C. et al. (2017). Global socioeconomic material stocks rise 23-fold over the 20th century and require half of annual resource use. *Proceedings of the National Academy of Sciences of the United States of America 114* (8): 1880–1885.
43. Krausmann, F., Lauk, C., Haas, W., and Wiedenhofer, D. (2018). From resource extraction to outflows of wastes and emissions: the socioeconomic metabolism of the global economy, 1900–2015. *Global Environmental Change 52*: 131–140.
44. Graedel, T.E. (2010). *Metal Stocks in Society: Scientific Synthesis*. Nairobi: UNEP.
45. AGGBUSINESS (2019). Global aggregates growth examined by GAIN convenor Jim O'Brien. https://www.aggbusiness.com/sections/market-reports/features/

global-aggregates-growth-examined-by-gain-convenor-jim-obrien (accessed 17 January 2020).

46. BASF (2019). Sand in short supply. *Creating Chemistry*. https://www1.basf.com/ magazine/issue-8/article.sand-in-short-supply.en.html (accessed 6 May 2020).

47. Torres, A., Brandt, J., Lear, K., and Liu, J. (2017). A looming tragedy of the sand commons. *Science 357* (6355): 970–971.

48. Augiseau, V. and Barles, S. (2017). Studying construction materials flows and stock: a review. *Resources, Conservation and Recycling 123*: 153–164.

49. Tabata, T., Morita, H., and Onishi, A. (2018). What is the quantity of consumer goods stocked in a Japanese household? Estimating potential disaster waste generation during floods. *Resources, Conservation and Recycling 133*: 86–98.

50. Cote, M., Poganietz, W.-R., and Schebek, L. (2015). Anthropogenic carbon stock dynamics of pulp and paper products in Germany. *Journal of Industrial Ecology 19* (3): 366–379.

51. Bringezu, S. and ConAccount Workshop, Wuppertal Institut für Klima, Umwelt, Energie (eds.) (1997). *Regional and National Material Flow Accounting: From Paradigm to Practice of Sustainability. Proceedings of the ConAccount Workshop, Leiden, The Netherlands*, Wuppertal spezial. Wuppertal: Wuppertal-Inst.

52. Machiels, L., Bernardo, E., and Jones, P.T. (2019). Enhanced landfill mining, the missing link to a circular economy 2.0? *Detritus 8*: 1.

53. FAO (2015). *Global Initiative on Food Loss and Waste Reduction*; Save Food: Global Initiative on Food Loss and Waste Reduction, 8. Rome: Food and Agriculture Organization of the United Nations.

54. Kaza, S., Yao, L., Bhada-Tata, P., and Van Woerden, F. (2018). *What a Waste 2.0 A Global Snapshot of Solid Waste Management to 2050*. World Bank Group.

55. Parfitt, J., Barthel, M., and Macnaughton, S. (2010). Food waste within food supply chains: quantification and potential for change to 2050. *Philosophical Transactions of the Royal Society B 365* (1554): 3065–3081.

56. Ricci-Jurgensen, M., Gilbert, J., and Ramola, A. (2019). *Global Assessment of Municipal Organic Waste Production and Recycling*, 29. Vienna, Austria: International Solid Waste Association ISWA.

57. Gilbert, J. (2015). *Circular Economy: Carbon, Nutrient and Soil*; ISWA Task Force on Resources, 40. Vienna, Austria: International Solid Waste Association ISWA.

58. Geyer, R., Jambeck, J.R., and Law, K.L. (2017). Production, use, and fate of all plastics ever made. *Science Advances 3* (7): e1700782.

59. Velis, C., Lerpiniere, D., and Tsakona, M. (2017). *How to Prevent Marine Plastic Litter: Now! An ISWA Facilitated Partnership to Prevent Marine Litter, with a Global Call to Action for Investing in Sustainable Waste and Resources Management Worldwide*; ISWA Marine Litter Task Force, 47. Vienna, Austria: International Solid Waste Association ISWA.

60. Vidal, J. (2020). The plastic polluters won 2019 – and we're running out of time to stop them. *The Guardian* (2 January 2020).

61. European Commission (2020). Plastic waste: environment. https://ec.europa.eu/environment/waste/plastic_waste.htm (accessed 6 January 2020).

62. Ellen MacArthur Foundation (2016). *The New Plastics Economy: Rethinking the Future of Plastics*; The New Plastics Economy, 61. Ellen MacArthur Foundation.

63. Ellen MacArthur Foundation (2019). First annual New Plastics Economy Global Commitment progress report published. https://www.ellenmacarthurfoundation.org/news/first-annual-new-plastics-economy-global-commitment-progress-report-published (accessed 6 January 2020).

64. Ambrose, J. (2019). War on plastic waste faces setback as cost of recycled material soars. *The Guardian* (13 October 2019).

65. Investopedia (2019). How much Coca-Cola spends on advertising. https://www.investopedia.com/articles/markets/081315/look-cocacolas-advertising-expenses.asp (accessed 6 January 2020).

66. Wienner-Bronner, D., Business, CNN (2019). Coca-Cola promised to help fix recycling. Here's how it plans to deliver. https://www.cnn.com/2019/03/14/business/coca-cola-recycling-grant/index.html (accessed 6 January 2020).

67. Lerner, S. (2019). Leaked audio reveals how Coca-Cola undermines plastic recycling efforts. *The Intercept* (18 October 2019).

68. Excell, C., Salcedo-La Viña, C., Worker, J., and Moses, E. (2018). *Legal Limits on Single-Use Plastics and Microplastics: A Global Review of National Laws and Regulations*, 118. United Nations Environmental Programme.

69. Godfrey, L. (2019). Turning off the plastic tap in developing countries needs more than a one size fits all solution. *ISWA Guest Blog* (11 February 2019).

70. Godfrey, L. (2019). Waste plastic, the challenge facing developing countries: ban it, change it, collect it? *Recycling 4* (1): 3.

71. Baldé, C.P., Forti, V., Gray, V. et al. (2017). *The Global E-Waste Monitor: 2017, Quantities, Flows and Resources*. Bonn/Geneva/Vienna: United Nations University (UNU), International Telecommunication Union (ITU) & International Solid Waste Association ISWA.

72. ISWA (2017). E-waste rises 8% by weight in 2 years, as incomes rise, prices fall. https://www.iswa.org/home/news/news-detail/browse/38/article/e-waste-rises-8-by-weight-in-2-years-as-incomes-rise-prices-fall/109/programmes (accessed 6 January 2020).

73. The Global E-Waste Partnership (2017). National legislation map. https://globalewaste.org/map (accessed 17 Febuary 2020).

74. PACE, E-Waste Coalition (2019). *A New Circular Vision for Electronics Time for a Global Reboot*, 24. Geneva: World Economic Forum.

75. Parajuly, K., Kuehr, R., Awasthi, A.K. et al. (2019). *Future E-Waste Scenarios*. StEP, UNU ViE-SCYCLE & UNEP IETC.

76. Natkunarajah, N., Scharf, M., and Scharf, P. (2015). Scenarios for the return of lithium-ion batteries out of electric cars for recycling. *Procedia CIRP 29*: 740–745.

77. Jamasmie, C. (2019). Recycled lithium batteries market to hit $6 billion by 2030: report. *MINING.com* (31 October 2019).

78. Kang, D.H.P., Chen, M., and Ogunseitan, O.A. (2013). Potential environmental and human health impacts of rechargeable lithium batteries in electronic waste. *Environmental Science and Technology 47* (10): 5495–5503.

79. Liu, P. and Barlow, C.Y. (2017). Wind turbine blade waste in 2050. *Waste Management 62*: 229–240.

80. Collier, C. and Ashwill, T. (2011). *Materials and Design Methods Look for the 100-m Blade*. Windpower Engineering Development. https://www.windpowerengineering.com/materials-and-design-methods-look-for-the-100-m-blade (accessed 6 May 2020).

81. Ramirez-Tejeda, K., Turcotte, D.A., and Pike, S. (2017). Unsustainable wind turbine blade disposal practices in the United States: a case for policy intervention and technological innovation. *New Solutions 26* (4): 581–598.

82. IRENA (2016). *End-of-Life Management: Solar Photovoltaic Panels*. International Renewable Energy Agency.

83. NC Clean Energy Technology Center (2017). *Health and Safety Impacts of Solar Photovoltaics*. NC State University.

84. IDC (2019). IDC forecasts steady double-digit growth for wearables as new capabilities and use cases expand the market opportunities. https://www.idc.com/getdoc.jsp?containerId=prUS44930019 (accessed 6 January 2020).

85. Cadence PCB Solutions (2018). Wearable technology: types of materials used. https://resources.pcb.cadence.com/blog/wearable-technology-types-of-materials-used-2 (accessed 30 April 2020).

86. Smith, C. (2017). *Will Wearable Technology Destroy Advances in Recycling?* GreenBiz.

87. Song, Q. and Li, J. (2015). A review on human health consequences of metals exposure to e-waste in China. *Environmental Pollution 196*: 450–461.

88. Ellen MacArthur Foundation (2017). Circular fashion – a new textiles economy: redesigning fashion's future. https://www.ellenmacarthurfoundation.org/publications/a-new-textiles-economy-redesigning-fashions-future (accessed 30 April 2020).

89. Yalcin Enis, I., Ozturk, M., and Sezgin, H. (2019). Risks and management of textile waste: the impact of embedded multinational enterprises. In: *Nanoscience and Biotechnology for Environmental Applications*, Environmental Chemistry for a Sustainable World, vol. *22* (eds. K.M. Gothandam, S. Ranjan, N. Dasgupta and E. Lichtfouse), 29–53. Springer Nature.

90. Chaffey, D. (2019). Ecommerce growth statistics: UK, US and worldwide forecasts. https://www.smartinsights.com/digital-marketing-strategy/online-retail-sales-growth (accessed 7 January 2020).

91. E-Commerce Packaging Market (2019). Growth, trends, and forecast (2020–2025). https://www.mordorintelligence.com/industry-reports/e-commerce-packaging-market (accessed 7 January 2020).

92. 1421 Consulting Group (2019). *Plastic Packaging Waste Recycling in China's E-Commerce Sector: A Market Outline and Opportunities for Dutch Companies*, 72. Beijing: The Embassy of the Kingdom of the Netherlands in Beijing.

93. Peters, A. (2018). Can online retail solve its packaging problem? https://www.fastcompany.com/40560641/can-online-retail-solve-its-packaging-problem (accessed 7 January 2020).

94. Bird, J. (2018). What a waste: online retail's big packaging problem. https://www.forbes.com/sites/jonbird1/2018/07/29/what-a-waste-online-retails-big-packaging-problem (accessed 7 January 2020).

95. Wessel, R. (2019). Wrap battle: how to tackle e-commerce packaging waste. http://www.delivered.dhl.com/en/articles/2019/04/tackle-ecommerce-packaging-waste.html (accessed 7 January 2020).

96. Silverstein, K. (2017). The trials of packaging for e-commerce. *Environment + Energy Leader* (22 February 2017).

97. Lansink, A. (2017). *Challenging Changes: Connecting Waste Hierarchy and Circular Economy*, 1e. Nijmegen: LEA.

98. Knauf, M. (2015). Waste hierarchy revisited: an evaluation of waste wood recycling in the context of EU energy policy and the European market. *Forest Policy and Economics 54*: 58–60.

99. DEFRA (2011). *Waste Hierarchy Evidence Summary*, 45. London: DEFRA.

100. Price, J.L. and Joseph, J.B. (2000). Demand management – a basis for waste policy: a critical review of the applicability of the waste hierarchy in terms of achieving sustainable waste management. *Sustainable Development 8* (2): 96–105.

101. JRC EC (2010). *Life Cycle Thinking and Assessment for Waste Management*. Brochure.

102. Schmidt, J.H., Holm, P., Merrild, A., and Christensen, P. (2007). Life cycle assessment of the waste hierarchy: a Danish case study on waste paper. *Waste Management 27* (11): 1519–1530.

103. Ekvall, T., Assefa, G., Björklund, A. et al. (2007). What life-cycle assessment does and does not do in assessments of waste management. *Waste Management 27* (8): 989–996.

104. Maalouf, A. and El-Fadel, M. (2018). Carbon footprint of integrated waste management systems with implications of food waste diversion into the wastewater stream. *Resources, Conservation and Recycling 133*: 263–277.

105. Cleary, J. (2009). Life cycle assessments of municipal solid waste management systems: a comparative analysis of selected peer-reviewed literature. *Environment International 35* (8): 1256–1266.

106. Clift, R., Doig, A., and Finnveden, G. (2000). The application of life cycle assessment to integrated solid waste management: part 1 – methodology. *Process Safety and Environmental Protection 78* (4): 279–287.

107. Maalouf, A. and El-Fadel, M. (2020). A novel software for optimizing emissions and carbon credit from solid waste and wastewater management. *Science of The Total Environment 714*: 136736.

108. Penney, S. (2018). Is waste hierarchy a misleading principle? *Wasteless Future* (14 February 2018). https://wastelessfuture.com/is-waste-hierarchy-a-misleading-principle (accessed 6 May 2020).

109. Lansink, A. (2018). Ad Lansink: waste hierarchy stimulates circular economy. *Wasteless Future* (12 March 2018). https://wastelessfuture.com/ad-lansink-waste-hierarchy-stimulates-circular-economy (accessed 6 May 2020).

110. Statista (2016). *Global Industry Forecasts*, 19. Hamburg.

111. Van Ewijk, S. and Stegemann, J.A. (2016). Limitations of the waste hierarchy for achieving absolute reductions in material throughput. *Journal of Cleaner Production 132*: 122–128.

112. Velis, C. (2015). Circular economy and global secondary material supply chains. *Waste Management and Research 33* (5): 389–391.

113. Iacovidou, E., Millward-Hopkins, J., Busch, J. et al. (2017). A pathway to circular economy: developing a conceptual framework for complex value assessment of resources recovered from waste. *Journal of Cleaner Production 168*: 1279–1288.

114. Goorhuis, M. (2015). *ISWA Key Issue Paper on Waste Prevention, Waste Minimization and Resource Management*, 14. Vienna, Austria: International Solid Waste Association ISWA.

115. Bartl, A. (2014). Moving from recycling to waste prevention: a review of barriers and enables. *Waste Management and Research 32* (9 Suppl): 3–18.

116. Potting, J., Hekkert, M., Worrell, E., and Hanemaaijer, A. (2017). *Circular Economy: Measuring Innovation in the Product Chain*, Policy Report 2544, 46. The Hague: PBL Netherlands Environmental Assessment Agency.

117. Dickinson, K. (2019). A new zero waste hierarchy for the circular economy. https://resource.co/article/new-zero-waste-hierarchy-circular-economy (accessed 9 January 2020).

118. ISWA Task Force on Resource Management (2015). *Circular Economy: Resources and Opportunities*. Vienna, Austria: International Solid Waste Association ISWA.

119. Mavropoulos, A. (2018). *ISWA: China's Ban on Recyclables: Beyond the Obvious*. International Solid Waste Association ISWA.

120. Velis, C. (2014). *Global Recycling Markets: Plastic Waste A Story for One Player – China*, Task Force on Globalization and Waste Management, 66. Vienna, Austria: International Solid Waste Association ISWA.

121. ISWA (2009). *Waste and Climate Change ISWA White Paper*, White Paper, 40. Vienna, Austria: International Solid Waste Association ISWA.

122. Defra, Resources & Waste Strategy Team (2018). *Our Waste, Our Resources: A Strategy for England*, 146. London: DEFRA.

123. Ragossnig, A.M. and Schneider, D.R. (2017). What is the right level of recycling of plastic waste? *Waste Management and Research 35* (2): 129–131.
124. Scharff, C. (2018). *The EU Circular Economy Package and the Circular Economy*, Coalition for Europe, 18. Circular Economy Coalition for Europe.
125. Anshassi, M., Laux, S., and Townsend, T.G. (2018). Replacing recycling rates with life-cycle metrics as government materials management targets. *Environmental Science and Technology 52* (11): 6544–6554.
126. Ricardo Energy & Environment (2018). *Why Wait? Weight Isn't Working: Smarter Measures for the Circular Economy*, 95. Environmental Services Association.
127. Velis, C.A. and Brunner, P.H. (2013). Recycling and resource efficiency: it is time for a change from quantity to quality. *Waste Management and Research 31* (6): 539–540.
128. Ervasti, I., Miranda, R., and Kauranen, I. (2016). A global, comprehensive review of literature related to paper recycling: a pressing need for a uniform system of terms and definitions. *Waste Management 48*: 64–71.
129. Graedel, T.E., Allwood, J., Birat, J.-P. et al. (2011). What do we know about metal recycling rates? *Journal of Industrial Ecology 15* (3): 355–366.
130. Van Ewijk, S., Stegemann, J.A., and Ekins, P. (2018). Global life cycle paper flows, recycling metrics, and material efficiency: global paper flows, recycling, material efficiency. *Journal of Industrial Ecology 22* (4): 686–693.
131. Park, J.Y. and Chertow, M.R. (2014). Establishing and testing the "reuse potential" indicator for managing wastes as resources. *Journal of Environmental Management 137*: 45–53.
132. van Ewijk, S., Park, J.Y., and Chertow, M.R. (2018). Quantifying the system-wide recovery potential of waste in the global paper life cycle. *Resources, Conservation and Recycling 134*: 48–60.
133. Fellner, J., Lederer, J., Scharff, C., and Laner, D. (2017). Present potentials and limitations of a circular economy with respect to primary raw material demand: present potentials and limitations of a circular economy. *Journal of Industrial Ecology 21* (3): 494–496.
134. Blok, K., Hoogzaad, J., Ramkumar, S. et al. (2016). *Implementing Circular Economy Globally Makes Paris Targets Achievable*, 18. Circle Economy: Ecofys.
135. Sturgis Associates (2010). *Redefining Zero: Carbon Profiling as a Solution to Whole Life Carbon Emission Measurement in Buildings (RICS)|Isurv*. RICS.
136. Monahan, J. and Powell, J.C. An embodied carbon and energy analysis of modern methods of construction in housing: a case study using a lifecycle assessment framework. *Energy and Buildings 43*: 179–188.
137. Challenge, C.C (2018). Circular construction challenge. circularconstructionchallenge.org (accessed 17 January 2020).
138. ISWA (2018). "Circular construction challenge – rethink waste" reveals six ideas to reduce waste in the built environment competing for the three winning spots on innovation fast track in 2019. https://www.iswa.org/home/news/news-detail/article/

circular-construction-challenge-rethink-waste-reveals-six-ideas-to-reduce-waste-in-the-built-e/109 (accessed 17 January 2020).

139. ISWA (2017). Barriers to sustainable resource management. https://www.iswa.org/index.php?id=1529 (accessed 17 January 2020).

140. Boston Consulting Group (2019). *Puls of the Fashion Industry, 2019 Update.* Global Fashion Agenda, Boston Consulting Group, and Sustainable Apparel Coalition.

141. Khusainova, G. (2019). Why the circular economy will not fix fashion's sustainability problem. https://www.forbes.com/sites/gulnazkhusainova/2019/06/12/why-the-circular-economy-will-not-fix-fashions-sustainability-problem (accessed 17 January 2020).

142. Lemille, A. (2019). For a true circular economy, we must redefine waste. https://www.weforum.org/agenda/2019/11/build-circular-economy-stop-recycling (accessed 17 January 2020).

143. Scheinberg, A. (2011). Value added: modes of sustainable recycling in the modernisation of waste management systems. Thesis, Wageningen University, Wageningen, NL.

144. Scheinberg, A. and Savain, R. (2015). *Valuing Informal Integration: Inclusive Recycling in North Africa and the Middle East.* Eschborn, Germany: Deutsche Gesellschaft für Internationale Zusammenarbeit (GIZ) GmbH.

145. Zink, T. and Geyer, R. (2019). Material recycling and the myth of landfill diversion. *Journal of Industrial Ecology 23* (3): 541–548.

146. Modaresi, R. and Muller, B. (2012). The role of automobiles for the future of aluminum recycling. *Environmental Science and Technology 46* (16): 8587–8594.

147. Haupt, M., Vadenbo, C., Zeltner, C., and Hellweg, S. (2017). Influence of input-scrap quality on the environmental impact of secondary steel production. *Journal of Industrial Ecology 21* (2): 391–401.

148. Cimpan, C., Maul, A., Wenzel, H., and Pretz, T. (2016). Techno-economic assessment of central sorting at material recovery facilities: the case of lightweight packaging waste. *Journal of Cleaner Production 112*: 4387–4397.

149. Geyer, R., Kuczenski, B., Zink, T., and Henderson, A. (2016). Common misconceptions about recycling: common misconceptions about recycling. *Journal of Industrial Ecology 20* (5): 1010–1017.

150. Mavropoulos, A. (2015). Circular economy needs more waste management than linear one! *Wasteless Future* (14 May 2015).

151. Cobo, S., Dominguez-Ramos, A., and Irabien, A. (2018). From linear to circular integrated waste management systems: a review of methodological approaches. *Resources, Conservation and Recycling 135*: 279–295.

152. Sandin, G. and Peters, G.M. (2018). Environmental impact of textile reuse and recycling: a review. *Journal of Cleaner Production 184*: 353–365.

153. Genovese, A., Acquaye, A.A., Figueroa, A., and Koh, S.C.L. (2017). Sustainable supply chain management and the transition towards a circular economy: evidence and some applications. *Omega 66*: 344–357.

154. Peng, W. and Pivato, A. (2019). Sustainable management of digestate from the organic fraction of municipal solid waste and food waste under the concepts of back to earth alternatives and circular economy. *Waste and Biomass Valorization 10* (2): 465–481.

155. Ljunggren Söderman, M., Eriksson, O., Björklund, A. et al. (2016). Integrated economic and environmental assessment of waste policy instruments. *Sustainability 8* (5): 411.

156. Massarutto, A., de Carli, A., and Graffi, M. (2010). *La Gestione Integrata Dei Rifiuti Urbani: Analisi Economica Di Scenari Alternativi*, 83. Bocconi University.

157. Massarutto, A., de Carli, A., and Graffi, M. (2011). Material and energy recovery in integrated waste management systems: a life-cycle costing approach. *Waste Management 31* (9–10): 2102–2111.

158. Crutzen, P.J. (2006). The "anthropocene". In: *Earth System Science in the Anthropocene* (eds. E. Ehlers and T. Krafft), 13–18. Berlin, Heidelberg: Springer.

159. Erickson, B. (2017). How many chemicals are in use today? *Chemical and Engineering News* (27 February 2017).

160. Brunner, P.H. and Kral, U. (2014). Final sinks as key elements for building a sustainable recycling society, *Sustainable Environment Research, 24* (6), 443–448.

161. European Environment Agency (2019). The European environment: state and outlook 2020: knowledge for transition to a sustainable Europe. https://www.eea.europa.eu/soer-2020 (accessed 30 April 2020).

162. Kuhn, E., Leumann, A., and Oetiker, D. (2019). *The Urban Mining of Construction Waste: the Canton of Zurich as the Materials Warehouse of the Future*, Mineralische Nebenprodukte und Abfälle 6, 11. Thomé-Kozmiensky Verlag GmbH.

163. Rubli, D.S., Wagner, R., Kuhn, D.E. et al. (2013). *Dynamische Modellierung der Asphalt-sowie PAK-Lager und Flüsse in den Strassen der Region St.Gallen, Thurgau, Zürich und Fürstentum Liechtenstein*, 43. Schlieren: Energie- und Ressourcen-Management GmbH.

164. Chen, S.-J., Ma, Y.-J., Wang, J. et al. (2009). Brominated flame retardants in children's toys: concentration, composition, and children's exposure and risk assessment. *Environmental Science and Technology 43* (11): 4200–4206.

165. Pivnenko, K., Eriksen, M.K., Martín-Fernández, J.A. et al. (2016). Recycling of plastic waste: presence of phthalates in plastics from households and industry. *Waste Management 54*: 44–52.

166. Pivnenko, K., Laner, D., and Astrup, T.F. (2016). Material cycles and chemicals: dynamic material flow analysis of contaminants in paper recycling. *Environmental Science and Technology 50* (22): 12302–12311.

167. Brunner, P.H. (2010). Clean cycles and safe final sinks. *Waste Management and Research 28* (7): 575–576.

168. Kral, U., Morf, L.S., Vyzinkarova, D., and Brunner, P.H. (2019). Cycles and sinks: two key elements of a circular economy. *Journal of Material Cycles and Waste Management 21* (1): 1–9.

169. Abdel-Shafy, H.I. and Mansour, M.S.M. (2016). A review on polycyclic aromatic hydrocarbons: source, environmental impact, effect on human health and remediation. *Egyptian Journal of Petroleum 25* (1): 107–123.

170. Chauan, A. and Singh, A. (2016). Healthcare waste management: a state-of-the-art literature review. *International Journal of Environment and Waste Management 18* (2).

171. Hulgaard, T. (2015). *Circular Economy: Energy and Fuels*, Task Force on Resource Management, 48. Vienna, Austria: International Solid Waste Association ISWA.

172. Arena, U. (2015). From waste-to-energy to waste-to-resources: the new role of thermal treatments of solid waste in the recycling society. *Waste Management 37*: 1–2.

173. Brunner, P.H. and Rechberger, H. (2015). Waste to energy: key element for sustainable waste management. *Waste Management 37*: 3–12.

174. Pomberger, R., Sarc, R., and Lorber, K.E. (2017). Dynamic visualisation of municipal waste management performance in the EU using ternary diagram method. *Waste Management 61*: 558–571.

175. Redazione. 2019. Amsterdam waste crisis continues as conditions set for AEB restructure. *+31mag* (30 July 2019).

176. DutchNews (2019). Dutch poised to halt importing waste due to crisis at AEB incinerator. *DutchNews.nl* (8 August 2019).

177. McGlone, C. Dutch waste import ban could lead to emergency landfilling, warns industry. http://www.endsreport.com/article/1593635?utm_source=website&utm_medium=social (accessed 20 January 2020).

178. Pasotti, E. (2010). Sorting through the trash: the waste management crisis in southern Italy. *South European Society and Politics 15* (2): 289–307.

179. Rosendal, R.M. (2016). *The Important Role of Landfills in the Circular Economy*, 10. Kalmar, Sweden: Linnaeus ECO-TECH 16.

180. Committee to Assess the Performance of Engineered Barriers, Board on Earth Sciences and Resources, Division on Earth and Life Studies (2007). *Assessment of the Performance of Engineered Waste Containment Barriers*. Washington, DC: National Academies Press.

181. Morris, J.W.F. and Barlaz, M.A. (2011). A performance-based system for the long-term management of municipal waste landfills. *Waste Management 31* (4): 649–662.

182. Burlakovs, J., Kriipsalu, M., Klavins, M. et al. (2017). Paradigms on landfill mining: from dump site scavenging to ecosystem services revitalization. *Resources, Conservation and Recycling 123*: 73–84.

183. Komilis, D.P., Ham, R.K., and Stegmann, R. (1999). The effect of municipal solid waste pretreatment on landfill behavior: a literature review. *Waste Management and Research 17* (1): 10–19.

184. ARCADIS. *The Efficient Functioning of Waste Markets in the European Union: Legislative and Policy Options*, 463. European Commission.

185. Gerbrandy, G.-J. (2012). I will be proposing a sort of schengen area for waste. www.euractiv.com (23 February 2012).

186. UNHCR (2019). Refugees, U. N. H. C. for Europe situation. https://www.unhcr.org/europe-emergency.html (accessed 24 Febuary 2020).

187. Ignasi Rifé (2019). Europa, F. C. Fundació Catalunya Europa. https://www.catalunyaeuropa.net/en/activitats/105/circular-economy-a-shift-in-the-productivity-model.html (accessed 24 Febuary 2020).
188. Stegeman, H. (2019). No circular economy without the human factor. https://www.triodos-im.com/articles/2019/no-circular-economy-without-the-human-factor (accessed 24 Febuary 2020).
189. UN ESCAP (2018). Closing the Loop. Unlocking the informal economy in an inclusive circular economy approach. *UNCC Bangkok* (19 November 2018). https://www.unescap.org/sites/default/files/Plastics_Nov19_ESCAP_PPT_0.pdf (accessed 6 May 2020).
190. Task Force on Globalization and Waste Management (2014). *Globalisation and Waste Management*, Task Force on Globalization and Waste Management, 36. Vienna, Austria: International Solid Waste Association ISWA.
191. Preston, F., Lehne, J., and Wellesley, L. (2019). *An Inclusive Circular Economy, Priorities for Developing Countries*, Research Paper, 82. London: Chatham House.
192. WBCSD (2016). *Informal Approaches Towards a Circular Economy: Learning from the Plastics Recycling Sector in India*. Geneva, Switzerland: World Business Council for Sustainable Development.
193. Hande, S. (2019). The informal waste sector: a solution to the recycling problem in developing countries. *Field Actions Science Reports the Journal of Field Actions*, No. Special Issue 19, 28–35. http://journals.openedition.org/factsreports/5143 (accessed 6 May 2020).
194. Gall, M., Wiener, M., Chagas de Oliveira, C. et al. (2020). Building a circular plastics economy with informal waste pickers: recyclate quality, business model, and societal impacts. *Resources, Conservation and Recycling 156*: 104685.
195. Tong, X., Wang, T., Chen, Y., and Wang, Y. (2018). Towards an inclusive circular economy: quantifying the spatial flows of e-waste through the informal sector in China. *Resources, Conservation and Recycling 135*: 163–171.
196. Velis, C. (2017). Waste pickers in global South: Informal recycling sector in a circular economy era. *Waste Management and Research 35* (4): 329–331.
197. Offenhuber, D. and Lee, D. (2012). Putting the informal on the map: tools for participatory waste management. *PDC '12: Proceedings of the 12th Participatory Design Conference: Exploratory Papers, Workshop Descriptions, Industry Cases – Volume 2*, 4. https://doi.org/10.1145/2348144.2348150.
198. S3IDF (2019). Informal workers: the front lines of enabling circular economies. https://medium.com/s3idf/informal-workers-the-front-lines-of-enabling-circular-economies-29a9e11e992f (accessed 24 Febuary 2020).
199. Neuwirth, R. (2012). *Stealth of Nations*. Penguin Random House.

Chapter 5

Waste Management 4.0

The world as we have created it is a process of our thinking. It cannot be changed without changing our thinking.

— Albert Einstein

Recommended Listening
Hoppipolla, **Sigur Rós**

Because this song feels like an optimistic journey that is constantly opening new doors to rooms that are almost the same – only at a higher level.

Recommended Viewing
The Biggest Little Farm (2018), **John Chester**

Because it is a beautiful documentary about rethinking how farming can be done and how a more diverse approach playing along with nature can help reduce the input and increase the yield.

Industry 4.0 and Circular Economy: Towards a Wasteless Future or a Wasteful Planet?, First Edition.
Antonis Mavropoulos and Anders Waage Nilsen.
© 2020 John Wiley & Sons Ltd. Published 2020 by John Wiley & Sons Ltd.

5.1 PERCEPTIONS AND REALITY

Are the executives of the waste management industry ready for Industry 4.0? Not really. In a 2017 survey by ISWA 1087 [1], participants from 97 countries around the globe were asked a series of questions about Industry 4.0 and possible impacts. Only 14% of the respondents consider that they know a lot about the fourth industrial revolution, while 57% consider that they know "something" and 29% consider that they know "little" or "nothing." The future may be here already, but if that is the case, the knowledge about it is very unevenly distributed.

5.1.1 Expectations for Industry 4.0

When Tim Berners-Lee suggested an idea he called "world wide web" back in 1991, the answer he got from his boss was "vague, but exciting…" [2]. Based on the response from the survey in waste management, many executives are having a similar feeling. Ninety-seven percent believe the waste industry will be affected. Half of these respondents expect a major impact, as the other half expect a moderate impact. This picture – a general anticipation of a coming transformation, but confusion about the actual impact and outcome – is confirmed by several other surveys on Industry 4.0 readiness across sectors.

Although the participants are not fully aware of the changes ahead, most of them believe the technology in the value chain will change in several ways over the next 10 years. Fully robotic waste sorting seems to be expected by 80% of the participants. Seventy-two percent of the participants expect fully robotic recycling plants – meaning "dark factories" without any human intervention. Among the other expected innovations are chatbots – digital assistants guiding citizens for waste prevention and sorting. Half of the respondents believe that driverless garbage trucks and robotic bins will be a reality in 2030. When asked about spending, panelists say mobile apps, new sensors, social media, big data, new materials, and digital utilities platforms are the top six priorities for investments.

Will Industry 4.0 stimulate the circular economy? Not necessarily. Around 82% of the respondents expect that the fourth industrial revolution will stimulate circular economy in consumer goods in some way, but just one-fourth of the participants are really optimistic and expect serious progress. One in five are rather pessimistic.

Regardless of the outcome, new materials, advanced sensors, and the Internet of Things (IoT) are expected to have the most important impact on the waste management industry. Surprisingly, although robots are already in use in the waste industry, their expected impact is considered rather low, and the expected impact by driverless cars is considered the least out of the 11 categories in the survey. Many of the participants expect a lot of impacts from social media and mobile applications, most possibly about waste prevention, reuse, and recycling activities.

How will the developments in the broader society shape the reality of waste management? A big majority (69%) anticipate that eco-design of consumer goods and circular value chains for plastics will be a reality before 2030. Sixty percent also expect a further significant reduction of the carbon dioxide emissions, and half of the respondents expect the elimination of huge dumpsites. The industry representatives are rather pessimistic about significant reduction of marine litter and reduction of e-waste.

Broadly speaking the industry is split into skeptics and optimists, where the majority have limited expectations. The waste management community is aware that change is coming. But do they have a realistic picture about the actual speed of technological transformation?

5.1.2 The Hype and the Surprise

Many of the technological trends driving Industry 4.0 follow exponential curves. They are easily overestimated and hyped in the beginning, but then, at some point, they take us by surprise. The technological progress fuels the force that Joseph Schumpeter described as "creative destruction" of long-standing arrangements and assumptions [3]. This erosion process of "industrial mutation" from within can be hard to identify for traditional business. At some time the existing economic structure collapses, incessantly creating a new one. This is a constant process that creates a substantial risk of misreading future developments. This gap between expected developments and actual outcomes is described by Singularity University [4] as "the exponential growth factor" (Figure 5.1).

But when will a specific solution start bending upwards in terms of productivity? Many new technologies make bold promises. How do you discern the hype from the commercially viable? Gartner, the world's leading research and advisory company, has created the hype cycle: "Gartner Hype Cycles provide a graphic representation of the maturity and adoption of technologies and applications, and how they are potentially relevant to solving real business problems and exploiting new opportunities. Gartner Hype Cycle methodology gives you a view of how a technology or application will evolve over time, providing a sound source of insight to manage its deployment within the context of your specific business goals" [5].

According to Gartner, technologies pass through five phases of the hype cycle (Figure 5.2):

1. Innovation Trigger: A potential technology breakthrough kicks things off. Early proof-of-concept stories and media interest trigger significant publicity. Often no usable products exist and commercial viability is unproven.

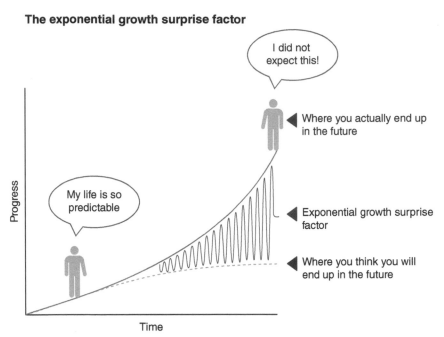

Figure 5.1 The exponential growth factor makes predictions uncertain.
Source: **From Ref. [4], © The Authors 2016, published by Singularity University under the CC BY-ND 4.0 license.**

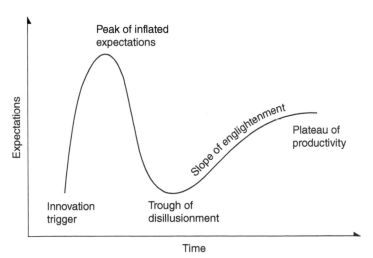

Figure 5.2 The Gartner hype cycle. Technologies pass through the five phases on the hype cycle.
Source: **From Ref. [5], with kind permission from Gartner Methodologies.**

Industry 4.0 and Circular Economy

2. Peak of Inflated Expectations: Early publicity produces a number of success stories – often accompanied by scores of failures. Some companies take action; many do not.

3. Trough of Disillusionment: Interest wanes as experiments and implementations fail to deliver. Producers of the technology shake out or fail. Investments continue only if the surviving providers improve their products to the satisfaction of early adopters.

4. Slope of Enlightenment: More instances of how the technology can benefit the enterprise start to crystallize and become more widely understood. Second- and third-generation products appear from technology providers. More enterprises fund pilots; conservative companies remain cautious.

5. Plateau of Productivity: Mainstream adoption starts to take off. Criteria for assessing provider viability are more clearly defined. The technology's broad market applicability and relevance are clearly paying off [5].

There are currently technologies and concepts at all stages of maturity in waste management. Many do underestimate how fast the technological development actually happens, as soon as proven concepts start to add substantial value. But it is also a tendency to overestimate the speed from the first proof of concept until the technology is widely adopted. A lot of the technical innovations expected in 2030, such as robotic waste sorting and chatbots, are already on the market. Others will be widely available within the next decade.

Another relevant question is whether there are sufficient incentives in place to put the technologies to use. "Even if the technology is available, many countries will not invest in it, but rather continue with manual labour. We still lack the standards and legislation that should be there first," says Toralf Igesund, head of research and development in BIR, an inter-municipality company in Norway. Others point to culture and mindset. "This is reflecting the fact that the solid waste management sector and especially the MSW management branch is a rather traditional thinking one, bound to tenders with local authorities and to medium/long term infrastructure investments, not so keen nor rapid in innovations and disruptive changes," says Marco Ricci, chair of the ISWA working group on Biological Treatment of Waste.

Regardless of what waste management executives think about their own pace of innovation, the surrounding world will change. This means that there will also be changes in the volume and quality of the waste streams. Industry 4.0 represents both good and bad news for the circular economy.

The good news: Sensors and a range of enabling technologies are becoming more applicable, due to both falling prices, increasing energy efficiency, and improved battery capacity. This means a higher sensory capacity within many industries and, theoretically at least, the possibility to follow closely the life cycle of any item sold. Service-based business models open up for recurring revenue, which give product designers a reason to optimize the design for a lasting product life. This is opposite from the linear ownership-based economy, where a short product life equals more revenue. Data from actual usage will help them minimize useless functions and materials, prevent damages, improve maintenance, and finally develop closed loops of clean materials.

From a technological point of view, reuse, recycling, and material recovery will become easier. "Zero waste" approaches for specific supply chains will become more realistic than ever. In additive manufacturing, already gaining ground in many industries, the design can optimize for less material, and the surplus material during the production can be reused. So, this phase of the supply chain can, in many cases, be really wasteless.

The bad news: The rebound effects, as described in Chapter 2, will probably change or increase consumption. And as we know, raw materials are not necessarily recyclables. In 3D printing a plastic powder (polyamides, thermoplastic elastomers, polyether ketone, polystyrene) is the most commonly used. Is it better or worse to make production of plastic products much easier (and with the potential to be fully decentralized in each and every household)?

A possible outcome is that waste will be eliminated from parts of the supply chain, but it will be relocated, in new forms, in other parts of it. If the result is more manageable waste streams, this will not happen by the market mechanisms alone. Even if the technological tools could be applied to create clean, closed materials loops, they will never be realized on a massive scale without eco-design becoming the norm.

The enormous positive potential relies upon societal constraints and the adoption of the extended producer responsibility by the major industries. The X factors defining the outcome will be policies, regulations, and industry standards. But these will not have any effect without a change on the business level in the full value circle – as well as the social demand for that change. The response needs to come fast. We can't manage exponential technologies with a slow linear evolution of the current governance models and industrial strategies.

5.1.3 Dilemmas Facing Waste Management

Even if there are external factors defining how Industry 4.0 will play out, each and every business will have to define their own roadmap in this unchartered territory. Leadership is needed in every organization, and the decision making will be always have to consider technological anticipations, organizational structure, skill set, existing software, and growth ambitions. Just as there are opportunities, there are multiple dilemmas.

One question is whether adding technology will add the most value upstream or downstream in the value chain. Should waste management companies invest in robotic sorting plants or in user communication to promote better source sorting? This dilemma faces many waste management operations right now. Do we overestimate the performance of robotic sorting and underestimate the cost? Should the consumer play a key role in a circular economy or be excluded from the emerging data-driven value chain all together?

Another dilemma is what the innovation process should focus on. Should we have a technical focus on the traditional waste management – handling a diverse stream of new composites and biomaterials? Or should the industry climb the waste pyramid, winning new position in the value chain to enable new business models. Who will orchestrate the closed loop value chains of the circular economy?

A third dilemma is related to speed. Should organizations be front runners that create the new solutions or should they wait and see? Industry 4.0 poses major modernization opportunity and also an opportunity to win new positions. A recent report predicts that by 2020 one-third of the top 20 firms, in every industry, will be seriously disrupted or even failed. But technological transformation is often clouded in hype. There is always a risk of picking the wrong horse or spending a lot of money on developing innovations that after a few years become a commodity available to anyone.

A fourth dilemma is about ambition and mindset. Should waste management leaders proactively try to reshape the industrial metabolism that they are an integral part of? Should waste management organizations experiment with more progressive and disruptive business models to unleash the creative destruction in manufacturing? Or should the industry be loyal to existing value chains and regulations? In the beginning, what feels the safest and least painful will always be to continue business as usual. But in the long run, although it may be a smart move to wait and see, it will make you fall behind.

5.1.4 An Ongoing Transformation

The digital revolution has already transformed waste management – especially for the last decade. Paper piles are replaced by databases. Letters are replaced by emails. The market for smart infrastructure is growing rapidly. Sensors, identity tags, and access-controlled containers are becoming a part of the everyday life of many waste management operations. Many technologies have already reached the "plateau of productivity." Over the last few years, a range of emerging services have been made available for the industry. Thanks to more (and better) data, new generations of software can be developed. And the software opens up for new ways of running the operation.

Thanks to global positioning system (GPS) trackers we now have fleet management systems – giving a real-time overview of the whereabouts of trucks and containers within a logistical operation. Mobile communication enables a more efficient feedback loop with both field workers and customers – in real time. Level sensors, measuring the waste amounts within bins or containers, open up for route optimization and remote monitoring. New algorithms can combine data about the fill level with data from online map services, including live data from traffic. Data is being used to generate better utilization of available vehicles.

Improved customer management systems and tracking of individual transactions open up for new pricing models – including dynamic fees, rewards, and incentives. Some waste management operations have established pay-as-you-throw (PAYT) programs. The enabler is IoT-enabled infrastructure and systematic tracking of events.

A very proving innovation during the last decade has been robotized sorting. The first fully automated waste sorting facility for municipal waste was opened in 2016, and several others are on its way. Robots, scanners, and optical recognition algorithms are slowly making their way into the industry.

Another interesting development is the concept of "industrial symbiosis," where waste handling facilities are colocated within a network of other industrial companies, and by-products like surplus heat, captured carbon, or material streams can be shared in real time. Industrial symbiosis is mimicking the symbiotic relationships found in nature and is a great example of how new collaborative structures are enabled within Industry 4.0.

These are all early examples of specific "cyber–physical systems," which will increasingly become a commodity in every waste management operation. First step to enable these systems is digitizing available information and capturing data that can easily and cheaply be measured. Next step is to start using the date as fuel for digitally enhanced processes and services. Where earlier practices were based on routine and schedules, the coming generation is not using just data for statistical or compliance purposes. In the operational systems of Industry 4.0, data will feed applications that make decisions on the fly – connected to machinery that receives and remotely executes instructions.

Several scientific experiments have validated that new kinds of "smart equipment" can increase performance within known processes in waste management. But the most important transformation probably happens indirectly, in the wider economic context creating the conditions around material streams and waste handling. Additive manufacturing is gaining ground, not only reducing surplus material in manufacturing but also adding new composites into the material stream. Access-based business models enable waste prevention in a range of product categories. Some companies are also creating their own closed-loop programs in collaboration with retailers, bypassing the traditional waste management value chain. Buildings are becoming material banks – designed to be deconstructed. Companies are increasingly assessing their environmental footprint – keeping track of the input and output of resources themselves.

A general new paradigm opens up – reshaping the way assets and resources are reused and recycled across industries, unlocking material value by providing transparency in reverse logistics and material stream. How will a waste management organization look if these trends evolve further and become the norm of the future? There are many possible philosophical answers to this important question. But most people in waste management are practical people. A good start may be to replace or upgrade some of the equipment.

5.2 HARDWARE IN WASTE MANAGEMENT

Based on the principles of Industry 4.0, as described in Chapter 2, what does the "sensing layer" of a cyber–physical waste management system look like? Generally in a smart factory network "machines are connected as a collaborative community" [6].

This machine community of waste management will be changing in the years to come, but many connected hardware concepts are already established and tested – waiting for a wider distribution. Traditional waste management is about collecting, transporting, and sorting physical materials. The mechanical equipment needed to collect waste across fractions are to a large degree "boxes" in different sizes – bins, containers, and compressors. To empty the trash from a small box into a larger box, you need a machine: the trash truck with a lift or crane can in many ways be considered a moving factory that increasingly can act autonomously. The basic concept of the receptacle and the collection vehicle has basically been the same for 150 years. There are long traditions of keeping some sort of log; the emptyings are counted; the volumes are, at least in aggregated form, measured or weighted.

In an Industry 4.0 environment, these measurements can be done in real time, more granularized with a way higher precision. Containers, inlets, lifts, and weights can easily be connected wirelessly to the Internet and retrofitted with machine-readable IDs, sensors, and smart locks.

In addition a range of cheap sensors can be installed to measure conditions and events. Many sensors and standard systems are already available on the market – and can even today be assembled easily to create customized solutions. They can, directly or indirectly, give important information about the customer behavior, logistical events along the value chain, and the condition within the container. Using basic technology, it is easy to measure security-critical data such as temperature, gas, light, and smoke. Physical movements can be registered by accelerometers and tilt sensors. Opening of hatches can be registered through touch and light sensors.

Even if sensory systems today may easily be connected to the internet through open build-it-yourself platforms like BeagleBoard [7] or Arduino [8], very few waste management organizations have put flexible systems like this to practical use. The Industry 4.0 solutions that have been commercialized are mainly trash trucks systems (lifts with weights and RFID scanners), smart connected underground containers, autonomous vacuum systems, and sorting plant technologies. There are also a range of single-purpose sensors (mainly level sensors and geo trackers) with associated software in the market. An important hardware device – probably the most important – will be the mobile phone in the pocket of the end user. Smartphones (and tablets) come with a range of capabilities that, if connected to the right software system, can both provide data and make the data available in useful ways – both for the end users and the people working within a waste management operation out in the field.

5.2.1 The Connected Devices of Waste Management

In this chapter we will go through some of the established hardware concepts available on the market (Figure 5.3).

5.2.1.1 Smartphones

A smartphone is a multipurpose mobile computing device with touch screen, replacing traditional single-purpose mobile phones. Thanks to hardware capabilities and advanced mobile operating systems, they can be used for a wide array of tasks alongside traditional phone functions such as voice calls and messaging. They are connected to the Internet through mobile broadband and include processor and a

Figure 5.3 Some of the key connected hardware components relevant to an Industry 4.0 operation in waste management. These can be used for different kinds of technical systems and business models.
Source: With kind permission from Nick Rigas, D-Waste.

range of sensors, such as proximity sensors, gyroscope, accelerometer, and geo-positioning capabilities. Smartphones support a range of wireless communications protocols, like WiFi and Bluetooth. Increasingly, payments can also be made through the phone. A smartphone can both be programmed to share information with different systems and subscribe to data – creating a feedback loop. The use of smartphones in waste management is still in its infancy, even if many waste management operations have started to experiment with personalized apps.

5.2.1.2 Connected Weights

Weight is an important metric used for monitoring of material streams, for compliance purposes, and for revenue/cost sharing in the marketplace. In a reverse logistics value network where material streams are aggregated, there will be several kinds of weight systems involved, ranging from very small weights integrated in smart inlets, lift weights mounted on bin collection trucks, and weighbridges for the trucks that check in and out from incineration or recycling plants. Technically the instruments that measure the weight are based on a sensor type called load cell. The load cell actually converts the physical force into electrical signals, creating data that can be transmitted in real time or related to specific events. Weight capabilities can come as stand-alone devices or be integrated with other solutions.

5.2.1.3 Container Level Sensors

A sensor using ultrasonic or infrared beams (dependent on waste type) can be placed in any bin or container to determine the fullness level. Automated monitoring of fill levels has been proven very useful for many purposes. Traditional trash pickup routes rely on weekly or biweekly routing based on maps. By using data from fill level sensors, these routes can be improved based on usage statistics, and unnecessary trips can be avoided. By applying the data to route optimization algorithm software, applications can create daily routes that optimize for personal usage, time, or climate footprint. Condition-based collection will be way more efficient than a schedule-based system. According to the sensor manufacturer Enevo, moving from fixed collection days to a more flexible model based on fill level data could provide up to 28% in cost savings [9]. An historical record of data from fill level sensors can also be used for planning and improving public infrastructure.

5.2.1.4 RFID Bin Tags and Scanners

RFID (acronym for radio-frequency identification) is the method of uniquely identifying items using radio waves. This has proven a robust solution for several purposes in waste management. An RFID system needs a tag, a reader, and an antenna. The reader sends a signal to the tag via the antenna, after which the tag responds with its unique information. Passive RFID tags (with no own power source) are a cheap way of giving individual bins a unique machine-readable identity. Tags can be attached to waste bins or containers and can be read by a scanner. This way all assets can be checked in and out, and all emptyings can be registered. Some garbage trucks are equipped with RFID readers and lift weights in combination, so you get to know the accurate weight of individual bins. This is data that can be used in a PAYT system [10].

5.2.1.5 Smart Locks

The waste being deposited in communal or private waste containers could be controlled by installing electronic locks that require some sort of user identification. In many systems users would need an electronic key – often a passive RFID card or tag – to get access and open the containers. When a valid access key is held over the RFID panel (or requested through the app), the container opens, and waste can be deposited. The advantages of access-controlled containers are real-time control with usage of containers, less unauthorized usage, and the ability to actually log transactions that can be used for dynamic pricing. In many PAYT systems the volume in the inlet is reduced to only fit one bag at the time – so every time the key is used and the hatch is open, it is considered a transaction.

5.2.1.6 GPS Trackers

A GPS tracker is a device that uses the GPS to track its movements and determine its location. The recorded location data can either be stored within the tracking unit (data logging) or transmitted to an external service through cellular, radio, or satellite. This allows the location to be displayed in external systems and used for different purposes. Waste management is to a large degree about moving things around. GPS trackers are already used to monitor trucks but can also be used to keep track of containers and other assets in a waste management operation. GPS trackers can be programmed to transmit their location at set intervals or only when specific events occur – for example, when a "geofence" (a defined geospatial area defined in a mapping software) is crossed.

5.2.1.7 Material Scanners

Near-infrared (NIR) spectroscopy is a great option for automated sorting and identification of materials – including different types of plastic. During the identification, light from an incandescent bulb is reflected on the material. A certain type can be identified by the reflectance of the light back to a receiver, which is sensitive to the NIR part of the spectrum. There are also other kinds of scanners, identifying color, density, and other material properties. Each material has a unique reflectance "fingerprint." By using machine learning algorithms, unknown plastic waste can be sorted with high accuracy, regardless of their color. Material sensors and scanners can be incorporated into mechanical sorting systems and different kinds of smart infrastructure. They also come in the form of handheld material scanners that can be a useful tool for speeding up and improving the quality of manual sorting.

5.2.1.8 Industrial Robots

Industrial robots are standardized robots, usually with one arm capable of movement on three or more axis, that can be programmed to do a lot of different work, including waste sorting. Research experiments show that industrial robots even without much training can do a great job when it comes to separating waste [11]. Standard robots have numerous sensors, such as computer vision, which enable them to monitor the waste stream. Thanks to machine learning and artificial intelligence (AI), the robot identifies the waste materials – and the robotic arm gets hold of the items. Various research projects have been involved in the development of robotic technologies for the sorting of radioactive waste or production in radiation-contaminated areas. Industrial robots for sale from leading manufacturers are not limited to working in laboratories. These machines can sort rubbish under various categories and even determine the appropriate method for disposal. An area where robots are considered especially interesting is in the area of construction and demolition waste, both because of limited capacity of dealing with big or heavy objects and because of the dust produced during processing (including asbestos). Robotic waste sorting will not be sufficient in all areas as a future solution. Especially in the area of municipal solid waste (MSW), in particular packaging waste, many manufacturers see the combined use of robotic technologies and optical sensor sorting systems with pneumatic discharge as the ideal solution for the future. In some cases, pre-shredding of the material up to a certain particle size is necessary.

5.2.1.9 Screening Machines

Mechanical screening is the practice of taking material and separating it into multiple grades or fractions by running it through physical processes. Increasingly there are also digital screening equipment with different scanners. This is a very important part of the downstream handling of waste streams. The machines use different techniques to achieve the highest possible sorting rates – such as vibration, gravity, and electrostatic force. A trommel screen, also known as a rotary screen, is a mechanical screening machine used to separate materials, mainly in the mineral and solid waste processing industries. Physical size separation is achieved as the feed material spirals down the rotating drum, where the undersized material smaller than the screen apertures passes through the screen, while the oversized material exits at the other end of the drum. Different kinds of screening equipment can be combined to achieve the best possible result in sorting plants.

5.2.2 Integrated Hardware Systems

Increasingly manufacturers integrate different hardware solutions into more complex, integrated products, often combining sensors, actuators, processors, and software solutions (Figure 5.4).

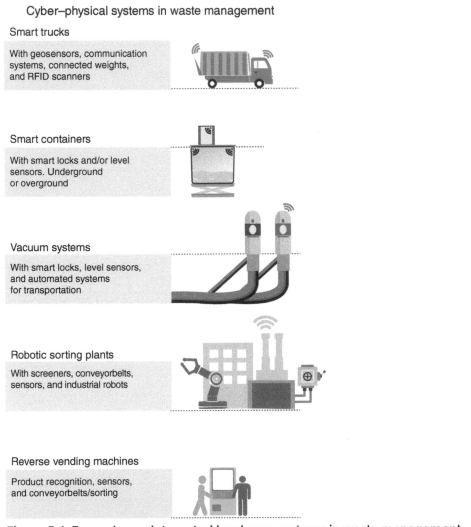

Cyber–physical systems in waste management

Smart trucks

With geosensors, communication systems, connected weights, and RFID scanners

Smart containers

With smart locks and/or level sensors. Underground or overground

Vacuum systems

With smart locks, level sensors, and automated systems for transportation

Robotic sorting plants

With screeners, conveyorbelts, sensors, and industrial robots

Reverse vending machines

Product recognition, sensors, and conveyorbelts/sorting

Figure 5.4 Examples on integrated hardware systems in waste management, where several different technical solutions are compiled into more complex hardware systems with associated software.
Source: With kind permission from Nick Rigas, D-Waste.

5.2.2.1 Reverse Vending Machines

An interesting and promising technology is the reverse vending machine, a smart robotic collection infrastructure often located at the retailer that can instantly count the number of containers returned, sort away ineligible containers, and pay out deposit refund to recyclers – much faster than is possible through manual human handling. The first reverse vending machines were originally made in 1972 by TOMRA to create a collection system for plastic bottles in a shopping center. Over the years the company has developed from a small, two-person endeavor to an international company spread over 55 countries, and the machines are now part of an incentive system. The technology evolved to be used in the collection of many different kinds of products and to support different incentive programs [12].

5.2.2.2 Underground Container Systems

In modern smart city environments, a popular solution is underground waste containers that can be fully equipped with both access control systems and level sensors, supported by associated software. These systems can be equipped with access-controlled inlets, level sensors, and technical surveillance. The administration can be done either a dedicated software or an application programming interface (API) that can be connected to other systems. Many underground containers, like the ones made by VConsyst [13], have an option for fill level measurement that gives a specific measurement of how full the containers are, enabling forecasting of emptying times for optimal planning.

5.2.2.3 Automated Vacuum Collection (AVAC)

The most complex infrastructure solution in waste management is automated vacuum waste collection systems. This system is also known as pneumatic refuse collection or automated vacuum collection (AVAC). The system was pioneered in the 1960s, designed by the Swedish corporation Envac AB [14]. Today there are several vendors that offer similar solutions, and there are a range of vacuum systems established in urban areas all over the world. The process begins with the deposit of trash into intake hatches, called portholes. These may be specialized for waste, recycling, or compost. A network of pneumatic tubes transports waste at high speed to a collection terminal. A control system automates the emptying and secures that only one fraction of material is traveling through the pipe at a time. The waste fractions are compacted and sealed in containers at the terminal. When the container is full, it is transported away and emptied. These systems can be equipped with access control systems, and a range of sensors and actuators can make them almost fully autonomous. The challenge with establishing systems is often the costly and difficult installation of the infrastructure [15].

5.2.2.4 Automated Waste Sorting Plants

Waste sorting plants can be more or less technically complex but generally comprise a series of sorting processes – often in combination with manual sorting (Figure 5.5). The process is supported by a variety of ancillary facilities designed to manage the process and maintain output quality [16]. At the plant, waste undergoes a series of procedures that refine the material stream. Specific materials that can be recycled are extracted, and materials suitable only for disposal are removed. Depending on the extent of the sorting processes, the output from the initial sorting can either go directly into manufacturing processes or into additional fine sorting. In a sorting plant there is a range of different systems interacting, including waste screeners, air separators, and different sensors. The sensors available can differentiate between plastics, identify materials based on color and atomic density, or sort different metals based on their conductivity.

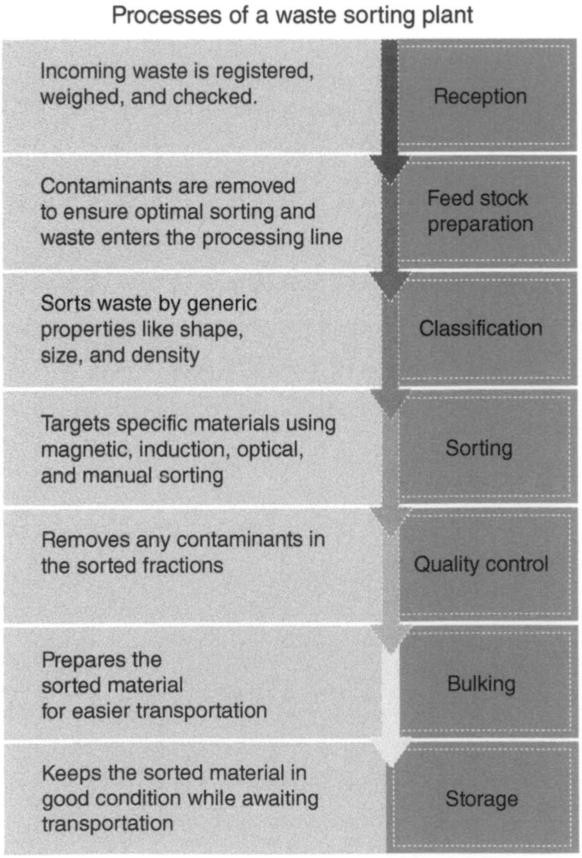

Processes of a waste sorting plant

Incoming waste is registered, weighed, and checked.	Reception
Contaminants are removed to ensure optimal sorting and waste enters the processing line	Feed stock preparation
Sorts waste by generic properties like shape, size, and density	Classification
Targets specific materials using magnetic, induction, optical, and manual sorting	Sorting
Removes any contaminants in the sorted fractions	Quality control
Prepares the sorted material for easier transportation	Bulking
Keeps the sorted material in good condition while awaiting transportation	Storage

Figure 5.5 The different processes of an automated sorting plant.
Source: **Based on Ref. [16], with kind permission from Nick Rigas, D-Waste.**

5.3 SOFTWARE IN WASTE MANAGEMENT

5.3.1 The Changing Landscape of Software Development

Describing the software for Industry 4.0 systems is not an easy task. A major shift in how software is designed and written is happening before our eyes. A lot of the new processes and business models are being reflected in new kinds of software. One driver is changing programming languages. There has generally been a revolution in how applications are being made over the last 10 years. Many of the existing "legacy systems" in waste management are designed to deal with single-purpose functions, not to serve the dynamic and interconnected requirements of Industry 4.0 – and cyber–physical systems. The transformation from old and static to open and dynamic software has proven quite challenging for many waste management companies. One of the reasons is probably (intentional or unintentional) vendor lock-in – when a system cannot be replaced or connected with other systems due to lack of compatibility between different field equipment, operating systems, or file formats. Another challenge is the procurement process.

The traditional labels put on different kinds of software tend to reflect the past, not the future paradigm. Software categories are increasingly outdated, and there are many hybrid software solutions and middleware, with features unknown just a few years ago. Before describing the different software systems, some general trends in software development should be noted.

5.3.1.1 From Servers to Cloud

There has been an important shift from local servers to cloud computing during the last decade. Due to high-capacity networks, low-cost computers, and storage devices, the concept of local servers and on-site data management is almost completely replaced by distributed remote storage. A range of different services are provided by the big companies like Alibaba, Amazon, Google, IBM, Microsoft, and Oracle. There are obvious advantages with the cloud compared with traditional local servers. It dramatically takes down the costs related to on-premise hardware and software management; it simplifies monitoring and reducing the need for extensive capacity planning. Instead, administrators can focus on other more important tasks. The key requirement for moving into the cloud is high-speed Internet from all sites and remote offices.

5.3.1.2 From Files to APIs

Modern software is mostly built on top of APIs. This basically means that the functions you see on the screen are separated from the data. APIs provide data in clean, programmatically accessible formats. No need to convert the data from one file format to another; no need to fix bad entries or load files into your local system [17]. An API makes the same data available for multiple systems and eases the integration between systems and stakeholders. "Today, a firm without application program interfaces (APIs) that allow software programs to interact with each other is like the internet without the World Wide Web," *Harvard Business Review* says in an article about the strategic benefits [18]. An API dramatically simplifies the way developers work; it makes it way easier to build rapid prototypes and enables easier sharing and collaboration. APIs can be access controlled, but by creating a public API parts of the data can be made available to every client or citizen, not just a select few. APIs can also help to support unanticipated future uses, as needs change.

5.3.1.3 From Monoliths to Modules

Generally software has gone from monoliths (complicated systems that do almost everything) to microservices (small independent applications that fit together like a puzzle). From building very complex software as a big chunk of increasingly messy code, modern programmers split the program into manageable, highly structured parts. Microservices make software easier to upgrade, as each service can be optimized or rewritten completely without affecting the functionality of the platform. This is also one of the reasons new programming languages, such as JavaScript, have gained so much popularity. As each service (notifications, customer handling, payment processing, etc.) is built and hosted independently, the software is both more stable and scalable – enabling a system to be built with the long-term future in mind [19]. This is good news for waste management operations that often struggles with complex interconnected software and challenges with outdated legacy systems.

5.3.1.4 From Centralized to Distributed Architecture

There is also a shift from centralized to distributed systems. A distributed system is a system whose components are located on different networked computers, communicating and coordinating their actions by passing messages to one another. The components interact in order to achieve a common goal. In Industry 4.0 systems some of the processing actually does not happen in the cloud, but at the edge – the processor physically placed nearby the equipment. This reduces response times and saves bandwidth. In waste management operations, the different equipment will often be dispersed within a big geographical area, so processing power locally at the site of the container or plant can be a key.

5.3.1.5 From Reactive to Predictive Software

Thanks to AI and machine learning, applications are going from "passive" (performing tasks according to protocol based on predefined orders) to more active and autonomous. For software developers and IT teams, AI offers an array of tantalizing possibilities for making applications faster, more scalable, and more efficient. AI is not yet implemented in many software solutions for waste collection and user communication, but it is being used for different purposes such as robots and waste stream monitoring in waste sorting plants. Access to AI capabilities is increasingly made available in the cloud, based on open "question-answering systems" and out-of-the-box algorithms such as IBM Watson [20] and Google Cloud AI [21].

5.3.2 Current Software in Waste Management

5.3.2.1 ERP Systems

Enterprise resource planning (ERP) is traditionally the mothership of systems within a corporation. It provides integrated management of main business processes, often in real time and integrated with other kinds of software and technology. The ERP system is often a suite of different applications that an organization can use to collect, store, manage, and interpret data from many different activities. Modern ERP systems track business resources such as cash, raw materials, and production capacity as well as keeping track of orders, purchase orders, and payroll. Data can be accessed, often in the form of reports, both to internal and external stakeholders. If there is a record of individual collection events, the systems might also be directly or indirectly integrated with hardware. Some platforms, like AMCS [22], have a "single platform approach" solving almost every task and process within an organization.

5.3.2.2 Customer Management Systems

Customer relationship management (CRM) is a software platform that handles a company's interaction with current (and potential) customers. It uses data analysis about customers' history with a company to improve business relationships with customers, specifically focusing on customer retention and ultimately driving sales growth. Due to growing complexity, a modern CRM system compiles data from a range of different communication channels, including social media, website, telephone, email, live chat, and more. In waste management a CRM system will often be based on businesses. In municipal waste management operations, the "customer" will often be the household – so the CRM system will get data from a registry of official addresses (public cadastre or a geographical information systems). Through the CRM approach and the systems used to facilitate it, businesses learn more about their target audiences and how to best cater to their needs. Many ERP systems will have CRM capabilities.

5.3.2.3 Project Management Software (PMS)

Project management software (PMS) has the capacity to help plan, organize, and manage tools for efficient handling of projects – both within an organization and between multiple parties. There is a broad variety of tools available within this category – supporting different project management frameworks and with functions such as estimation and planning, scheduling, cost, collaboration, communication, decision making, time management, and documentation.

5.3.2.4 Data Lakes and Data Warehouses

With increasing amounts of data, it is important to handle the information. There are two established concepts in use currently, either a data lake (unstructured raw data) or the data warehouses (processed, structured data). Both systems can pull together data from many different sources within an organization – and the data can be used for reporting and analysis. The reports created from complex queries within a data warehouse can be used to make business decisions. Some data warehouse software is delivered with dynamic dashboard tools that help visualize the data.

5.3.2.5 Fleet Management Systems

Any company involved in waste management will likely operate a variety of specialized vehicles, from refuse trucks to vacuum tankers and road sweepers. There is a range of software solutions that is made to give an overview of all the vehicles and other assets involved in daily operations. By subscribing to data from geo trackers and field equipment, administration can gain real-time visibility over their portfolio, drivers can get driving routes or notifications, and the logistical system can be optimized.

5.3.2.6 Route Planning Software

There are many specialized software solutions for creating routes based on parameters, like level sensors or requested container collections. This kind of software will help waste collectors use less vehicles, staff, and fuel. Systems may also provide better overall service to customers, due to less missed bins. In route planning software, an optimization algorithm will often be a key feature. The system helps in finding the most efficient route for your collection points. These systems can often be a combination of automated routes and manual routes – depending on your business requirements. The route optimizer will base suggestions on a geographical data source and can take start and finish location, collection points, road types, time of day, and vehicle capacity into account.

5.3.2.7 Field Reporting Software

In operations with many field workers and drivers, there will be a need to coordinate the operations. There are a range of different in-cab (mobile) solutions in waste management. This operates in real time and gives a complete overview for the employee, with back-office integration. This way schedules can be updated, orders can be communicated, and feedback can be provided back to the headquarter. Thanks to the sensors in mobile devices like smartphones and tablets, field applications can also report on position automatically. In many modern trash trucks, which can be considered a movable collection machine, the system is integrated to give continuous feedback on different events.

5.3.2.8 Customer Apps

Several apps are currently in use to communicate easily with end users. They can be used not only to provide useful services such as recycling guidance for materials and localize recycling centers but also to aggregate relevant news. The RecycleNation app in the United States is an example of these functions [23]. As the records of waste collections are increasingly being logged in real-time customer apps, they can also be used to provide feedback to customers on recycling behavior. Informational strategies have proven efficient in stimulation to recycling behavior in households [24]. There is obviously a great potential for further innovation in creating new incentives and feedback systems involving the end user. A very interesting development is the Alipay-developed mobile app that is designed to help Shanghai citizens navigate the eastern city's tough new trash-sorting regulations [25]. The mini-programs help citizens to understand which bin to use for the waste by looking up the object they want to throw away or simply by taking a photo of it. This is also an example of the potential in Industry 4.0. The waste image identification uses AI to help customers sort correctly as the first step of an integrated industrial recycling process.

5.3.2.9 Marketplace Software

One of the interesting developments in recent years is the development of digital platforms that exchange information between customers, producers, and service providers. Online booking of container systems for the disposal of waste have become established on the Austrian and German waste market and are expanding to new markets. Platforms like WasteBox [26] and Rubicon [27] offer private individuals or construction companies to book waste disposal orders for different types of waste through an app. The size of the container can be selected and ordered online and delivered to the chosen location and picked up again if necessary. Inspired by similar digital solutions in other industries, platforms like Schrott 24 [28] provide information on the possibility of selling scrap metal.

5.3.3 Dealing with Complexity and Creating Higher-Level Systems

As you add a higher number of data sources and more software tools into any organizations, the technological complexity and vulnerability will increase. Traditionally data exchange requires special integrations. Many processes need manual input or human data conversion. To streamline sharing of data and code is a major challenge in the transition from traditional to Industry 4.0-type systems. Ideally, based on the trends described earlier in this chapter, data should be managed independent from applications, as data usually will be reused and repurposed across software solutions (Table 5.1). To create a 4.0-type technology, a portfolio should be developed and planned based on actual tasks and processes. Each process will typically require the combination of a hardware and software solutions that serve a specific purpose within the organization (Figure 5.6).

There are many new categories of software that in different ways addresses the challenge of interoperability, enabling solutions that works end-to-end and across value chains. Waste management organizations that start digitizing from scratch do have an advantage that they can build a new structure from the ground up, without having to deal with old systems and legacy data.

5.3.3.1 Connectivity/Integration Platforms

In Industry 4.0 systems, a multitude of connected devices and software components need to be connected to a shared stream of data. To enable the flow of information between hardware and software, there is a new generation of middleware – IoT platforms or "connectivity platforms." They may be technically oriented and generalized like MuleSoft [29] or more industry specific like WasteIQ [30]. The difference from traditional integrations is that these platforms serve as a bridge – a middleware connecting hardware to the cloud and vice versa. These platforms combine flexible connectivity options, security mechanisms, and broad processing powers to reduce complexity. These platforms are often made with developers in mind. They may be built using microservices and ready-to-use features that greatly speed up development of applications. Platforms have a value at a technical level – as they simplify data collection from remote devices and the flow of data all the way to applications. Often they can process raw data to enable business critical processes or make the data available through an API. Connectivity platforms can thus easily be integrated with third-party applications and reduce the complexities and technical dependencies between software and hardware installations in heterogeneous technical environments.

5.3.3.2 Cloud Communications Platforms

As tools and technologies continue to evolve, so will the way business gets done. Most businesses now use multiple channels to manage their workflow, sell their products, interact with customers, and create seamless collaboration between different parts of their own organizations. As interaction happens in real time, often in different channels, there is a need for systems that handles communications in real time. Cloud communication platforms are built for developers to programmatically make and receive phone calls, send and receive text messages, handle chatbots, and perform

Table 5.1 Examples of different new business functions in an Industry 4.0 operational environment.

Business functions in Waste Management Industry 4.0

Planning	Development	Administration	Information	Equipment	Maintenance	External
Project planning	Product life cycle	Customer management	Customer apps	Remote monitoring	Predictive maintenance	Automated compliance
Predictive analytics	Collaborative engineering	Inventory management	Customer notification	Container monitoring	Virtual twins	API monitoring
Route planning	Team collaboration	Business intelligence	Chatbots	Geotracking	Field communication	Smart contacts
3D product modeling	SCRUM			Access management		

Source: With kind permission from Nick Rigas, D-Waste.

Most of them will need to be organized as end-to-end solutions – where different systems and devices must be connected to support the cyber–physical process.

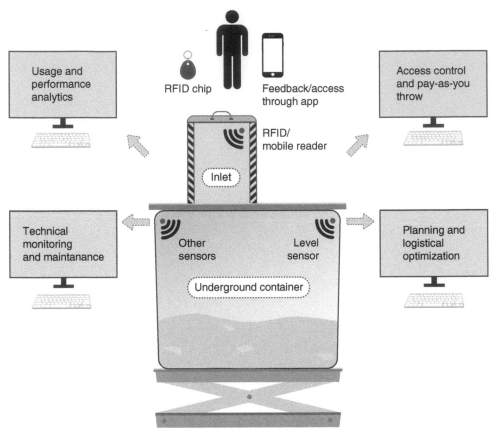

Figure 5.6 The relationship between software and hardware. A connected underground container with access control at inlet can be used for multiple purposes. It can create a new business model (pay-as-you throw) and feedback to customers based on data. Due to a level sensor, the container can become a part of a data-driven logistical model (collection based on waste level). By using other sensors and status monitoring, the maintenance can be done based on actual events/conditions – not scheduled inspections or customer complaints.
Source: With kind permission from Nick Rigas, D-Waste.

other communication functions using different APIs, sometimes including AI. They are usually provided on a platform-as-a-service (PaaS) model [31]. Similar to IoT platforms (that is mainly centered around communication with machines), cloud communication platforms can work independently from software systems – but the focus is communication with and between humans.

5.3.3.3 Smart Contract Frameworks

Contracts are still mainly in the form of written documents, but the management of judicial agreements is about to change. Many of the terms and conditions in a formal agreement can today be built into code – so-called smart contracts. There are several platforms enabling smart contracts, most of which have their origin in the different blockchain framework, especially Ethereum. Solidity – a new programming language – is developed to reflect contractual logics in applications [32]. This may seem distant future, but it is actually already being implemented – also by public bodies. By implementing the Decree on Development of Digital Economy, Belarus became the first ever country to legalize smart contracts [33]. Making the contract a dynamic application instead of a static document has major implications. The clauses in a smart contract may be made partially or fully self-executing and self-enforcing, the security can be superior to traditional contract law, and transaction costs associated with contracting can be substantially reduced.

5.3.3.4 Code Repositories

In a world run by software, management code is increasingly important. Many organizations use a repository hosting service, sort of like a cloud for code, both to keep their code projects at hand and to share parts of the technology with others. Repository services such as GitHub or Bitbucket support a variety of different programming languages and keep track of the various changes made to every iteration. They have extensive collaboration features that let programmers work together on projects from anywhere. Open-source projects, where code is made available for anyone, are usually shared through a code repository. There are also specialized platforms for open API specification and documentation in the cloud, such as Swagger [34]. These make documentation readily available for external developers, making integrating different systems easier.

5.4 SELECTED CASE STUDIES

5.4.1 Developing a Platform to Simplify Access Management Across Different Technical Solutions

Sometimes politically imposed reforms create challenges for waste management. Can such forced transformations be used to create new opportunities? BIR, the waste management company owned by nine municipalities in the Bergen region of Norway, started their journey towards a data-driven operation in 2009, when a PAYT fee was introduced to reward recycling. This was politically imposed and represented a huge transformation for the organization. From a flat fee to a data-driven system where fee is calculated for every household based on amount of mixed solid waste, is it even possible?

To make this work the ICT department looked for a solution that could work across technical solutions. In Bergen they have a complex underground vacuum system in the inner city, and there are several different container solutions. Many of the inlets were already outfitted with smart locks – supporting RFID keys. The majority of households were served by traditional collection of waste bins using trash trucks.

Can data from different smart bin systems and smart lock systems be combined to create a unified stream of data? What about the access management, and can a small administration handle this for a full city? How can the cost of integration be reduced? How can transactions be integrated into one unified system – with data in a consolidated format?

The questions were many. In the new system, the data would turn business critical, being used to calculate the fee for each individual household. There was no general connectivity platform on the market serving the purpose of integrating different access-controlled solutions. What was needed was a system that served as a bridge between the hardware and software, normalizing data. The system needed to be open. Most software are tied to a specific IoT solution. To manage IDs in a consistent way in a complex modern city is impossible if this should be done using a range of different incompatible vendor systems.

The ICT department in BIR took on a big challenge: they started building an application themselves. To connect the established customer management software with the different infrastructure systems out in the field was not an easy task. Many vendors became involved in the process. The customer system needed a big upgrade. A new technical standard was developed for the different smart lock systems to exchange data seamlessly with the system.

The waste management company decided to use RFID for their customer identification. Bins were outfitted with passive RFID tags, and households with access to underground solutions got their own RFID keys – issued by the waste management company. This way, all households got a specific ID. Every bag that entered into the system could be tracked to someone responsible, and an individual cost could be estimated.

The integrations needed to go both ways. From the customer system you needed to issue RFID keys and update the infrastructure. And from the infrastructure you needed to send data about usage from individual users back to the CRM system. The data was business critical and needed to be accurate as it was used to calculate PAYT fee for every household.

The system was designed to create an incentive for recycling. Households pay more for mixed waste than for recyclables. The results were impressive. Recycling rates peaked. Improved data quality increased operational efficiency. As other cities wanted a similar solution and the data was business critical for the operation, BIR decided to collaborate with a private technology company to establish a software company: WasteIQ.

With a new development team on board, the system was reprogrammed to run in the cloud and offered as a software-as-a-service (SaaS) platform to serve multiple cities. As the former project was converted into a product, the technical team developed WasteGraph – a universal information model that could be used by any waste management operation. The integrations with the vendors were simplified, and a system of certified adapters was introduced to reduce the cost of adding new infrastructure to the system.

The openness of the system through an API is a key feature in the new platform. This makes it very easy for third parties and external software solutions to exchange data from the platform. WasteIQ collects, checks, and structures the data. Data from WasteIQ is currently being used by CRM systems, ERP systems, data warehouses, and business intelligence solutions.

Every time someone opens a hatch of a smart bin or a bin is emptied, an event is logged in the database. The system supports level sensors and GPS trackers, so the data can also be used for fleet management and route optimization software.

The platform offers a solution to the problem of legacy systems with data formats that cannot be read by other systems. WasteIQ "liberates" data and creates a richer context that combines geographical areas, IDs, system types, and events. The platform runs automated checks in the background to ensure high data quality. Onboarding of new equipment is easy with no lock-in to specific software.

The key lesson from WasteIQ is that the combination of industry know-how and external resources can be a key to develop new solutions and that a publicly owned municipal waste management organization actually can become a co-founder of a software company.

5.4.2 Using Bin Sensors to Decrease Overflowing Trash Cans with 80%

Can the use of IoT sensors improve cleanliness on the streets in a big American city? And can the data from such sensors be used in improving the infrastructure? On an annual basis, San Francisco Public Works was receiving 10 000 service requests for overflowing trash cans. The city wanted to find a solution that could tackle this issue effectively, decrease the number of annual complaints, and ensure cleaner streets. The department turned to technology, partnering with Nordsense in a pilot project to deploy sensors in 48 cans – initially to monitor trash levels in real time.

The goal with the project was to provide a better and cleaner experience for community members. The municipality wanted to detect service needs before community members complain. The city wanted to reduce the number of collections necessary and subsequently reduce cost. By analyzing the data they also wanted to optimize the placement of bins.

The results clearly show the benefits of smart bin infrastructure. They reached 80% decrease in overflowing trash cans. There was a 66% decrease in street cleaning service requests. Due to highly successful pilot projects, San Francisco Public Works will now expand the collaboration with Nordsense. In the near future a total of 1000 cans will be equipped with sensors.

Many cities across the world have "zero waste" ambitions and are turning to technological innovation to help transform the waste collection industry. In addition to ensuring a cleaner environment, the data gathered from the sensors has enormous potential to help cities predict waste generation activity in specific areas, fine-tune the routing of collection trucks, and even help to increase recycling rates.

"The pilot program showed us firsthand the benefits of the sensor technology to improve cleanliness on our city's streets and we're looking forward to implementing the expansion." San Francisco Public Works Director Mohammed Nuru says, "The partnership with Nordsense will allow us to use real-time data to better service the public trash cans and deploy resources more strategically" [35].

5.4.3 Creating Symbiosis-Like Industrial Parks to Curb Emissions and Increase Resource Productivity

On the cutting edge of innovative environmental strategy is industrial symbiosis. The concept emerged two decades ago, when researchers observed that waste and energy usage from industrial operations drastically could be decreased by using the waste of one firm as the input of another. To make this system work, you need sharing infrastructure such as steam or water processes. Sharing infrastructure and knowledge among diverse local industries creates a competitive edge, new collaborations, and reduced costs.

In many ways this is the perfect example of applied Industry 4.0. It is about capturing more value by creating new integrated value chains. The exchange of resources between companies requires monitoring and higher-level control systems, new business models, and collaborative mindset that cross organizational boundaries.

Over the last decade, China has taken the lead in this development. The 2008 Circular Economy Promotion Law in China demanded that factories and industries embedded themselves in networks of recycling. More than 200 eco-industrial parks are now approved by the Chinese government.

But industrial symbiosis is also an example of how ideas develop and spread. The cradle of industrial symbiosis is considered to be the eco-industrial park in Kalundborg, Denmark. Here a large ecosystem of colocated companies has gradually evolved organically [36]. The network started forming almost by a coincidence in 1961. Today more than 30 different kinds of by-products are shared among the tenants. The symbiosis is becoming a differentiator and a strategy. And it works. Data on the performance of the Danish plant indicates a $15 million in annual savings across the different companies in the ecosystem. They save both on reduced need for transportation and from synergies of sharing of infrastructure and services. Big savings also stem from the fact that waste that elsewhere would be treated as a loss, which now can generate revenue.

The Chinese have learned from these experiences and have been utilizing innovations developed in Germany, Japan, and the United States. The concept is being rolled out at a rapid pace and on a massive scale and is seen as a tool to make the circular economy principles a reality.

One example mentioned in a GreenBiz online article [37] is the Rizhao eco-industrial park, also known as REDA. The park was established in 1991 initiated by companies in the park hoping to get a tax break on their reduction of solid waste. The exchanges of resources and the sharing of infrastructure soon became incorporated in the core business model of its firms. This aligned the government strategy to promote reuse and energy efficiency, as well as reduce resource dependence on other countries. EDA was soon given status as a model for eco-industrial parks. The development has continued with new subsystems and "neighborhoods" developing. Today there are 31 by-products being exchanged between local firms. A fertilizer factory gets the by-product of sugar from a nearby beer brewery. A paper and pulp plant receives scrap wood from a nearby wood factory and in turn provides sludge for fertilizer, fly ash for a cement plant, waste hot water for an aquaculture mill, and so on.

The environmental benefits of the flow of energy and materials are obvious. According to GreenBiz 71 446 tons of white sludge were used as a substitute for calcium carbonate in the citric acid and cement factories in 2011. The nearby cement and building material factories were supplied with over 66 000 tons of fly ash and 20 000 tons of green mud as raw materials. Thanks to a combination of symbiosis and cleaner production practices, 98% of the industrial solid waste in the park is recycled.

Industrial symbiosis is already a part of the training of engineering students at universities. Industrial symbiosis principles are an interesting roadmap for future production models. Both in China and elsewhere, more industry parks are planned. If the industrial parks of tomorrow will be diversified and developed around resources, that would be good news for the circular economy.

5.4.4 Enabling a Fully Automated Waste Processing Facility

Many have been dreaming of a fully automated mixed waste processing facility. Due to the high cost of labor, the Norwegian MSW hauling company Romerike Avfallsforedling (RoAF) decided to turn the concept into a reality. The plant opened in 2016 in the village of Skedsmokorset, just outside of Oslo. Originally it was built to process household and food waste from 10 municipalities in Norway, with a population of roughly 53 000 people. But the operations turned out so successful RoAF completed a "major expansion" at the facility to increase capacity and open its doors to other waste management companies and municipalities interested in delivering waste and recyclables to the plant.

When the waste arrives at the plant, it's first fed onto a conveyor, which delivers the waste into the sorting plant. Green bags of food waste are sorted from the rest of the material stream and taken to a local anaerobic digestion facility, where they are turned into biogas. The gas is used to fuel RoAF's waste collection trucks. Residual waste and recyclable materials pass through a screening drum and NIR optical sorters. In the separation process, the material is separated by size in the screening drum and then into five types of plastic, mixed paper, metals, and residual waste. The recyclable materials such as plastics, metals, and mixed paper are recovered at the facility for further recycling, while the waste is incinerated and used for nearby heating and electricity applications.

Only two employees are needed on-site to monitor the machines. What did they do? An article in *Waste Today* [38] gives an insight into the process. The municipality wanted to push the limits and needed a technology provider. They put out a bid to design and construct the facility. The bid went to Stadler Anlagenbau. Even if Stadler got a challenge that had never been done before, it then managed to construct the plant in only three months.

The plant includes a variety of processing equipment, including 145 conveyors, 16 NIR optical sorters, two drum screens, one vibrating screen, a star screen, a shredder, two bag openers, two ballistic separators, and an eddy current. All the technology is well known in waste processing, but the trick was to put it together in a new way.

The quality and purity of the recycled fraction is high. The technical equipment in the production line is one of the reasons for this. Collaboration was a key to achieving this. Adler collaborated with TOMRA, known for optical sorters. By combining their own equipment with specialized equipment from other vendors, they got the best possible output.

Another reason for high-quality output is that the residential waste and recycling collection program was redesigned before the plant went operational. ROAF asked households to put food waste in bright green bags for food waste collection. The color increased the ability of optical sorters to identify the bagged food from the rest of the waste. This way food can be sorted out together with fiber early in the process. This makes downstream sorting easier and removes odor.

Five thousand tons of recyclables are recovered annually, including 2,500 tons of high-quality plastics, including polyethylene terephthalate (PET), polyethylene (PE), polypropylene (PP), and film grades. The material recovery rates at the automated facility are comparable with recovery rates at single-stream material recovery facilities elsewhere in the world.

There is already a high level of automation in waste processing facilities, contributing to offset challenging commodity prices and take down the high operating costs. New technology is increasingly being adopted at waste and recycling facilities across the world. This can be seen from the amount of manual labor needed to operate an average processing plant. According to Adler standard staffing rates used to be at least 40 sorters in a 40-ton/h system. This number of people can easily be reduced to less than 10.

> The issue is not the technology or programming, but it takes people at the plant thousands of hours to figure out how to get it set up right. Manufacturers making the robotic sorters are just getting enough hours in real-life applications to get the kinks worked out. It's definitely improving, but it's not as quick as everybody would like it to be.

5.5 THE VALUE OF OPENNESS

The shift from digitalization to Industry 4.0 implies different systems are connected more seamlessly. Data is reused across different applications, and increasingly the software should also be able to subscribe to data from hardware and communicate with the connected devices. Ideally the business critical data generated at the hardware level will be used in real time – both for analytics and automated processes.

The first wave of digitalization in waste management created a lot of "single-purpose" applications, based on proprietary solutions. Data exchange is often based on complex integrations, with low frequency. There is a lack of standards for the specific processes of "reverse logistics" that creates challenges. Bad data quality and costly integrations are a problem especially as legacy systems without APIs need to interact. Many of the general trends in software development are addressing the challenge of interoperability – the ability for different systems and data sources to work together. In Industry 4.0 seamlessness and connectivity should be an ideal.

Open source and open standards are expected to play a large role in the next big wave of evolution in the tech world. If software is published as open-source license, anyone can freely use, change, and share software in modified or unmodified form. Promoters often emphasize the value of four "freedoms": the freedom to use it, study it, improve it, and share it. As the code evolves continuously and constantly, so does the quality of the software. Current systems in waste management are dominated by proprietary software and lack of open specifications.

Open standards are often initiated by enthusiasts and slowly turning into wider industry initiatives. The reason for standards is to make it easier to combine different systems. Open APIs enable data sharing across stakeholders and technical solutions. The open-source community is a large and passionate society that enables open collaboration and sharing, which results in added value to existing products and creates new ones. Embracing an open-source mindset is an invitation for innovation and enables organizations to break free from proprietary vendor systems. The enterprise customer avoids being locked in and can access innovation across the community.

The openness and portability of systems is an important consideration in the procurement process. Choosing open systems is choosing flexibility but can also mean supporting a sharing culture that will help spread technology and reduce the development costs for the industry as a whole. In Table 5.2 we listed different concepts of openness.

Table 5.2 Understanding openness.

Concept	Explanation
Open licenses	A license grants permissions and states restrictions; it specifies what can and cannot be done with a work. If a license is open, this means that the work can be accessed, redistributed, and reused with few or no restrictions. By publishing content through an open license, we do not have to ask permission – but we should give credits according to the rules specified by the license – often defined by Creative Commons [39]
Open-source software	Open-source license is a license for computer software that allows source code to be used, modified, and shared under defined terms [40]. The free software movement was initiated in 1983 and has since then developed a system for sharing open-source code within a system of open licenses that has been very influential on the development of software
Open-source hardware	Even if open source started with software, the same principles are increasingly used within the creation of hardware solutions and physical artifacts. This is relevant to Industry 4.0. The term usually means that information about the hardware is easily discerned so that others can make it – coupling it closely to the maker movement [41]
Open data	Open data is data that can be freely used, reused, and redistributed by anyone. It must be available in a convenient and modifiable form, at no more than a reasonable reproduction cost, preferably by downloading over the Internet. The data must also be available. Data must be provided under terms that permit reuse and redistribution including the intermixing with other data sets. Everyone must be able to use, reuse, and redistribute with no discrimination against fields of endeavor or against persons or groups
Open standards	An open standard is a document that provides requirements, specifications, guidelines, or characteristics [42]. It can be used to ensure that materials, products, processes, and services are fit for their purpose. Standards are developed through a consensus process, and at the highest level it is recognized to be an international standard
Open API	Open application programming interfaces (APIs) provide developers and others with access to a data set, a software, or web service. Restrictions might include the necessity to register with the service providing the API. Open APIs are often used to make open data available in a programmatic form – licensed to be use and republish as they wish, without restrictions from copyright, patents, or other mechanisms of control. Open APIs are based on an open standards
Open access	Open access is a publishing model that makes research information, both publications and research data, available at no cost [43]. This is opposed to the traditional subscription model in which readers have access to scholarly information by paying a subscription (usually via libraries)

Different solutions and the degree of openness in technological systems and solutions.

This is also a challenge that can be addressed by national and international associations. Even if the events and processes in waste management are quite similar across businesses and countries, there is generally a lack of shared standards in waste management software. There are a few national initiatives such as the British KnoWaste standard [44] and the Dutch STOSAG standard [45]. But other industries have developed way more complex models that enable a digital representation of physical and functional processes.

An inspirational example is Building Information Model (BIM) [46] – a shared knowledge resource for information about a facility forming a reliable basis for decisions during its life cycle. With BIM any building can be defined as existing from earliest conception to demolition. The early concepts of the BIM standard were developed by pioneers in the 1970s. Gradually this has gained momentum and transformed both the software tools, business models and processes in the construction industry. Today it gives architects, engineers, and construction professionals the ability to plan, design, and manage building projects more efficiently. It is growing in popularity and has created the need for an international framework that allows the industry to work together across projects and national borders. Can a similar process be imagined in waste management?

Key Take-Outs of Chapter 5

Key take-out	Why it is important
Executives in waste management have different opinions about the technological progress and how it will affect the industry in the near future. The respondents ranging from optimism on behalf of the circular economy to pessimism regarding e-waste and marine litter	There is a need for the waste management sector to prepare themselves for more uncertainty and to take action to create the transformations they want to see in the industry
New technologies follow a cycle of hyped expectations, disillusionment, and productivity. But many enabling technologies are now mature and about to scale and be applied into new business models	Big changes can be expected in the years ahead within the waste management industry and more importantly in the wider economy. Some technologies are probably still hyped and will not give the expected short-termed transformations
The waste management industry faces many dilemmas – given the expected transformations. As technology evolves the big X factor is political reform, regulatory constraints, and industry standards	Leadership and willingness to change are important. To give direction to the technological development in society, it is important not only to innovate and experiment on a practical level but also to work on a political and regulatory level

Key take-out	Why it is important
Many inspiring examples show how new technology can lead to improved efficiency, resource quality, and recycling rates in waste management. But the technological solutions have yet to be spread widely	Sharing and standardizing on a global industry level is an important challenge that can ease the transition for many waste management organizations
There is a range of new hardware solutions on the market that can be applied by most waste management organizations. A lot of the equipment can be used to upgrade and retrofit existing equipment. Marginal costs are falling, and some systems are open source	Taking the first steps towards Industry 4.0 practices does not necessarily require very expensive investments. Small experiments can lead to more cost-effective, transparent, and efficient waste management
To get full value out of digitizing hardware, smart software is needed, and business processes need to be redesigned end to end. A new generation of open cloud-based software enables more holistic digital systems	Waste management can be complex but can be simplified by embracing open standards and connectivity platforms. A new operational paradigm requires rethinking the digital architecture and systems portfolio
Interoperability and openness are a key consideration when investing in both hardware and software. The systems should have open APIs to enable exchange of data between different systems	Open APIs should be considered as important criteria in procurement. The technology vendors in the industry should generally open up to enable a more diverse technical ecosystem
There is generally a lack of domain-specific technical standards in waste management, especially related to the handling of data. Open data and open-source code can help spread the best practices faster	The waste industry globally should collaborate to create standards and open technology projects for both hardware and software – to speed up and democratize the access to new technology

REFERENCES

1. ISWA (2017). *The Impact of the 4th Industrial Revolution on the Waste Management Sector*. ISWA.
2. cern.info.ch (1989). Tim Berners-Lee's proposal. http://info.cern.ch/Proposal.html (accessed 1 February 2020).
3. Schumpeter, J.A. (1934). *The Theory of Economic Development: An Inquiry into Profits, Capital, Credit, Interest, and the Business Cycle*. Transaction Publishers. 255 p.

4. Berman, A.E., Dorrier, J., and Hill, D.J. (2016). How to think exponentially and better predict the future. *Singularity Hub*. https://singularityhub.com/2016/04/05/how-to-think-exponentially-and-better-predict-the-future (accessed 12 February 2020).
5. Gartner (2018). Hype cycle research methodology. https://www.gartner.com/en/research/methodologies/gartner-hype-cycle (accessed 12 February 2020).
6. Lee, J., Kao, H.-A., and Yang, S. (2014). Service innovation and smart analytics for Industry 4.0 and big data environment. *Procedia CIRP 16*: 3–8.
7. BeagleBoard. Community supported open hardware computers for making. beagleboard.org (access 1 February 2020).
8. Arduino. Home. https://www.arduino.cc (accessed 1 February 2020).
9. Enevo (2018). How technology can cut your waste management costs. https://www.enevo.com/blog/how-technology-can-cut-your-waste-management-costs-197 (accessed 30 January 2020).
10. Saleh, D., Salova, M., Bulbena, B., and Loderus, T. (2019). User ID for municipal waste collection in high-density contexts. User ID for municipal waste collection in high-density contexts. https://ent.cat/wp-content/uploads/2019/03/User-identification-for-municipal-waste-collection1.pdf (accessed 28 January 2020).
11. Barrero, N., Galvis, D., and Martinez, C. (2018). Industrial robots for waste separation tasks: an approach to Industry 4.0 in Colombia. *The 9th International Conference on Production Research-Americas 2018*. http://dx.doi.org (accessed 22 January 2020).
12. WIPO (2010). Recycling for the future. https://www.wipo.int/ipadvantage/en/details.jsp?id=2589 (accessed 1 February 2020).
13. VConsyst. Home. https://vconsyst.com/en (accessed 3 February 2020).
14. Envac. Smart waste handling for sustainable cities, hospitals and airports. www.envacgroup.com (accessed 3 February 2020).
15. Igesund, T. (2018). Guest Blog | BIR Net. Underground waste handling in medieval city centre, Bergen. https://www.iswa.org/home/news/news-detail/article/guest-blog-bir-net-underground-waste-handling-in-medieval-city-centre-bergen-1/109 (accessed 3 February 2020).
16. ISWA (2018). Waste sorting plants: extracting value from waste. https://www.iswa.org/home/news/news-detail/article/waste-sorting-plants-extracting-value-from-waste/109 (accessed 12 February 2020).
17. UC Berkely. Why use an API? | Engineering and integration services. https://integration-services.berkeley.edu/using-apis/why-use-api (accessed 29 January 2020).
18. HBR (2015). The strategic value of APIs. *Harvard Business Review*. https://hbr.org/2015/01/the-strategic-value-of-apis (accessed 29 January 2020).
19. Skelia (2018). 5 Major benefits of microservice architecture: Skelia. https://skelia.com/articles/5-major-benefits-microservice-architecture (accessed 29 January 2020).

20. IBM. IBM Watson. https://www.ibm.com/watson (accessed 3 February 2020).
21. Google Cloud. Cloud AI. https://cloud.google.com/products/ai (accessed 3 February 2020).
22. AMCS. https://www.amcsgroup.com/home (accessed 27 January 2020).
23. RecycleNation (2017). Search. Find. Recycle. recyclenation.com (accessed 1 February 2020).
24. Van der Werff, E., Vrieling, L., Van Zuijlen, B., and Worrell, E. (2019). Waste minimization by households: a unique informational strategy in the Netherlands. *Resources, Conservation and Recycling 144*: 256–266.
25. TechNode (2019). Alipay helps Shanghai citizens sort trash as authorities get tough. https://technode.com/2019/07/01/waste-sorting-alipay (accessed 1 February 2020).
26. Wastebox. Main page. https://www.wastebox.biz/en (accessed 27 January 2020).
27. RUBICON (2019). Waste container options: commercial dumpsters, totes, and carts. Rubicon Global: Waste, Recycling, and Smart City Technology Company. https://www.rubiconglobal.com/waste-recycling-equipment (accessed 27 January 2020).
28. Schrott 24 (2017). Online Altmetall verkaufen | Aktuelle Altmetallpreise. www.schrott24.de (accessed 27 January 2020).
29. MuleSoft. Integration platform for connecting SaaS and enterprise applications. www.mulesoft.com (accessed 3 February 2020).
30. WasteIQ. wasteiq.com. www.wasteiq.com (accessed 3 February 2020).
31. CEQUENS. Why should you invest in a communication platform? https://www.cequens.com/story-hub/why-should-you-invest-in-a-communication-platform (accessed 23 February 2020).
32. Solidity. Solidity 0.5.1 documentation. https://solidity.readthedocs.io/en/v0.5.1/# (accessed 3 February 2020).
33. Emerging Europe (2017). ICT Given Huge Boost in Belarus. News, Intelligence, Community. News, Intelligence, Community. https://emerging-europe.com/business/ict-given-huge-boost-in-belarus (accessed 3 February 2020).
34. Swagger. The best APIs are built with Swagger tools. swagger.io (accessed 3 February 2020).
35. Maile, K., Toto, B.D., Cottom, T., and Redling, A. San Francisco partners with Nordsense to monitor bins. *Waste Today*. https://www.wastetodaymagazine.com/article/san-francisco-partners-with-nordsense-to-monitor-bins (accessed 22 January 2020).
36. Jacobsen, N.B. (2008). Industrial symbiosis in Kalundborg, Denmark: a quantitative assessment of economic and environmental aspects. *Journal of Industrial Ecology 10*: 239–255.
37. Childress, L. (2017). Lessons from China's industrial symbiosis leadership. *GreenBiz*. https://www.greenbiz.com/article/lessons-chinas-industrial-symbiosis-leadership (accessed 16 January 2020).

38. Maile, K., Redling, A., and Redling, A. (2019). Inside the world's first fully automated mixed waste processing facility. *Waste Today*. https://www.wastetodaymagazine.com/article/romerike-avfallsforedling-roaf-automated-mixed-waste-facility (accessed 22 January 2020).

39. Creative Commons (2002). When we share, everyone wins. creativecommons.org (accessed 15 February 2020).

40. OpenSource (2007). Open source definition | Open source initiative. https://opensource.org/osd (accessed 15 February 2020).

41. Bajarin, T. (2014). Why the maker movement is important to America's future. *Time*. https://time.com/104210/maker-faire-maker-movement (accessed 15 February 2020).

42. Opensource.com (2010). What are open standards? https://opensource.com/resources/what-are-open-standards (accessed 15 February 2020).

43. Openaccess. What is open access? https://www.openaccess.nl/en/what-is-open-access (accessed 16 February 2020).

44. OpenDataManchester (2018). OpenDataManchester/KnoWaste. *GitHub*. https://github.com/OpenDataManchester/KnoWaste (accessed 1 February 2020).

45. STOSAG (2015). Welkom bij STOSAG. http://stosag.nl/nl/home (accessed 1 February 2020).

46. Autodesk. What is BIM | Building information modeling. https://www.autodesk.com/solutions/bim (accessed 9 February 2020).

Chapter 6

Towards the Digitalization of the Waste Industry

In thinking about the microeconomics of information technology – or, of biotechnology, new materials, and other developing "generic" and systemic technologies – one often is prone to suffer from a kind of "telescopic vision": the possible future appears both closer at hand and more vivid than the necessary intervening, temporally more proximate events on the path leading to that destination.

— Paul A. David, Department of Economics Stanford University (in 1989)

Recommended Listening
Quiet Is the New Loud, **Kings of Convenience**

Because it is thoughtful, relaxing and acoustic - the simplicity makes every chord more important.

Industry 4.0 and Circular Economy: Towards a Wasteless Future or a Wasteful Planet?, First Edition.
Antonis Mavropoulos and Anders Waage Nilsen.
© 2020 John Wiley & Sons Ltd. Published 2020 by John Wiley & Sons Ltd.

6.1 FROM WASTE MANAGEMENT TO RESOURCE INNOVATION

Change is always hard work. Organizations evolve over time to serve a purpose in a certain way. When the way is winding, structures and processes need to follow. This can be quite painful and frustrating. If you talk to almost any waste management organization anywhere in the world, you will soon get to learn about the difficulties of meeting regulatory requirements, keeping customers happy, and implementing new systems. Transforming from one paradigm to another can feel almost impossible – especially on a management level.

Compared with many other industries, waste management is still a highly mechanical yet complex process with lots of manual labor. Limited digitalization and lack of transparency in the value chain mean there is huge potential for both short-term and long-term gains. Many companies struggle with old legacy systems, incompatible information, and manual data handling. Even improving current practices requires a clear long-term vision and a new set of management tools. Let's first have an overview of lessons learned and experience gained.

As seen in Chapter 5, there are plenty of new technological tools that can be used in reinventing waste management. Self-driving cars, autonomous sorting plants, route optimization software, reverse vending machines, and data-driven pay-as-you-throw (PAYT) programs are all improvements to the existing waste management practices. All over the world there are many interesting initiatives and proof of concepts on how to organize operations more efficiently. According to an ISWA report on Industry 4.0 in waste management [1], the new technology "opens new ways to prevent, reduce and even eliminate waste from specific sectors and streams, to advance resource recovery, to achieve high standards of treatment and disposal, to substantially reduce pollution and environmental impacts. At the same time, it provides new tools to stimulate stakeholders' interaction, awareness and citizens' participation, to apply the 'polluter pays' and the 'extended producer responsibility' principle on a global scale, to make the governance of cities (including waste management) more inclusive and participatory, and to reduce or eliminate 'dirty – dangerous – difficult' jobs and improve occupational safety and working conditions."

6.1.1 A Fundamental Shift

The transformation of the waste management industry represents more than an improvement of the known business models. Together, the circular economy and Industry 4.0 represent a journey into the unknown, a makeover that will and should transform not only the processes but also the identity of people working with waste management. "To absolutely decouple growth from resource use, we must change the way we produce, market, consume and trade, and the way we deal with waste," the EU commission said as they launched the first draft of a strategy to decouple economic growth from natural resource use in February 2020 [2].

Big words – coming from a political institution that will define the regulatory framework in which waste management operates. We are facing a paradigm shift, not an incremental evolution. Leaders in waste management are, if they like it or not, leaders within a revolution. They can either follow or contribute to shaping the outcome. They will have to reshape culture, organizational structures, and business models – using technology as an enabling tool for deep and conceptual transformation. Siloed, territorial businesses will be challenged by collaborative alliances. Waste handling in this paradigm becomes something more than just collecting containers and sorting the contents. The economic and material responsibilities will be redistributed in new and yet unknown ways. The integrated value loops enabled by Industry 4.0 principles increasingly also involve interaction with consumers and the secondary users of the raw material (Figure 6.1). Operations that used to be based on schedules and routine are changing to data-driven optimization in real time. Work experience will, at least partly, be replaced by algorithms generating work orders. A new relationship between humans and machines will develop. As the policy makers raise the bar, the industry needs to perform better than ever before but also reimage their own processes and role in an extended and circular value chains.

Exciting things happen across the globe, driven by (often resourceful) pioneers and early adopters. Established companies, start-ups, and research teams are making great progress to enable the new systems and best practices of an Industry 4.0 world. Many political reforms are introduced to improve recycling rates and the quality of the recycled waste streams. Many countries are closing landfills. Import bans on low-quality waste in several Asian countries requires Western

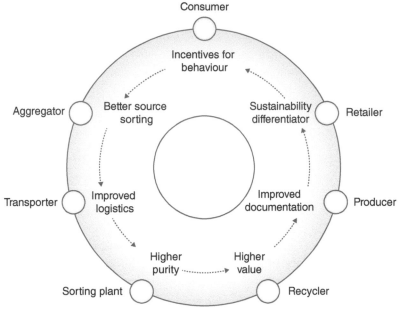

Value creation in an integrated circular value chain

Figure 6.1 An integrated circular value loop, where Industry 4.0-type innovations can improve resource efficiency for every stakeholder at different steps. In a holistic model like this, the materials can be tracked, and costs and rewards can be distributed according to the contributions in the process. The requirement is open data sharing within a trusted, transparent business ecosystem.

Source: With kind permission from Nick Rigas, D-Waste.

countries to increase the industrial capacity and deeply rethink their own systems for material handling. Many municipalities establish economic incentives and compensation mechanisms to increase profitability of recycling [3]. There is a lot of ongoing experimentation to create new frameworks, platforms, processes, and experiences that can be shared with others. And there is hope this sharing will span borders – unlocking new opportunities for developed developing countries that face very real waste management challenges like weak collection services, uncontrolled dumping, and open burning.

There are many anecdotal proofs that the industry is responding to regulatory ambitions, but the big systemic challenge – decoupling growth from waste – is still unresolved. Will it even be possible?

Innovative new approaches will require leaders with the ability to think outside the box and unleash the potential of new technology. Organizations dealing with the handling of materials across the globe will have to develop the ability to innovate and experiment. As we move from a linear to a circular economy, the waste management will go from "downstream" to center in the economy. They should actively search for a new and more defining role in the value chains they take part in. The recipes that work should be shared and scaled fast. International collaboration is increasingly important, also the peer-to-peer collaboration between leaders across nations and regions. The political level may define the metrics, but it is the industry itself that will have to come up with the actual answers. This revolution starts from within. Leadership has never been more important.

6.1.2 Risks and Challenges

There is no safe route. Many risks and challenges are associated with the transformation to Industry 4.0 principles. A study of supply chain management [4] summarizes several key lessons that leaders should be aware of:

- The economic risks are obvious. There are high costs associated with new equipment, training of staff, and redesigning organizational structures. The return on investment may be unclear and hard to calculate. Changing existing business models can be risky, especially for early adaptors trying to win new positions by offering services in new ways.

- Creating a data-driven organization always comes with privacy concerns, both related to customers and partners. There may be reluctance related to distrust, fear of surveillance, or a general skepticism towards data collection. IT security issues are greatly aggravated by the inherent need to open up systems. When machine-to-machine (M2M) communication becomes critical for the operation, reliability, stability, and latency becomes very important.

- Within any organization there will always be certain resistance towards change. Internally, new and digitized organizational practices may be considered a threat. The IT department may, sometimes with good reason, protect existing solutions. The blue-collar workers can feel threatened by increasing automation. Many employees may be happy with their roles and work environment and feel a loyalty to the siloed departments they are currently a part of.

- Insufficient skills within the organization are another obvious challenge. Industry 4.0 requires knowledge about new tools and methodologies. Lack of adequate competence to expedite the transition towards the fourth industrial revolution poses a highly relevant threat in most organizations. Lack of top management commitment and curiosity has also proven to be a problem.

- Lack of openness and industry-wide standards can also create complexity and difficulties in achieving the goals set for the organizations. There are many existing software solutions that are based on proprietary data formats – or even lock-in strategies – where business critical data is trapped in legacy systems.

Another risk is the lack of industry regulations, as well as unclear or shifting legal and political frameworks. What if you play your bets on the wrong horse? Many of the pioneering waste management companies that invested early in sensors, access systems, and advanced sorting plants do still struggle with proprietary software and lack of universal systems and standards for data exchange. Poor data quality can have a strong, negative impact on the predicting capabilities of machine learning and curtails the overall value of the data gathered [3]. Being a late bloomer is not necessarily a bad thing.

A fresh start can actually be a privilege.

6.1.3 From Specific to Holistic Problem Solving

To successfully implement Industry 4.0 in any organization, there are several different areas that need to change – in a coordinated fashion. Most enterprises can be considered complex systems with many subsystems. Stamina and consistency over time and a clear vision are required on the management level. Many of these changes will have to be initiated from within. The people doing the job out in the field or in the back office must be involved in the process.

The shift from specific problem solving to holistic, integrated models that involves multiple stakeholder increases the number of dependencies. It can be critical for success to address the different challenges in the right order. Sometimes you will need to introduce costly digital infrastructure to get the data needed for new and sometimes unproven business models. Upgrading the technical equipment is only the first step in a long-term plan for a more automated operation with higher capacity and flexibility.

The strategies need to be foresightful and take into consideration both the emerging trends and different possible outcomes. Humans tend to predict the future based on linear expectations. The strategies should not only take into consideration the exponential nature of the current technology shift but also consider the tendency to overvalue new technology early in the hype cycle.

To conquer new land you need to prepare for the unknown. Planning is important, but there is always a lot of "dark matter" that can surprise, challenge, or open new opportunities. Waste management has for a long time been sheltered from the disruptive changes that have completely altered other sectors. If you look into the music industry, media, and finance, you find a possible blueprint for the changes that can be expected in waste management. New players may enter the market and challenge the position of incumbents, both on higher and lower levels in the value chain. Many try to fix the problem of industry fragmentation by creating new digital platforms on top of existing products or services. Experience from other sectors shows that profitability in existing news organizations was deeply affected when global platforms like Apple, Facebook, and Google created new structures on a higher level in the value chain. Many sectors have also been disrupted by new players that build a competitive advantage (often lower marginal costs) by building products or services for underserved markets. Using the right methodologies and strategies, the unchartered territories can be explored and utilized also by existing organizations [5].

Who should have the power of definition in the increasingly important resource industry? This is a key question to consider. An important keyword is relationships. In a world of new possibilities and challenges, leaders need to build new alliances both internally and externally. The need for data sharing calls for standardization, legal structures, shared infrastructure, and open innovation frameworks between several companies. The resource efficiency imperative of the circular economy calls for new strategic metrics. The need for fast innovation calls for new processes and professional development programs [6]. There is no quick fix or standard recipe. Someone needs to show the way.

6.2 LEADERSHIP AND MANAGEMENT

The combined effect of circularity and Industry 4.0 is likely to undermine the foundation for existing business models, and challenge established cultural conventions. Leaders need to navigate a landscape that is shifting before their eyes. To succeed within a rapidly transforming industrial paradigm, organizations need to increase flexibility, speed up decision making, reshape processes, and open up for new impulses and partnerships. User habits in flux requires the ability to listen and understand customer needs and then change and adapt, connect, and share – fast and efficient. The dynamic processes of Industry 4.0 makes a new interaction between hardware, software, and humans possible. The ability to work across disciplines and organizational silos becomes an increasingly important differentiator.

Industry 4.0 is way more than a technical challenge. It requires a clear vision that utilizes the bright minds, courage, and skills of people – enhanced and supported by data. The right employees need the right mandate to drive operational and cultural change. The organization should appropriately structure their departments to align the analytics capability with their overall business strategy. Transforming the insights into business value and a competitive advantage requires information management systems capable of making sense of the increasing volumes of data [7].

To be thriving in an Industry 4.0 world, many existing organizations need to go through a deeply transformational process in the coming years. According to economist Jeremy Rifkin, we are entering a collaborative age. "Its completion will signal the end of a 200-year commercial saga characterised by industrious thinking, entrepreneurial markets and mass workforces, and the beginning of a new era marked by collaborative behaviour, social networks and boutique professional and technical workforces," he wrote in 2011 [8].

He argues that over the next decades, conventional, centralized business operations will be increasingly subsumed by the "distributed" business practices.

Waste management has for decades been swimming against the current in the linear economy. The challenge now is not to swim harder but to, somehow, turn the stream. The green vision was at some point idealism. But as we have seen in both Europe and China, the underlying drivers and motivation are now fear – fear of lacking resources, environmental consequences, and climate change – and the need to create new jobs in a new reality. The motivation for circular economy is now realism. The job at hand is very serious, and the expectations are high.

One of the challenges is where to put the efforts. One of the challenges is to use innovation directionally – as a strategy to win new positions. Being a leader in an organization undergoing transformation will typically lead to a wide range of dispersed initiatives. The innovation ambition matrix [9] is a tool to create a manageable portfolio of projects and find a balance between efforts between different levels of scope and ambition (Figure 6.2). The core projects are projects

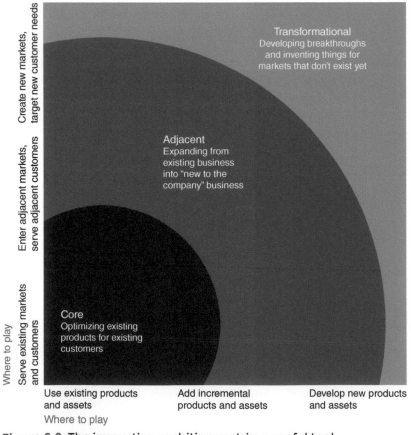

Figure 6.2 The innovation ambition matrix, a useful tool.
Source: **From Ref. [9], with kind permission from Harvard Business Publishing.**

to optimize your current products for current customers. A software upgrade to make administration more efficient is a typical example. Adjacent projects add new features to your existing business, like adding a new option or feature with an existing service. Transformational projects aim to create breakthroughs for markets that do not currently exist.

The challenge for leaders is at least a quadruple:

- There is a need to support new and sustainable business models that include the reverse handling of materials.

- Traditional static models must be morphed into dynamic systems that use data not only for records but also for immediate response.

- The corporation-centric logic needs to be replaced by distributed value propositions that spans long value chains.

- And, maybe the most difficult: The siloed, territorial mindset reflected in traditional competitive management theory must be replaced by a mindset that embraces interdependence and collaboration.

Leaders must not only transform individual business but also take part in the redesign and modernization of the surrounding industry. There is also a need to challenge conventions, routines, and mindsets. Do the organizational structure reflect the challenges of the past or the needs of tomorrow? Do we have the diversity and background in our own staff making us capable of adapting to a more customer-centric business model? Are we doing enough fun stuff to attract young talent?

There will often be a need for organizational upgrading of both specific operational processes, company visions, and tactical dispositions. Information must be digitized. Systems, at all stages of product development and service life, must be orchestrated. The entire organization, from product development and purchasing through manufacturing, logistics, and service, should increasingly be integrated to move in the direction of Industry 4.0 principles [10]. As digital services are increasingly personalized, the new "smart factories" often include a new form of involvement with customers. Logistical operations will have to establish feedback loops and work more systematically to understand the needs beyond the company borders.

Balance sheets could look very different in a future circular economy. Business leaders need to adapt to this new reality. Material value, resource productivity, and asset utilization will become more important key metrics. Performance and sharing models will change customers' and suppliers' need for capital – and possibly disrupt capital markets. This could disrupt several financial aspects of businesses, including capital structure, accounting, and valuation [11].

Meeting the goals and targets of this new reality calls for unprecedented transformations. New people to realize their leadership potential shape systems and move beyond business as usual approaches. The new management skills need to strategically align projects and initiatives and transcend polarizing discussions. Waste management needs to build the capacity to design and implement projects that, over time, will shift systems and cultures.

So where do the journey start?

6.2.1 Long-Term Vision or Short-Term Pragmatism?

The first move for many leaders may be personal: to really recognize the magnitude of the earthquake themselves. The shift ahead is structural. It is a long term. It is systemic. Management is not only about incremental improvement but also about facilitating change at a deeper level. The world that most organizations are designed to function within is changing – in terms of material streams, regulatory requirements, and value chains. To respond strategically calls for leaders with long-term vision and short-term pragmatism. To enable new value chains of the circular economy, we need openness – reflected in both machines (technically) and humans (socially). Fast and secure data sharing requires high levels of documentation and strict process. At the same time a culture for open innovation and experimentation must be fostered. But there are also many devils in the details. The changes need to be specific.

A lot of the work related to handling waste happens with manual labor. Most of the infrastructure in waste management today is still unconnected. To become "IoT-ready" is about both access to and usage of data. Digitizing selected small parts of the operation is a first step for most organizations. This is more than just placing some sensors out in the field. For the data to be useful, you will need software that can deal with information in a flexible manner and start creating digital improvements of specific processes.

Taking the first steps of a digital journey can feel overwhelming. But leaders must stop dreaming and start doing. What is the first steps? In a report from Forrester [12], they focus on the first 100 days of a digital transformation. It is not a question of "finding a vendor and buying a technology or using some open source libraries," according to co-author Jeff Hammond. The initial challenges are organizational and cultural, not technological. The report breaks down the get-going challenge in six steps:

1. Get the top management on board.
   You need understanding and buy-in from the people that will actually need to implement future changes and support projects, processes, and teams mandated to challenge or transform business as usual. Create a team to define the objectives for the transformation. This includes both the CEO and different top-level managers in roles across the organization.

2. Identify pathfinder projects that build momentum.
   You need small wins, early. If a process drags out, people tend to think that transformation is worthless or unrealistic. Start by creating a minimum viable product or fixing a small part of a process or a customer journey. This can be to create a digital dashboard monitoring the operation or to fix very specific time-consuming tasks that can be automated. The point is to empower internal changemaker and provide an important nudge to the people sitting on the fence. It will also allow for more freedom in future digital projects and business innovation experiments.

3. Communicate early, involve everyone.
   Buy-in is important, and a high degree of involvement in innovation projects is important. The ongoing initiatives and results should communicate early and often. The wider organization need to understand the digital strategy – and how the projects can improve the daily life within the operation.

4. Shift your firm's relationship with digital.
   As digital transformation takes time, you need to use the first stage to do the necessary groundwork. Enterprise architects should be involved to ensure that the infrastructure is ready. Changing the digital customer experiences is way more than a tech project, in also how employees work. Sometimes a new digital service can be implemented fast, and initiatives that are "bolted on" can be used tactically to change processes and culture.

5. Introduce agility by hiring new people.
   To create a new culture for innovation, new skills and experiences should be embedded into the existing organization – to challenge existing ways of thinking. The right mix of talent is important. This may mean hiring mobile developers, design facilitators, or data scientists to work alongside internal employees with knowledge about the industry and internal processes.

6. Create space for the innovation team.
   According to Forrester, colocation of the team is important to get them in productive state. Collaboration on a day-to-day basis is hard for people that are physically separated. Some companies even move some of their team out of the regular office building (even to another city) to create attract new talent.

Only by experiencing the value of digital systems, the organization can be convinced to replace old habits with data-driven decisions and user involvement. In waste management, an industry that is often based on experience and expertise, this first step is important. The point is not to collect the data, but understand how a structured information model and digital enablers can create new value. This will build acceptance within the organizations on the key metrics future innovations should seek to improve.

6.2.2 Leadership Styles and Strategies

Is Industry 4.0 a people or technology challenge? What kind of leadership style is required to create a thriving organization in an Industry 4.0 business environment? New services will have to be developed for new markets. Production process will be modernized in a continuous process of upgrading. The new products and services will be within a framework of new forms of cooperation and collaborative structures. One approach to the leadership style is to define a matrix of leadership styles based on orientation towards people and/or technology [13]. According to this simplified framework, there are four different leadership styles (Figure 6.3):

- The "freshman leader" is concerned with an ability to focus on traditional manufacturing structures with the primary focus on the finalized product; it has no employee focus, customer needs of minor interest, same for emerging technologies.

- The "social leader" style refers to an ability to create a friendly atmosphere for employees without regard to innovation and technology. Supporting employees is important.

- The "technology leader" is related to an ability to determine how new technology can be used to deliver enhanced value; it has a strong focus on innovation and a low employee focus.

- The "4.0 digital leader" focuses on ability to understand how technology impacts people and the organizational model is aligned with human nature.

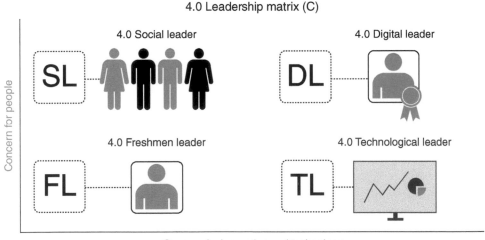

Figure 6.3 The 4.0 leadership matrix – the different leadership styles for Industry 4.0.
Source: From Ref. [13], © The Authors, published by Canadian Institute for Knowledge Development under the CC-BY 4.0 license, adapted by Nick Rigas, D-Waste.

Competence and culture is an important differentiator for any traditional organization transforming into an Industry 4.0 organization. An Industry 4.0 leader needs to be hands on in how new technology can be used to deliver enhanced value. But equally important is the ability to shape a collaborative culture, a workplace that attracts talent and services that are playing along with human nature.

Given the challenges described earlier in Section 6.1, a fast, team-oriented, and cooperative leadership style is preferred, with a strong focus on innovation. The "4.0 digital leader" leadership style is probably the most productive in many traditional organizations that need a deep transformational change.

This is of course a very simplified model. In a world of diverse challenges, there will also be a need for diverse leadership. There is no one best leadership style in all situations, and there will certainly be a need for leadership also on lower executive levels. But it is partly confirmed by other research in the field of transformational leadership. According to research among executives by Deloitte, four key characteristics define successful leaders in the age of Industry 4.0: A commitment to do good, data-driven decision making, a longer-term vision of technology, and an aggressive willingness to develop the workforce. Leaders with these attributes appear to be better prepared to survive and thrive in times of rapid economic changes. In the same survey they find that four personality types, with associated strategies, seem to make the most progress in transforming industry businesses.

The first is what they call "social supers" – who are oriented towards impact and societal initiatives. This is a leader personality that wants to "do well by doing good," aligning profit and purpose, creating new revenue streams through conscious products or services.

The second is labeled "data-driven decisives" – applying a methodical, data-focused approach, with bold decision making. This is a confident leader, with a clear vision, often investing heavily in technology to disrupt their markets.

The third type is given the name "disruption drivers" – who understand that investment in new innovations is required for growth. They often invest in technologies with a focus on changing their markets, helping their business capitalize on the opportunities associated with Industry 4.0.

The last category is called "talent champions" – who work hard to build the right skill set within their organizations and invest in training their employees for the future of work.

All four leadership personas characteristics are attuned to using Industry 4.0 technologies in an ethical manner focusing on impact rather the short-term financial returns. They work methodically in setting the strategies, and their companies follow clearly defined processes.

6.2.3 Choosing the Right Innovation Methodology

In a world of deep transformation, far-sighted, creative innovative solutions are required. A simple, linear plan does not help if you do not know what you are planning for. In an uncertain business environment, the problem-solving capabilities of your organization is a key differentiator. Opportunities and possibilities for change must be explored, systematically. This requires training and the right frameworks. There are many different methodologies and techniques that can be used for this purpose. We emphasize three high-level approaches that are developed software development: Agile, design thinking, and lean start-up. Together they represent almost unlimited opportunities for combination and exploitation [14].

6.2.3.1 Design Thinking

Broad, sometimes complex questions are the baseline for design thinking. It focuses on qualitative and not on quantitative insights. Very central to this process is understanding the needs and requirements of the user. Design thinking was described and the term coined by Tim Brown at IDEO [15]. The team will need to challenge themselves to step into the users' shoes, leaving aside any prior assumptions regarding the specific problem and user. Through interviews and observations in their natural context, the creative teams gradually get a shared and deeper understanding, a necessary baseline for the idea generation phase. The process uses "divergent thinking" to come up with as many solutions as possible for giving answer to a specific problem before using more critical "convergent thinking" to narrow down the options and find the most optimal solution. In divergent stage it is important to open up for wild, outside-the-box ideas and not be constrained by conventions and critical thinking. As ideas are further developed and turned into simple prototypes, a more critical

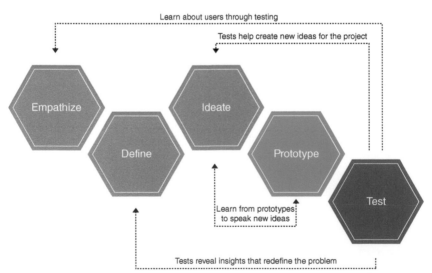

Figure 6.4 The stages of a design thinking process with feedback loops.
Source: From Ref. [16], with kind permission from Teo Yu Siang, Interaction Design Foundation.

perspective must be applied, bringing together considerations like user desirability, financial viability, and technical feasibility. The prototypes are used to validate ideas, involving the end user. The user feedback is used to iteratively improve ideas (Figure 6.4).

6.2.3.2 Lean Start-Up

In the lean start-up methodology, described by Eric Ries in the *The Lean Startup* [17], the team will often already have a good understanding of the problem. It is a technique used to develop new businesses and products under conditions of extreme uncertainty (Figure 6.5). In some ways it can be considered a framework for failing (and learning) fast, before implementing a new product or service broadly. Understanding the problem does not imply that your solution will actually work. Therefore everything in this process is strictly being considered a hypothesis that needs validation. Instead of theorizing, the team will turn ideas into a "minimum viable product" (MVP), measuring how customers respond and learning whether to further develop the idea. There are many assumptions that can be tested within one idea, and these nuts and bolts should be tested if identified and specific experiments designed to validate them. Through a process of testing and rebuilding based on learning, the solutions are gradually being validated. Lean start-up uses qualitative insights early on, but as the product is becoming more sophisticated, actionable quantitative data is being used to measure how effectively the solution addresses the problem [18].

6.2.3.3 Agile

Agile was developed by a group of software developers and is basically a set of shared values, ideas, and approaches to software development, described in the Agile Manifesto [19] and corresponding principles. The key message can be summarized in four principles:

- Individuals and interactions over processes and tools.
- Working software over comprehensive documentation.
- Customer collaboration over contract negotiation.
- Responding to change over following a plan.

In a way it is just as much a general work process based on a certain philosophy as a specific methodology. But to work Agile, there are some methodologies that should be followed. One is SCRUM [20], the most widely used process framework for software development. This means that you create a product backlog, a list of features that should be built, in prioritized order. The development is organized in a two-week sprint, where only these specific features are in scope. The progress is reported, and prioritization is discussed in recurring meetings. At the end of each sprint, the team reviews and demonstrates the results, discussing future steps in the process.

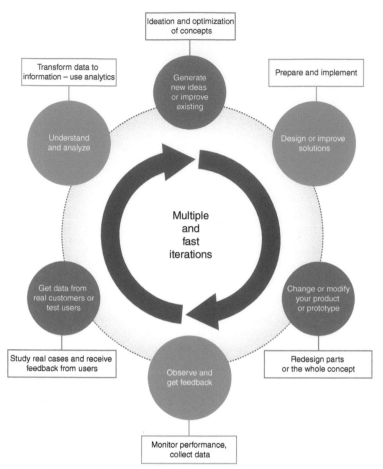

Figure 6.5 The build–measure–learn loop is a process for fast validation of ideas in several iterations

Source: **Adapted From Ref. [17], with kind permission from Nick Rigas, D-Waste.**

All these methodologies can be combined, given the challenge at hand. Insights developed in a design thinking process can be further validated in lean start-up methodology. And when the specifications for the final product is described in detail, the actual product can be developed using the Agile principles. Working with innovation does not necessarily take very much time. Actually time can be used creatively as a constraint that fuels the process with creative energy. Google Ventures has developed their own methodology, the design sprint method – where a team get to exchange knowledge, develop ideas, build prototypes, and test a concept within five days [21]. During this process you will learn, ideate, decide, and get feedback from real users during the span of a workweek [22]. Given the short time frame, design sprints focus only on part of the solution, but it's an excellent way to validate really quickly whether you're on the right track or not.

6.2.4 Building Industrial Ecosystems

Developing a circular business model requires a broader set of capabilities. In the waste management industry, there will be a new skill set needed in design, engineering, logistics, and procurement. The choice exists between developing such capabilities in-house, working with partners or outsourcing them – leading to a number of difficult decision points for organizations but also to the possible emergence of new service sectors [11]. To understand and harness the new opportunities, companies need to rethink recruitment and training strategies and create a culture of interdisciplinary exploration.

As described throughout this book, we generally need to go from a "zero sum" business approach to a collaborative mindset focused on synergies and complementarities. To find the sweet spot between collaboration and competition is not easy. Generally most organizations in linear value chains are territorial, and there is a need to open up the processes. A culture of sharing will help spread the best available solutions and also "cross-pollinate" ideas to stimulate innovation. The concept of "open innovation" is defined as "a distributed innovation process based on purposively managed knowledge flows across organizational boundaries, using pecuniary and non-pecuniary mechanisms in line with the organization's business model" [23].

The basic idea is to manage the inflows and outflows of knowledge across the boundary of a firm to get access to otherwise inaccessible knowledge and new commercialization paths. External resources can be used strategically to boost internal innovation activities, and the unutilized internal knowledge from inside the firm can be used as input in similar processes in other external organizations. In the open innovation model, these flows are directed strategically, and the mechanisms are designed. Thus, what was unspecified and unmanageable can now be a strategic tool.

The inflows of knowledge can be fostered by a range of mechanisms, such as scouting, university research programs, funding start-up companies in one's industry, collaborating with suppliers and customers, and utilizing nondisclosure agreements [24]. New digital tools and platforms also open up for concepts like crowdsourcing, online competitions and tournaments, and communities.

The outflows of knowledge require organizations to allow unused ideas and underutilized assets to create value outside the organization. Interactions in this type of open innovation can involve either selling or revealing knowledge [25]. The mechanisms for managing outflows include out-licensing IP and technology, donating IP and technology (open source), spin-outs, corporate venture capital, corporate incubators, joint ventures, and alliances [26]. Sometimes a company can choose to become a supplier to or a customer of a new initiative, instead of executing the initiative internally.

A third type links outflows and inflows, creating a mutual collaboration between two or more partners to collaboratively develop or commercialize an innovation. Mechanisms for this coupled type of open innovation include strategic alliances, joint ventures, consortia, networks, ecosystems, and platforms, all involving complementary partners [27].

In the integrated value chains of Industry 4.0, there is a horizontal flow of information from the end user over the whole corporation to the point of the supplier. Similarly there is also a vertical integration specifying the intelligent cross-linking and digitalization within the different hierarchical levels of a value chain. Customer communication and manufacturing are all parts of the same process. A general challenge in an Industry 4.0 world is to become a part of a collaborative environment where businesses, universities, and innovators converge to solve problems in an open environment.

But open innovation will not automatically lead to synergies for everyone.

A key element of Industry 4.0 is creating a "higher-level system" on top of technical subsystems and processes, as described in Chapter 2. Going from linear value chains to circular value networks requires new collaborative structures to be established, both inside an organization and with existing and new partners. In waste management that span the whole reverse logistical value chain, this cross-organizational collaboration is especially important – as life cycle and resource productivity require the ability need to monitor and control material flows across multiple stakeholders. Citizens will also play an increasingly important role, both in the form of consumers and recyclers. They are not only customers, but they also contribute to the process and may also be rewarded for their contributions.

6.2.5 Dealing with Complexity

Complexity is a major concern in most companies, deeply affecting the ability to get things done. To cope with the messy problems in a rational way, it is important for leaders to identify the different drivers behind the mess. There will always be institutional manifestations of complexity: the organizational structure, operations within different geographical regions, regulatory requirements, a portfolio of technical systems, and number of workers. At the same time there will also be a lot of individual complexity, facing employees on all levels – confusing roles, poor processes, and inconsistent routines. For leaders this difference is important to bear in mind. There are always problems that can be solved internally and increase both satisfaction and productivity, as well as external problems defined by the political or industrial context. Parts of the complexity cannot be solved easily – or takes a lot of effort to fix.

While charting the landscape of complexity, it can be useful to bear in mind the different categories of complexity, as suggested by the Harvard scientists:

- *Intrinsic complexity* is created by the political and industrial context: Industry regulations and law. To change these conditions is not always possible, and if so it requires industry-level engagement and long-term perspective.

- *Inherent complexity* follows from actual business processes and can only be changed by changing (exiting from) parts of the current operations.

- *Designed complexity* results from active choices in the business model: customer base, product offering, and channels. It can be removed by changing strategy, but this does not necessarily add value.

- *Unnecessary complexity* is the result of the actual processes in place and the needs of the organization. These misalignments need to be identified and can easily be handled.

The key insight is that complexity is a natural part of any business operation, and not necessarily a bad thing. According to research by McKinsey [28], failing to tackle individual complexity can be financially costly. By clarifying roles, refining key processes, and developing appropriate capabilities among the employees, it is possible to create more room to increase institutional complexity. This way capacity for dealing with a shifting business environment and potential scaling increases.

Growing institutional complexity can actually be an indicator of increasing success. This recognition by leaders is a great way to start working with the kinds of complexity that actually hinders productivity. Removing the unnecessary mess frees up time for employees to handle stream of difficult questions that naturally follows from running any business operation. To identify the low-hanging fruits requires the involvement of employees. A general curiosity and openness is of course important, but structured interviews, employee surveys, and focus groups can also be used to gather information – both on the intensity of complexity and insights into the underlying drivers. Organization charts and work instructions can also be a good source of information on role duplication and lack of responsibility.

6.2.6 Political Leadership

Waste management is a team effort and highly regulated. In Chapter 2 we established how the big wave of waste in society really emerged as technology liberated society from the constraint of limited resource availability. The big challenge now is to create constraints political and judicial and at the same time create space and incentives for innovation that leads in the direction of sustainability.

According to ISWA report on Industry 4.0, governance is the key to unlock the benefits and restrict the problems. New principles, protocols, rules, and policies are needed to accelerate the positive and inclusive impacts of these technologies while minimizing or eliminating their negative consequences. IND 4.0 makes necessary a revolution in governance in all the levels involved: international cooperation, national, regional, municipal, and corporate. This requires a new role for the private sector and academia working alongside public officials to provide expertise on the technologies they are developing, as well as their applications and potential consequences.

There are many unresolved questions for policy makers. Will addressing environmental breakdown require a direct downscaling of economic production and consumption in the wealthiest countries? The "Decoupling debunked" report from the European Environmental Bureau [29] summarizes research on resource productivity across different resource types and is quite clear in the conclusion: they advocate "complementing efficiency-oriented policies with sufficiency policies, with a shift in priority and emphasis from the former to the latter even though both have a role to play." From this perspective, it appears urgent for policy makers to pay more attention to and support the developing diversity of alternatives to green growth.

Policy making for the circular and digitized world requires a tailored approach to problem-solving and setting goals specific to a region. A general challenge is creating an appropriate regulatory environment around a technological innovation that ensures the trust of stakeholders while encouraging innovation [30]. Generally technologies, and the entrepreneurs using them for business model innovations, are accelerating away from the capabilities of existing legal infrastructure to manage risks effectively. This development provokes alternative ways of providing a more effective regulatory system.

"Matter can change meta. And meta can change matter," says strategic designer Dan Hill [5]. Sometimes specific projects can challenge the regulatory framework – in interesting ways. And sometimes changing regulatory framework can stimulate interesting projects. An important side of political leadership is to listen to feedback from the industry and be curious on how to and actively remove barriers and support new and disruptive models. Sometimes a local innovation experiment can be a practical and feasible solution to large-scale challenges, in contrast to reforms made by multiple stakeholders without technical knowledge or a unifying vision. Successful local models can also be scaled up on national on an international level – by becoming a shared baseline. Designers often talk of "design by committee," when a solution ends up as a compromise between viewpoints of the participants, often with needless complexity and logical flaws and lack of sophistication. Political measures often tend to have these characteristics.

Related to Industry 4.0 more specifically, a key challenge for policy makers lies in stimulating open innovation while ensuring data security and generating trust for those who are directly and indirectly linked through intelligent assets. The report "Growth Within: A Circular Economy Vision for a Competitive Europe" suggests instruments within six categories for policy making [31]:

1. Information and awareness. To change the ingrained behavior patterns of individuals and companies, there is a need to redesign the curriculums in schools, universities, and continuing education. On a policy level one should also focus on broad or specific information campaigns and increase the general understanding

of the economic potential, the limits of the linear economy and alternative, circular ways of doing business.

2. Collaboration platforms. Policy makers should incentivize businesses to overcome interoperability hurdles by fostering cross-industry collaboration, including providing safe sharing environments. The ambition should be to stimulate multiple stakeholders to come together to solve existing problems. Government support should be given to industrial symbiosis arrangements, R&D collaborations between academic and industry players, and demonstrator platforms.

3. Business support schemes. Government should provide support such as grants, capital injections, and financial guarantees, as well as provide levers of technical support, advice, training, and the demonstration of best practices. This can be done in many ways, including brokering traditional investment through public–private partnerships to using more innovative solutions including crowdfunding. Sufficient funding should be given research and education that explores the relationship between circular economy and new technology.

4. Public procurement and infrastructure. The public sector can step in to provide purchasing power to enable a circular economy driven by IoT and other Industry 4.0 components. Being forward-thinking public sector can provide an important proof of promising technology and help drive up interest and reduce prices. There may be lacking incentives in the private sector due to entrenched customs and habits among companies or citizens.

5. Regulatory frameworks. Government regulations can be used actively to stimulate circular economy activities enabled by IoT technologies. It can also discourage noncircular practices. Given differences in the speed between the development of new policies and new technology, it is practically impossible for policy makers to implement effective policy targeted at specific possibilities. According to the report policy makers should focus on creating a secure, enabling environment. For example, an IoT-enabled information infrastructure could make it easier to differentiate between resources and "waste" in cross-border trading, which would in turn help governments control and monitor resource flows.

6. Fiscal frameworks. These policy instruments include different types of taxes and government subsidies. They can be applied to the market to either encourage circular economy activities-enabled technology or business models or discourage noncircular activities. For example, getting rid of (sometimes) indirect subsidies for raw material extraction or fossil fuels could further incentivize performance models. Shifting the taxation structure from labor to natural resource use could further incentivize circular practices as well as potentially create hundreds of thousands of jobs.

6.3 EXPLORATION VERSUS EXPLOITATION

Waste management organizations do a very important job on a day-to-day basis. To do this job properly is critical both to businesses and society. Without management of the material streams in the economy, everything will collapse. This is a heavy responsibility. Many leaders will naturally focus on improving existing business operations in incremental steps to increase efficiency, not explore new opportunities with uncertain return on investment. In the book *The Innovation Paradox* [32] describes how the very processes and structures responsible for established companies' enduring success prevent them from developing breakthroughs. Improving step by step is a recipe for growth, but will not lead to major innovations. A management objective should also be to foster a culture and organizational structure that balances innovation with improvement of established processes, partnerships, and business models.

6.3.1 Playing with Both Hands

These are common management considerations in any technological transition. Established companies will have an existing organization and business processes to take care of. At the same time they will have to innovate and transform. According to Gartner research on innovation management [33], inconsistent funding, slow product development, and even overinnovating can lead to breakdown in the R&D processes. Many companies overspend on incremental improvements at the expense of the future. And nearly 80% of innovation leaders reported pressure to scale down or kill breakthrough projects in favor of more immediate priorities.

What is the best strategy to deal with this dilemma?

According to scientist at Harvard business school, the solution is becoming an "ambidextrous organization" [34]. Ambidextrous means being able to use the right and left hands equally well. The most successful companies have developed capabilities of both exploiting the present and exploring the future. The research shows that these companies share an important characteristic: They split the organizations. To be able to experiment with new ideas and processes, they establish new, exploratory units from their traditional, exploitative ones. They consciously allow new structures and cultures to emerge, but importantly, they keep maintaining tight links across units at the higher executive levels.

This structure allows cross-fertilization among units but prevents cross-contamination. Tight management coordination enables the sharing of important resources from the traditional units – cash, talent, expertise, and customers. But the organizational separation ensures that the new business units are not constrained by the sometimes strong forces of tradition and business as usual.

According to Harvard scientist, this setup provides "a practical and proven model for forward-looking executives seeking to pioneer radical or disruptive innovations while pursuing incremental gains."

This is important insights when preparing to become an Industry 4.0 organization: "A business does not have to escape its past, to renew itself for the future," the Harvard scientists conclude.

6.3.2 The Three Types of Innovation

Traditional leaders should note the importance of keeping a variety of innovation efforts (Figure 6.6). One challenge is to improve through *incremental innovations*: small improvements in existing processes and operations – in the form of deliver more customer value or higher efficiency.

Can routes or maintenance routines be optimized? Can manual processes be automated? Can paper documents be turned into structured data? These are all important questions that should be asked on a daily basis.

But in the transition there will also be a need to make *architectural innovations*. There are technologies or processes that should be replaced or fundamentally changed. If you move all your software into the cloud and buy them on a platform-as-a-service model, do you even need your existing ICT department? Could parts of customer management be replaced by an automated notification system?

The third kind of innovation is discontinuous innovations – radically new ways of doing business that could change the role of the company in the value chain, sometimes rendering old products or ways of working obsolete. In innovation theory there is a lot of talk about disruption, when new ways of doing business undermines the existing value chains. Regular examples are how Kodak was challenged by digital photography and how Nokia, the cellphone manufacturer, was totally overrun by the smartphones and touch screens.

An important question to keep in mind: Who are the targets for your innovation experiments? Some may be aimed at a known group: the current customers. But

Three types of innovation

	Incremental innovations Small improvements in existing products and operations	Architectural innovations Technological or process advances to fundamentally change an element of the business	Discontinous innovations Radical advances that alter the competition in the industry
New customers			
Existing customers			

Figure 6.6 Three types of innovation according to Harvard business review. In this matrix innovation initiatives can be mapped according to the types of innovation and the customer groups you want to address
Source: From Ref. [9], with kind permission from Harvard Business Publishing.

others can be aimed at existing markets that are not currently considered a strategic segment. Can a waste management operation serve a manufacturer with the logistics of a closed-loop program? In a new paradigm here may also be a third category: entirely new markets that are yet to be clearly defined.

6.3.3 Resource Optimization: The Forgotten Perspective

In the mainstream management literature, "productivity" basically means economic efficiency within a linear economy. There is generally little consideration given the resource footprint of products, services, and related business models. The reason is simple: as long as exploiting finite resources is profitable, it will financially outcompete more sustainable solutions. Sustainability tends to be the icing on the cake, not the primary reason for innovation. There is obviously a need to develop better methodologies for innovation in the material throughput in different product–service systems.

The best suited framework taking material footprint into consideration is eco-design. This is a set of consideration for producers that want to meet their customers' needs while using the minimum levels of resources and reducing the impact on the environment and society. It involves designing or redesigning products, services, processes, or systems to avoid or repair damage to the environment, society, and the economy. At the heart of eco-design, you will find ten core environmental considerations:

- Using materials with less environmental impact.
- Using fewer materials overall in the manufacture of products.
- Using fewer resources during the manufacturing process.
- Producing less pollution and waste.
- Reducing the environmental impacts of distributing products.
- Ensuring that products use fewer resources when they are used by end customers.
- Ensuring that products cause less waste and pollution when in use.
- Optimizing the function of products and ensuring the most suitable service life.
- Making reuse and recycling easier.
- Reducing the environmental impact of disposal.

The challenge is lack of knowledge about the capabilities and constraints in handling waste. Most business developers, consultants, and product designer have no or little experience with pollution handling, waste sorting, recycling, and disposal. This lack of competence is an important opportunity for the organizations working with waste management on a daily basis. Garbage is the direct result of highly effective business models and product strategies aimed to create more value. But on a material level, the value is often reduced dramatically – due to the fact that the resources are mixed up. What started a journey as a pure and valuable raw material ends up in a landfill in degraded form – without secondary value. In all the innovative processes needed for the decoupling of waste from growth, recycling expertise and garbage-handling experience constitute a unique competence. The practical, technical, and chemical insight is a "underutilized" competence, an outflow that should be used externally, especially in the design stage of physical products and services.

To create mutual collaborations between waste handlers and producers is especially important as nanomaterials, and customized composites become more common. A great example is the collaboration between WindEurope, the European Chemical Industry Council (Cefic), and the European Composites Industry Association (EUCIA) and their initiative to advance novel ways of recycling wind turbine blades [35]. These blades are made up of a composite material that boosts the performance of wind turbines by allowing lighter and longer blades, but they represent a huge waste problem. Today, 2.5 million tons of composite material are in use in the wind energy sector. Around 130 000 wind turbines are operational in Europe, and during the next 5 years, 12 000 of them will be decommissioned. To deal with this problem, the consortium will look into different techniques both cement coprocessing, mechanical recycling, solvolysis, and pyrolysis. Experience gained from wind turbine recycling can be transferred to other markets to enhance the overall sustainability of composites.

The resource perspective can also serve as a fresh and innovative approach to the development of new policies. A Natural Resource Strategy for Finland from 2009 [36] is a rare and inspiring example on how resource optimization and open innovation methodologies can be combined to design new measures and political perspectives – centered around ecosystems and the intelligent and sustainable utilization of natural resources. In this strategy commercial and productive exploitative view on natural resources is supplemented by an environmental impact perspective. It takes into account the economy, production, and the environment and their links to social and societal factors.

6.4 FROM DIGITALIZATION TO NEW BUSINESS MODELS

The driving question behind innovation is how to create, capture, and deliver value in a more efficient way. Business models are "stories that explain how enterprises work," says Joan Magretta in the article "Why Business Models Matter" [37]. This story should answer the question of who the customer is and what the customer is trying to achieve. It should explain the economic logic: How can you deliver something of value to customers and make them pay more than the cost of creating this value?

When the technology, value chains, or regulatory conditions change, the business models will often have to respond. A context in constant transformation creates pockets of new threats – and opportunity. Redesigning business models can be used as a tool to find new customers, new problems to solve, and new partners to collaborate with.

To seize opportunities in existing and new markets, the organizations need to adapt more quickly and embrace new processes faster. A general trend in Industry 4.0 business model design is "everything as a service": new business models are driven by the use of smart data for offering new services – the functionality and accessibility of products, not the tangible product - not products in itself [38].

The ability to design, test, and implement new business models will be increasingly important in a marketplace with shifting supply and demand patterns, changing commodity prices, and new regulatory requirements. The ability to rethink processes and implement new systems fast becomes a competitive advantage. IoT and real-time data enable organizations to optimize for environmental and economic performance – in a world of shifting business dynamics and regulatory requirements.

It is important that leaders in the resource management end of the economy develop the ability to work strategically with business models. There is no natural law stating that Industry 4.0 will lead to more sustainability. There are many threats embedded into the digitized and integrated value chains. Established business models will become more efficient and customer oriented due to the increased connectivity and analytical abilities. One such business model is the concept of mass customization: design and production that can be dynamically adapted to each specific customer order or continuous changes in market preferences. Seen from a waste management perspective, this represents more complexity in the waste stream and decreased recycling efficiency. Another example is the use of artificial intelligence in the design of custom composites with product-specific characteristics. Sounds great from an engineering perspective, but composites are already hard to recycle. A third example is nanomaterials – expected to become a huge future game changer in many industries. But preliminary research suggests that some of these materials can have catastrophic implications for natural ecosystems. And how should the nano-stuff be dealt with as the products turn into waste? Maybe there is a business opportunity hidden in these problems?

6.4.1 Sustainability as a Driver of Value

In traditional innovation the focus has been either to create more value or to capture more of the value you are already creating to increase profits of your operation. In a circular economy the resource perspective becomes a part of this equation. Based on the assumptions of changing user preferences, more transparency, and regulatory reform, the footprint becomes business critical. Can you create more value but use less resources?

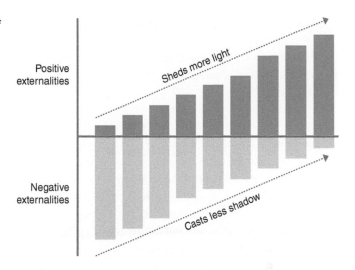

Figure 6.7 The effects of sustainability efforts.
Source: **From Ref. [39], with kind permission from Springer.**

Or, a question highly relevant to every leader working with waste handling, can you solve the problem of "externalities" for other companies more efficiently?

When the constraints of nature and the enablers of technology are combined, it opens up for a range of new business models. And these business models can be optimized either to create more positive externalities or to reduce the negative impacts of existing value chains (Figure 6.7). An interesting perspective is considering every business to have a shadow side and sunny side in relation to sustainability. A new business model should ideally shed more light or cast less shadow. The shadow leads to neutrality, but the trajectory of casting light is in some sense endless – there is no limit to how comprehensive a company's positive externalities might become.

There are many examples on how Industry 4.0 can enable upcycling (Figure 6.8) and more sustainable new ways of creating value, with less footprint.

One example is the emergence of *product–service systems* [40]. By bundling together products and services for customers, a product's life cycle becomes more holistic. Tesla is a well-known example – the car gets software updates that expand the functionality continuously. Another example is sharing services, where physical products can be accessed by multiple users. Going from products to product–service systems increases the interaction between a company and the customer. The relationship is lifelong, not limited to the moment of purchase. These new systems open up for new and recurring revenue streams.

Another example is *mass personalization* – the customer is involved in the designing of the end product. Direct customer input can enable companies to produce more customized products with shorter cycle-times and lower costs than those associated with standardization [41]. In this model the producer and the customer collaborate to create a product and share the new value in a novel way.

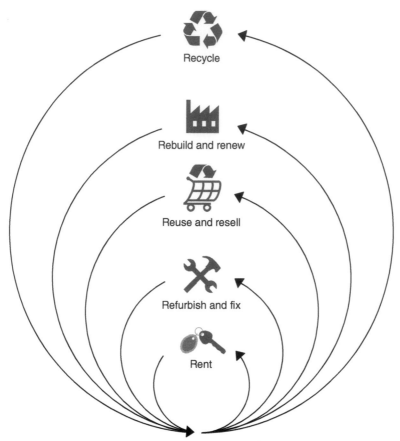

Figure 6.8 Different types of upcycling in circular business models.
Source: From Ref. [39], with kind permission from Springer.

A third example is *predictive maintenance*. Is there an alternative to fixing machines when they break or putting up costly inspection schedules to keep them from breaking? Yes. In Industry 4.0 the combination of sensors, cloud services, and smart algorithms can be used to get early warnings. Machine learning can be used to improve predictive capabilities, and service personnel can be used to prevent problems – not fix them. This way "uptime" can be delivered as a service, based on a highly and distributed cost-efficient business model [42].

A fourth example is *dynamic pricing systems*. In an Industry 4.0 where every transaction and event is logged and the data is shared across stakeholders, the pricing can be based on actual usage rather than flat fees. A very relevant case of this is PAYT systems in waste management. In a PAYT system the end user pays for the actual amount of waste generated. Different price levels for different fractions create an incentive for separation at source of recyclable waste and for reduction of mixed waste

[43]. The business model transformation is not only a passive response – it is being used as a tool to change user behavior in a way that add value further downstream in the value chain.

6.4.2 The Elements of a Business Model

To start exploring new business models, established players can learn a lot from the methodologies of the start-up movement. One such methodology is a well-proven framework for design of business models: the Business Model Canvas (Figure 6.9). This template was developed by Alexander Osterwalder [45] and visualizes the different elements of any new or existing business model. It has been highly influential – especially in the start-up movement. In this framework the business model is created by nine different building blocks. In a modified version [46] of the original canvas, eco-social costs and benefit are added. To make the walkthrough of the model relevant, we will use an example: a PAYT business model for households and business in a publicly owned waste management company.

1.   *Partner network:* In most business models you are dependent on one or more buyer–supplier relationships or collaborations to make the model work. This can also be the relationship between public and private sector. In the PAYT case one key partner could be the municipality that collects the fee from end user through taxation – based on the data from the waste management company.

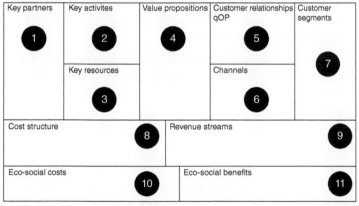

Figure 6.9 In a business model canvas, any business model can be broken down into a set of different interdependent parts. The sustainable business model canvas is based on the business model canvas developed by Alexander Oster-walder/Strategyzer.
Source: Adapted from Ref. [44] by Nick Rigas, D-Waste.

Other partners could be the provider of the physical equipment needed and a digital agency that develops the app.

2. *Key activities:* The most important activities in executing a company's value proposition. An example for a waste management company establishing a PAYT would be to create a source sorting infrastructure, a connected waste collection that identifies customer and communication systems for end user – enabling an effective feedback loop.

3. *Key resources:* These are resources necessary to create value for the customer and sustain and support the business. Key resources can be human, financial, physical, and intellectual. In the PAYT system you would need access-controlled infrastructure and/or bin collection system with RFID tags. In an advanced system you would probably also need a mobile app for personalized feedback.

4. *Value proposition:* This is the collection of products and services a business offers to meet the needs of its customers. It describes the value created for the user by solving specific problems. In this use case it may be "providing a fair and motivating system that rewards recycling behaviour through differentiated fraction pricing, giving feedback in a friendly and transparent way."

5. *Customer relationships:* To ensure the survival and success of any businesses, companies must identify the type of relationship they want to create with their customer segments. Various forms of customer relationships can be based on transactions, personal assistance, automated processes or self service platforms. This may be an assistive interaction with employees, self-service, and a community. In the PAYT example there will probably be both infrequent contact with employees and automated feedback through a mobile app and a monthly invoice. The customer should be treated as a co-creator of value in the value chain – getting access to the final outcome.

6. *Channels:* A company can deliver value propositions to its targeted customers through different channels. An organization can reach its clients through its own channels (store front) through partner channels (major distributors). In the PAYT case the information is provided to the end users through letters, email, and public information channels. If a personalized app is established, this will probably be a very important channel.

7. *Customer segments:* This describes which customers the business models try to serve. Various sets of customers can be segmented based on their different needs and attributes. Some business models will target a mass market, but other niche models may address very specific customer segments. In platform business models that creates network effects between different customers, you may describe the different users of the platform. In the PAYT case the users would probably be households and businesses within the operational area.

8. *Cost structure:* Here you describe how the activity affects the cost. There may be fixed costs, such as infrastructure and variable cost, such as volumes and quality of the collected waste and associated logistics. In the PAYT case the cost will be associated with running the infrastructure and handling of the waste streams. If the recycling rates improve, cost will fall. So the system should see that the increased source sorting quality adds more value than the additional costs of running a more complex system.

9. *Revenue streams:* There are multiple ways to generate a revenue stream. You could base it on sales, subscriptions, licensing, or provisions – or a combination. In the PAYT example there could be a baseline fee and differentiated fee where you only pay for the mixed waste – and the sorted fractions go for free.

10. *Eco-social costs:* There will often be environmental costs of the business model. In this section you can map out the nonrenewable materials in the "key resources" section and the usage of resources in key activities. In the PAYT operation, this will typically be the resources of the embedded electronics, the fuel in collection trucks, and the energy consumption in streaming services. Here you can also discuss how these costs can be reduced. One example is replacing fossil fuel truck with electrical trucks.

11. *Eco-social benefits:* Any business model should optimize for ecological and social benefits. In the PAYT case the obvious benefit is higher sorting quality and a tax discount given to those citizens taking responsibility. From a sustainability perspective, it is important that the benefits are surpassing the costs.

6.4.3 Business Model Experiments: The RESTART Approach

In their book *RESTART* (open access), Sveinung Jørgensen and Lars Jacob Pedersen have created a practical framework for sustainable business model innovation – based on controlled experiments [39]. They are concerned with including the sustainability perspective into business development simply because companies are a really important part of the problem. Companies have played a major role in the development of our sustainability crisis. It is difficult to imagine that these problems can be solved without their efforts.

They consider sustainable business model innovation a tool to reduce the negative influence on society and environment but also to find profitable ways to exploit the business opportunities that arise because of those problems (Figure 6.10).

RESTART is an acronym of seven letters that correspond with seven features of more sustainable business models. The seven features are contrasted with their opposites, all of which are arguably characteristics of business as usual. In this way,

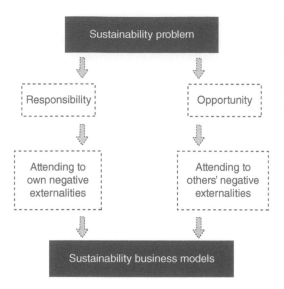

Figure 6.10 Two ways to a sustainable business model. Taking responsibility or seeking opportunity.
Source: From Ref. [39], with kind permission from Springer).

the framework highlights seven main changes that can make business models smarter and more sustainable:

REDESIGN rather than standstill

EXPERIMENTATION rather than turnaround

SERVICE LOGIC rather than product logic

THE CIRCULAR rather than the linear economy

ALLIANCES rather than solo runs

RESULTS rather than indulgences

THREE-DIMENSIONALITY rather than one-dimensionality

The RESTART approach is based on three basic assumptions:

- They believe businesses in general will need to shift their business models more often. This necessitates a more structured innovation methodology.

- They say that to lower the footprint business strategies should emphasize services rather than products (or functionality rather than ownership). They also stress the importance of collaboration rather than single companies competing in isolation.

- They believe sustainable business models need a new framework of governance and control – including richer performance metrics that include environmental measures.

From this baseline they developed their seven-step program that reflects new ways to create, deliver, and capture value by taking into consideration sustainability issues and technological capabilities as drivers for business model innovation.

1. *Redesign:* Step one is to start thinking of business models as something that can and should be consciously and continuously challenged and reconceptualized. This is a design process in line with the innovation methodologies described earlier in Section 6.3 and subsections.

2. *Experimentation:* Step two is to start learning. In a shifting environment there is no given answer. You need to identify and analyze what works and what does not. Rather than putting all eggs in the same basket, one should rather conduct a range of small-scale experiments on the way to a new business model. This is something that should be embedded as a part of the business culture.

3. *Service Logic:* Step three is rethinking the business model. Shifting the focus from the product alone, and innovating on the associated services, will in many cases lead to a lower resource footprint. This implies that value creation and value delivery are oriented towards giving the customer access to what he or she needs, rather than offering it in the form of a product based on ownership.

4. *Circularity:* Step four is identifying the actual footprint. There is no company in the world that does not use energy, water, and other natural resources or that do not generate to some extent excess resources and waste from their operations. To become more sustainable in a way that is compatible with financial performance, the ideal of the *circular economy* should be the leading star while designing the way resources are acquired, processed, used, and ultimately reused.

5. *Alliances:* Step five is rethinking alliances and partnerships. Solutions of the type that promotes circulation and service thinking will often require that businesses enter into collaborations with other organizations that may enable them to create and deliver value in this way. This enables exploitation of the complementarity between companies and even facilitates manufacturing, logistics, or other processes that involve smarter use of resources than if companies do not cooperate.

6. *Results:* In order to set the right goals and to prioritize the right efforts that can promote sustainability and profitability, it is essential to emphasize the right results. This involves identifying key externalities and material sustainability concerns that are critical for corporate strategy and operations.

7. *Three-Dimensionality:* To succeed in achieving these goals, the entire organization must be designed in a way that reflects social, environmental, and economic objectives. This requires new governance structures reflected in organizational design, leadership, and management control systems.

To solve the sustainability problem collectively, and to do so in a manner that allows for continued profitable growth, we need a comprehensive set of big and small RESTARTs across all industries. While the specific solutions in each industry will be different, these seven components can inform new ways of creating, delivering, and capturing value in all sorts of organizations.

6.5 DEMOCRATIZING TECHNOLOGY

How can the evolving technology and the possibilities created from Industry 4.0 become accessible to everyone? There is currently a big technology gap between the developed countries and developing countries. It is the under-regulated countries of the world that face the most serious challenges related to handling of waste. We live on the same planet; we are connected through the same globalized value chains and material streams. It is in everybody's interest that everybody is empowered to step up the game. Without massive scaling of new business models and solutions, the transition will be too slow.

Democratizing technology should not be considered philanthropy – but ecosystem strategy. Creating collaborative communities around technological progress and business development is closely aligned with the challenge of going from linear, siloed waste handling to circular and networked resource management. Leadership in an Industry 4.0 world requires individual companies to find a new and productive relevancy within dynamic networks of multiple stakeholders interconnected by technology. These networks need to be created – on a local, regional, and global level – to speed up the transition and scale new systems and solutions. We need a diverse kaleidoscope of perspectives to come up with the important ideas, and we need a lot of field experimentation to validate that they are actually working.

Luckily the creativity and brain power is evenly distributed across the globe. We share the same challenges, even if we operate in Scandinavia, West Africa, and Latin America: creating circular systems, incentivizing consumers, promoting eco-design, and enabling closed-loop programs. To avoid reinventing the wheel several times, we need to form better systems for knowledge transfer – networks that share results and insights. This will help the integrated value chains and the circular systems of Industry 4.0 to form and evolve faster.

6.5.1 The Benefits of Openness

The spreading of knowledge, ideas, and skills is the fuel that runs any technological revolution. The degree of openness is an important condition for both the innovating rates and the proliferation of new inventions. Actually, some research show that the earliest innovation in the British industrial revolution happened through "non-institutionalized and thus non-formalized diffusion, contagion and adaptation mechanisms for technical and technological innovations." According to this research the drivers were weakly connected manual workers, inventors, and entrepreneurs that were not scientists, but had practical experience [47].

Science is developed in an open and connected global community (even if much of the scientific papers are unavailable due to expensive subscription programs). Software has flourished as a result of social creativity. "Practicing the norms of openness and information sharing in a peer-production setting can result in the creation of complex technological products that approach, and sometimes rival, the scope and quality of similar products produced by proprietary efforts," the scientist say in the paper "The Value of Openness in Scientific Problem Solving" [48].

The challenge is of course that in a knowledge economy, knowledge is an important competitive advantage. The mechanism to gain and maintain power has always been protecting and fencing in the competencies and tools that make the difference – and selling the advantage of the invention to the customers willing to pay the most. In earlier times you had guilds and priesthoods protecting their knowledge, but the modern concepts of IPR law is closely related to the dynamics of the industrial revolutions. For engineers the 1883 Paris Convention [49] was groundbreaking – an international agreement that inventors could protect their innovations even if they were being used in other countries. Three years later, the Berne Convention [50] led to protection on an international level for all forms of written expression – including drawings, operas, and songs.

There are of course many legitimate reasons to keep parts of the technology, data, and skills behind closed doors. But experience from other industries show that openness creates a lot of added value – in the form of speed of implementation, security, cost, and flexibility. If openness becomes the norm within an industry, there will be network effects, creating a more dynamic and interesting business environment – including new start-ups that challenge the conventions.

What can we share that is not business critical but will create value for others? And what did others share that can be valuable to us? These are important questions for any industry leader to ask, both because it is simply smart and because open source and open licenses will also make important inventions available to anyone. Open APIs will invite new players to innovate on top of existing solutions. It is the first step in creating a more open ecosystem.

The open design principles, as described in Chapter 5, can reduce costs, increase the speed of implementation, and invite more skill into the process of innovation. Research also shows that involving stakeholders produce more innovative solutions than designers consulting users through traditional means [51]. Importantly, open design is also used in efforts to create solutions for the common good, sometimes where funding or commercial interest is lacking. It is also a way of bringing down the general costs of technologies or reducing the environmental footprint. Even if the open-source methodology started out in software, it is expanding to physical objects (especially through 3D models), hardware, and educational resources.

Waste management is a public service that buy a large amount of services from the private sector. Procurement should generally emphasize openness in technology – and all technical solutions should readily come with well-documented APIs. The use of open-source software can have several advantages to proprietary software for important long-term projects. As the source code publicly distributed, users relying on the software can be sure their tools won't disappear or fall into disrepair even if the original creators stop working on them.

6.5.2 Creating and Promoting Shared Standards

The easiest and least controversial way to democratize technology is to embrace existing open technical standards. A standard is norm or requirement for a repeatable technical task, described in a "formal document that establishes uniform engineering or technical criteria, methods, processes, and practices" [52].

The problem, or maybe the possibility, in waste management is that there are few global standards established for Industry 4.0 systems. The standards for many of the needed end-to-end functions need to be developed. Usually the standardization follows a long and winding path from an informal initiative to an internationally acknowledged standard. This is not, and should not be, a harmonious process. Experience from the development of software standards is that it is usually a harsh battle over a tough question: Who should have the power of definition within the industry?

In waste management this question is important and interesting. The whole industry is politically regulated and involves governmental services, private companies, and technology providers. There are many different cultures and local traditions regarding waste. A society based with a system of informal waste pickers and local sorting will need other solutions than a Western society heavily reliant on centralized robotic sorting. We need different templates that can be adapted to a local context. The key point is connectivity, flexibility, and simplicity.

A technical standard can be initiated privately or locally, within a corporation, regulatory body, or project. It can also be developed by groups such as national or international industry associations. Standards organizations often have more diverse input and usually develop voluntary standards. Standards can become mandatory if they are included in legislation or is a requirement in a procurement process. There will be negotiations on every level of the step from an informal, local initiative to an international standard. Often there are competing standards, with different advantages, that seek to be merged.

Industry 4.0 and the related business models need a range of codes and standards. In a formalized process, national and local government bodies, industry organizations, and individual companies should be involved. The importance of people at practical and technical levels should be stressed; standards often need to deal with the devil in the details.

Governments play a very important role in resource management. Public procurements can be used to push openness and standardization forward. When buying software or hardware solutions, the openness of data and licensing should be clearly defined. To push the development further, the solutions they ask for should be rooted in clearly defined needs within the organization. There is no need to be taken hostage by the features available in current technologies.

6.5.3 Global Collaboration for the Common Good?

Closing the technology gap is an important global challenge that can only be solved through global collaboration. In a research article on waste mismanagement in developing countries [53], Navarro Ferronato and Vincenzo Torretta conclude that common projects should be introduced at a global level in order to reduce the environmental contamination and health issues due to waste open dumping and burning. How should these common projects be organized?

Even for an industry working with physical material streams, there are many important lessons to be learned from the open design community. Popular open systems, designed to be shared, will by nature become more transparent and well documented as they are the result of a broad collaboration between independent developers. When anyone can view and modify a solution, errors are spotted and corrected, and systems continuously improved. A piece of open-source software can be upgraded and built upon without asking for permission from original authors. This is a very unfamiliar though for most traditional business managers, but very rational: there are not only advantages but also costs related to technology. The value of the communal effort often outweighs the value of owning and controlling (and having the responsibility for) technology.

The community of users and developers that form around important projects is especially valuable when the landscape is shifting. People who both use and contribute the product discuss, suggest, and work together – across borders. Yochai Benkler has named this the "commons-based peer production" [54]. This approach, routed in ideas of "community momentum" as a value driver, stands in stark contrast to the approaches of the more territorial industrial economy, where control, patents, and copyrights are considered key assets.

This dilemma is emblematic for the ongoing paradigm shift of Industry 4.0. The Internet has made it technically easy to share ideas and knowledge within peer-to-peer networks, but many valuable existing business models are still rooted in the closed system of production.

An inspiring example on the power of combining open source, open standards, free licenses, and international collaboration is the Global Digital Library (GDL). This is a platform to distribute and share free early-grade reading resources in languages that children use and understand [55]. It has been developed to increase the availability of high-quality reading resources in underserved languages worldwide – to be used in education. All the educational resources are published under open licenses and available on web or mobile and for print. It can be used by ministries of education, school managers, teachers, donor agencies and their implementing partners, international and national nongovernmental organizations (NGOs), local publishers, digital distributors and content providers, and households in developing countries.

The whole project is based on contributions from users – similar to open-source projects. This way the content pool will expand over time through discovery and sharing of more existing quality content, translations, and localizations of the platform's content, as well as additions of newly created content. The GDL project was initiated All Children Reading: A Grand Challenge for Development (ACR) in 2014. The technology is being developed by the Norwegian Digital Learning Arena (NDLA), based on their publicly funded open-source digital infrastructure.

Even if the initial aim was to translate books from widely used languages to underserved languages, the opposite is now starting to happen. Books from African countries have been translated into Western languages and are now used in education in Europe. The creative community of contributors is growing, and the synergies go both ways.

This project could be a possible template for how the waste industry could create systems for sharing, learning, and collaboration using digital platforms and sharing tools. The project taps into existing networks and involves multiple stakeholders, including resourceful individuals, businesses, NGOs, and government funding agencies – very similar to the alliance trying to solve the problem of waste.

Could a global initiative be developed to promote new operational models in waste management? Could new software and hardware solutions be distributed based on open licenses and open design principles? Can smart waste management solutions be invented in the developing world and be exported to Western societies? Probably, yes. The principles of the circular economy should also be applied to technological inventions. But for ideas and solutions to spread across the plant, the door leading to the circular future needs to be open.

Key Take-Outs of Chapter 6

Key take-outs	Why is it important?
A paradigm shift is currently happening in waste management – due to the imperative of a circular economy and the emergence of new technology-driven business models.	A revolution is about to happen. Leaders will have to take on a new role and start transforming their organizations.
There are many risks and opportunities associated with this paradigm shift. The transformation requires organizations that can both improve existing operations and find new ways of creating value. In this process there is a lot to learn from similar shifts in other sectors.	Waste management leaders need to be updated on technological change yet focus on the people: preparing the organizations for deep changes and recruiting the skills needed for Industry 4.0-type operations are important keys.
With the transformation follows new leadership requirements. Both a long-term vision and short-term pragmatism are needed. There are several different innovation methodologies that can be applied to challenge business as usual.	Leadership in transformational processes can be challenging and requires a new way of operating – across established siloes. Innovation is a team sport that requires involvement and training.
Research from across industries shows the importance of innovation portfolio management. Projects should be highly prioritized and search to both improve existing operations and create new business models.	Disruption from within is hard. Leaders need to find the balance between exploiting current business opportunity and at the same time search for the new opportunities.
There are multiple innovation methodologies that stimulates idea creation and user involvement. These frameworks may speed up the process from problem statement to validated solution.	A more organized way of working continuous and systematic with innovation is important as the business environment is shifting faster. Some of these techniques are really efficient.
The territorial logic of the linear economy is challenged by industrial ecosystems and a more collaborative mindset.	Industry 4.0 requires a focus on synergies and alliances – and conscious management of the inflows and outflows of knowledge across the boundary of a firm.
In a more open, technology-driven, and changing business environment, complexity will naturally increase. Not all these conditions can be managed.	Leaders should focus on reducing the unnecessary complexity and the complexity that follows by design.

Key take-outs	Why is it important?
Political leadership is very important in waste – to accelerate the positive and inclusive impacts of technologies and minimizing or eliminating their negative consequences.	We need conscious political leaders that focus both on creating constraints and creating new possibilities in for circular resource management and the use of technological innovations.
There is currently an ongoing revolution in the design of business models based on the idea of reducing negative externalities and enhancing positive outcomes in an environmental perspective.	Leaders in resource management should take part in the value loop as the new circular business models evolve. Reducing material footprint or enabling closed-loop programs will reduce the waste volumes.
A new mindset and theory around business models have evolved – where eco-social benefits and costs are parts of a more holistic approach based on both taking responsibility and exploring opportunity.	Leaders should challenge existing management models and get involved in building momentum for the emerging green business movement.
For the innovation to really create an impact, it needs to scale globally. The gap between developed and developing countries is a challenge. Openness in technology and a global community around knowledge transfer is important to make Industry 4.0 solutions available for everyone.	Leaders should embrace open source and open standards and take part in the work to create standards that make systems work more seamlessly to support the collaborative business processes of Industry 4.0.

REFERENCES

1. ISWA (2019). ISWA president launches ground-breaking new report preparing the waste management industry for the 4th Industrial Revolution. https://www.iswa.org/home/news/news-detail/article/iswa-president-launches-ground-breaking-new-report-preparing-the-waste-management-industry-for-the-4/109 (accessed 7 February 2020).
2. Simon, F. (2020). LEAK: EU's new circular economy plan aims to halve waste by 2030. https://www.euractiv.com/section/circular-economy/news/leak-eus-new-circular-economy-plan-aims-to-halve-waste-by-2030 (accessed 5 February 2020).
3. Kannangara, M., Dua, R., Ahmadi, L., and Bensebaa, F. (2018). Modeling and prediction of regional municipal solid waste generation and diversion in Canada using machine learning approaches. *Waste Management 74*: 3–15.

4. Birkel, H.S. and Hartmann, E. (2019). Impact of IoT challenges and risks for SCM. *Supply Chain Management: An International Journal 24*: 39–61.

5. Hill, D. (2012). *Dark Matter and Trojan Horses: A Strategic Design Vocabulary*. Strelka Press. 181 p.

6. Kagermann, H., Helbig, J., Hellinger, A., and Wahlster, W. (2013). Recommendations for Implementing the Strategic Initiative INDUSTRIE 4.0: Securing the Future of German Manufacturing Industry. Final Report of the Industrie 4.0 Working Group. 112 p.

7. Information Resources Management Association (2015). *Business Intelligence: Concepts, Methodologies, Tools, and Applications: Concepts, Methodologies, Tools, and Applications*. IGI Global. 2284 p.

8. Rifkin, J. (2011). *The Third Industrial Revolution: How Lateral Power Is Transforming Energy, the Economy, and the World*. St. Martin's Press. 304 p.

9. Harvard Business Review (2012). Managing your innovation portfolio. https://hbr.org/2012/05/managing-your-innovation-portfolio (accessed 8 February 2020).

10. Dinkelmann, M., Siegert, J., and Bauernhansl, T. (2014). Change management through learning factories. In: *Enabling Manufacturing Competitiveness and Economic Sustainability* (ed. M. Zaeh), 395–399. Cham: Springer https://doi.org/10.1007/978-3-319-02054-9_67.

11. Ellen MacArthur Foundation and World Economic Forum (2016). Intelligent assets: unlocking the circular economy potential. https://www.ellenmacarthurfoundation.org/publications/intelligent-assets (accessed 7 February 2020).

12. Forrester (2018). Digital transformation: your first 100 days. https://www.forrester.com/report/Digital+Transformation+Your+First+100+Days/-/E-RES143673# (accessed 9 February 2020).

13. Oberer, B. and Erkollar, A. (2018). Leadership 4.0: digital leaders in the age of Industry 4.0. *International Journal of Organizational Leadership 7*: 404–412.

14. Gerber, N. (2018). What is innovation? A beginner's guide into different models, terminologies and methodologies. Medium. https://medium.com/@niklausgerber/exploring-innovation-a-beginners-guide-into-different-models-terminologies-and-methodologies-6f20c20fcee4 (accessed 6 February 2020).

15. IDEO (2018). Design thinking. designthinking.ideo.com (accessed 6 February 2020).

16. The Interaction Design Foundation. UX design courses & global UX community. www.interaction-design.org. (accessed 17 February 2020).

17. Ries, E. (2011). *The Lean Startup: How Today's Entrepreneurs Use Continuous Innovation to Create Radically Successful Businesses*. Currency. 336 p.

18. Koen, P. (2015). Lean Startup in Large Enterprises Using Human-Centered Design Thinking: A New Approach for Developing Transformational and Disruptive Innovations (27 January 2015). Howe School Research Paper No. 2015-46. SSRN. https://ssrn.com/abstract=2566916 or http://dx.doi.org/10.2139/ssrn.2556187.

19. Agile Alliance (2015). Agile manifesto for software development. https://www.agilealliance.org/agile101/the-agile-manifesto (accessed 6 February 2020).

20. Schwaber, K. and Sutherland, J. (2017) The Scrum guide. https://www. scrumguides.org/scrum-guide.html (accessed 4 April 2020).
21. GV. The design sprint. www.gv.com/sprint (accessed 13 February 2020).
22. Banfield, R., Todd Lombardo, C., and Wax, T. (2015). *Design Sprint: A Practical Guidebook for Building Great Digital Products*. O'Reilly Media, Inc. 272 p.
23. Chesbrough, H. and Bogers, M. (2014). Explicating open innovation. In: *New Frontiers in Open Innovation* (eds. H. Chesbrough, W. Vanhaverbeke and J. West), 3–28. Oxford: Oxford University Press.
24. Chesbrough, H.W. (2003). A better way to innovate. *Harvard Business Review 81* (7): 12–13, 115.
25. Dahlander, L. and Gann, D. How open is innovation? *Creating Wealth from Knowledge* https://doi.org/10.4337/9781848441248.00009.
26. Chesbrough, H.W. and Garman, A.R. (2009). How open innovation can help you cope in lean times. *Harvard Business Review 87* (12): 68–76, 128.
27. Enkel, E., Gassmann, O., and Chesbrough, H. (2009). Open R&D and open innovation: exploring the phenomenon. *R&D Management 39*: 311–316.
28. Heywood, S., Spungin, J., and Turnbull, D. (2007). Cracking the complexity code. *McKinsey Quarterly 2* (2): 85–95.
29. The European Environmental Bureau (EEB) (2019). Decoupling debunked: evidence and arguments against green growth as a sole strategy for sustainability. https:// eeb.org/library/decoupling-debunked (accessed 5 February 2020).
30. World Economic Forum (2016). Home. http://www3.weforum.org/docs/ WEF_Intelligent_Assets_Unlocking_the_Cricular_Economy.pdf (accessed 4 February 2020).
31. McKinsey (2015). Growth within: a circular economy vision for a competitive Europe. https://www.mckinsey.com/business-functions/sustainability/our-insights/growth-within-a-circular-economy-vision-for-a-competitive-europe (accessed 4 February 2020).
32. Davila, T. and Epstein, M. (2014). *The Innovation Paradox: Why Good Businesses Kill Breakthroughs and How They Can Change*. Berrett-Koehler Publishers. 240 p.
33. Wiles, J. (2018). How and why to keep innovation management on track: smarter with Gartner. https://www.gartner.com/smarterwithgartner/how-and-why-to-keep-innovation-management-on-track (accessed 6 February 2020).
34. Harvard Business Review (2004). The ambidextrous organization. https://hbr. org/2004/04/the-ambidextrous-organization (accessed 30 January 2020).
35. The Maritime Executive (2019). New project to advance wind turbine blade recycling. https://www.maritime-executive.com/article/new-project-to-advance-wind-turbine-blade-recycling (accessed 6 February 2020).
36. Sitra. A natural resource strategy for Finland. https://www.sitra.fi/en/publications/ natural-resource-strategy-finland-0 (accessed 6 February 2020).

37. Magretta, J. (2002). Why business models matter. *Harvard Business Review 80* (5): 86–92, 133.
38. Stock, T. and Seliger, G. (2016). Opportunities of sustainable manufacturing in Industry 4.0. *Procedia CIRP 40*: 536–541.
39. Jørgensen, S. and Pedersen, L.J.T. (2018). *RESTART Sustainable Business Model Innovation*. Springer. 253 p.
40. Marilungo, E., Papetti, A., Germani, M., and Peruzzini, M. (2017). From PSS to CPS Design: A Real Industrial Use Case Toward Industry 4.0. *Procedia CIRP 64*: 357–362.
41. Wang, Y., Ma, H.-S., Yang, J.-H., and Wang, K.-S. (2017). Industry 4.0: a way from mass customization to mass personalization production. *Advances in Manufacturing 5*: 311–320.
42. Li, Z., Wang, K., and He, Y. (2016). Industry 4.0: potentials for predictive maintenance. *Proceedings of the 6th International Workshop of Advanced Manufacturing and Automation*. https://doi.org/10.2991/iwama-16.2016.8.
43. Operate Foundation (2019). Pay as you throw (PAYT) toolkit for European cities. Urban Agenda for the EU.
44. CASE (2017). Sustainable business model canvas. https://www.case-ka.eu/wp/wp-content/uploads/2017/05/SustainableBusinessModelCanvas_highresolution.jpg (accessed 16 February 2020).
45. Osterwalder, A. and Pigneur, Y. (2013). *Business Model Generation: A Handbook for Visionaries, Game Changers, and Challengers with Wlyetxc Set*. Wiley.
46. CASE (2017). Sustainable business model canvas. https://www.case-ka.eu/index.html%3Fp=2174.html (accessed 9 February 2020).
47. Nuvolari, A. (2004). Collective invention during the British Industrial Revolution: the case of the Cornish pumping engine. *Cambridge Journal of Economics 28*: 347–363.
48. Lakhani, K.R., Jeppesen, L.B., Lohse, P.A. et al. (2007). *The Value of Openness in Scientific Problem Solving*. Harvard Business School WP. 58 p.
49. WIPO (1883). Summary of the Paris convention for the protection of industrial property. https://www.wipo.int/treaties/en/ip/paris/summary_paris.html (accessed 14 February 2020).
50. WIPO (1886). Summary of the Berne convention for the protection of literary and artistic works. https://www.wipo.int/treaties/en/ip/berne/summary_berne.html (accessed 4 February 2020).
51. Mitchell, V., Ross, T., May, A. et al. (2015). Empirical investigation of the impact of using co-design methods when generating proposals for sustainable travel solutions. https://repository.lboro.ac.uk/articles/Empirical_investigation_of_the_impact_of_using_co-design_methods_when_generating_proposals_for_sustainable_travel_solutions/9348323/files/16957391.pdf (accessed 4 February 2020).

52. Gergetz, D. (2014). LibGuides: mechanical engineering resources: codes and standards. https://libguides.msoe.edu/me/codes (accessed 15 February 2020).
53. Ferronato, N. and Torretta, V. (2019). Waste mismanagement in developing countries: a review of global issues. *International Journal of Environmental Research and Public Health 16* (6): 1060.
54. Open Design Now (2011). remko. Into the Open/John Thackara. http://opendesignnow.org/index.html%3Fp=403.html#fn-403-2 (accessed 4 February 2020).
55. Global Digital Library (2020). Global Digital Library. https://digitallibrary.io/en/books/category/library (accessed 15 February 2020).

Chapter 7
The Rise of a New Science

We grow in direct proportion to the amount of chaos we can sustain and dissipate
— Ilya Prigogine

Recommended Listening
The Dark Side of the Moon, **Pink Floyd**

Because this emblematic masterpiece will stimulate the right vibes to understand the double edge of urban informatics.

Recommended Viewing
The Rise of Surveillance Capitalism, **https://youtu.be/ 2s4Y-uZG5zk**

Because the author **Naomi Klein and Harvard Business School professor Shoshana Zuboff, author of** *The Age of Surveillance Capitalism: The Fight for a Human Future at the New Frontier of Power,* **discuss in depth the unprecedented form of power called "surveillance capitalism" and the quest by corporations to predict and control our behavior.**

Industry 4.0 and Circular Economy: Towards a Wasteless Future or a Wasteful Planet?, First Edition.
Antonis Mavropoulos and Anders Waage Nilsen.

In this chapter we are going to discuss the rise of new science as a result of the convergence of big data and artificial intelligence with the efforts to apply complexity models and principles for understanding cities. More specifically we are going to discuss how the availability of huge data sets that are generated by citizens, mobile apps, the Internet of Things (IoT), and other types of sensors create the basis for understanding urban centers in a more comprehensive and integrated way. Until few years ago our simulations for cities were based mostly on assumptions and rather static models that were not capable to capture the complex and thick nature of cities. The result was that we were dealing with problems like waste management, energy, water, transportation, and food in a rather isolated way, and then, we were trying to identify potential linkages and interfaces. The rise of big data sets and analytics provides a new opportunity that can drive the development of a new science, the science of cities. But, to utilize this new opportunity, there is a need to develop a new approach regarding our understanding about cities, a dynamic approach that will involve complexity as an integral part of cities and not as a negative externality. However, the rise of this new science creates difficult problems and challenges that should be managed for successful applications. In addition, it becomes clear that the same tools that can advance sustainability in cities can also be used as an excuse to cut services and abandon neighborhoods and citizens that lack participation in decision making. As it is the case with all major technological innovations, they can be used either to improve human societies or to worsen them. Once again, the future is going to be shaped based on our own choices and intentions.

7.1 THE RISE OF URBAN INFORMATICS

Urban informatics [1] is the exploration and understanding of urban systems for dynamic resource management, knowledge discovery and understanding of urban patterns and dynamics, urban engagement, civic participation, urban planning and policy analysis. Urban informatics is using the new big data sources. What are these new sources? In brief we are talking for the huge data sets that are generated by citizens, mobile apps, the IoT, and other types of sensors as they are emerging from technological, institutional, social, and business innovations. These data sets are now combined with the traditional ones like population census and surveys, maps of the energy, water and transportation networks, statistics about traffic and passengers, income distribution, and health conditions. The concept of urban informatics is the natural evolution of the concept of smart cities in the era of the IoT and the digitalization of everything.

It's good to have an idea of the relevant magnitudes of the new big data sets. It has been assessed that from 2005 to 2020 the digital universe will grow by a factor of 300, from 130 to 40 000 exabytes (EB) (40 trillion gigabytes (GB)) [2]. Another report from the International Data Corporation (IDC) forecasts that only the data generated by 41.6 billion objects and people connected to the IoT such as smart fridges, cars, or smart speakers combined with the focus of the 5G revolution will generate 79.4 zettabytes (ZB) of data in 2025 [3]. Trying to make these numbers better understood, let's give some examples. One gigabyte is, more or less, the size of a digital movie at TV quality. One exabyte is equal to 1 billion GB. Five exabytes is the size of all the words ever spoken by human beings [4]. One zettabyte is simply 1000 EB. Cisco estimated that in 2016 we have passed 1 ZB in total annual Internet traffic, that is, all data that we have uploaded and shared on the World Wide Web, most of it being file sharing [5]. On a more practical level [6]:

- The average US customer uses 1.8 GB of data per month on his or her cell phone plan.
- Walmart processes one million customer transactions per hour.
- Amazon sells 600 items per second.
- On average, each person who uses email receives 88 emails per day and send 34. That adds up to more than 200 billion emails each day.
- MasterCard processes 74 billion transactions per year.

There are three crucial details regarding the new big data sets. First, although the traditional data sets continue to play a crucial role, it seems that there is an increasing difficulty to achieve high participation rates in urban surveys and their representativeness is undermined by the increasing marginalization phenomena. In addition, these surveys are not dynamic at all, as usually they are repeated every two to five years, which is a substantial time period for a dynamic city that grows with 4–7% in population annually, as it is the case in many emerging megacities. Second, the emerging big data sources belong to different private and public sector stakeholders; thus their utilization is not something that will happen automatically. Third, the big data sets relevant to sensors, smartphones, and mobile applications are usually serving a rather narrow point of view of the urban life, based on the interest of the service provider; thus these data sets are becoming really important only when they are interconnected and combined in a way that represents a more holistic view of the urban metabolism.

However, in principle, assuming that these data sets are available and accessible for integration and acquisition using big data analytics, working on them can provide important new tools to answer difficult questions like the following: What is the city's environmental and energy footprint and how it changes in time, even in real time? How much pollution is generated and how is it spatially and temporarily distributed on an hourly or daily basis? How energy, water, waste, and food systems are integrated in the urban metabolism? What are the interlinkages between urban mobility, resource consumption, and pollution? How the different socioeconomic patterns of a city contribute to its sustainability?

7.1.1 New Answers to Old Questions, but Not for Free

We can put many similar questions, but the meaning is the same: combining the new emerging big data sets with traditional ones, we are capable to monitor, understand, model, and forecast the urban metabolism in a way that was never possible before. The potential of urban informatics [1] research and applications is expected to be huge in four areas: (i) improved strategies for dynamic urban resource management, (ii) theoretical insights and knowledge discovery of urban patterns and processes, (iii) strategies for urban engagement and civic participation, and (iv) innovations in urban management and planning and policy analysis. Even more, as we will explain later, this combination of old and new data sources with big data analytics and artificial intelligence can set the foundation for the rise of a new science, the science of cities.

However, this will not happen easily and certainly not automatically. One of the reasons concerns the barriers relevant to ownership and accessibility of the data sets. The real challenge is like this: a huge amount of data is being generated from different data sources, such as smartphones, computers, sensors, cameras, global positioning systems, social networking sites, commercial transactions, and games. For sure this amount of data, or better the size of all these data sets, will continuously grow, making the question of efficient data storage and processing a very difficult one [7]. If we are not going to be able to transform and integrate these data sets to meaningful information, in a holistic and integrated way, then big data can be easily turned to big waste of money, from the point of view of city governance and sustainability.

But there is another important problem. The concept of smart cities and big data has been widely (and rightly) critiqued for promoting economic development while giving low attention to environmental and social concerns, putting in practice a rather narrow and technocratic perspective that reduces sustainability to a series of "technical and economic fixes" while obscuring its political implications [8]. It was discovered [9] that in the discussion about smart cities, there is an obsession on modern technologies and "smartness," while urban sustainability frameworks that contain a large number of indicators measuring environmental sustainability are ignored in most of the cases. However, it must be very clear that the concept of smart sustainable and big data manageable cities has a real positive potential that goes beyond any narrow political and corporate interest. How, for whom, with what funds, where, and when it will be implemented is a matter of political and social debate. As it was mentioned [10], "Making urban data widely available, developing a city-scale 'Internet of Things' and mobilising smart and digital technologies has the potential to enhance social well-being, empower communities, reveal previously hidden urban environmental processes, enable resource and skills sharing, include citizens in co-creative governance processes, generate novel solutions to mundane urban problems, and underpin new models for more efficient use of idle assets. But this also requires the careful application of ICT technologies to avoid empowering corporate interests within urban visioning and management and further excluding those already marginalised by prevailing technocratic and entrepreneurial forms of urban governance. The varieties of smart-sustainable agendas emerging on the ground in hundreds of cities around the world occupy a wide range of positions between (and beyond) these critiques and promises. Social equity and environmental sustainability are not a-priori absent or de-facto present in technological designs of smart city initiatives, but have to be made, nurtured and maintained, as they materialise in particular places."

A crucial term in this discussion is the content of the term big data, in an urban context. Michael Batty, one of the most influential thinkers in urban complexity and its management, says that he prefers this definition: big data is "any data that cannot fit into an Excel spreadsheet" (Excel 2019 can have up to 1 048 576 rows and 16 384 columns, in total 17.179 billion cells – however these are the theoretical values that are subject to restrictions like the specific software and the memory available [11]). Speaking about big data and city planning [12], Batty mentions "I define big data with respect to its size but pay particular attention to the fact that the data I am referring to is urban data, that is, data for cities that are invariably tagged to space and time. I argue that this sort of data are largely being streamed from sensors, and this represents a sea change in the kinds of data that we have about what happens where and when in cities. I describe how the growth of big data is shifting the emphasis from longer term strategic planning to short-term thinking about how cities function and can be managed, although with the possibility that over much longer periods of time, this kind of big data will become a source for information about every time horizon." There are three important points here. First, it's the integration of data with location and time in a single package that allows any data set to be plotted easily in the desired time and space dimensions. Second, the "sea change" of the data availability is rather literal. The real situation is better described if we use a metaphor based on oceans: we continuously generate oceans of data that currently are transformed to very few islands of information. Third, the shift from long term to short term and real time deserves a better explanation. Planning for waste management or recycling as an example, you use as a short term a 3–5-year period, and as long term a 15–20. Planning for transportation networks has a short term of 5 years and a long term of 20–50 years. This is how traditionally city planners are working visioning and prioritizing interventions based on the big long-term picture. However, taking recycling as an example, we have in many cities real-time monitoring of the content of bins and continuously optimized routing for vehicles based on the level of fullness of each bin [13]. This allows a much more granulated understanding of the spatial and temporal patterns of recycling, moving our approach from the city or the neighborhood in months and years to the level of a single bin in minutes.

7.1.2 Citizens' Engagement

Of particular interest is a specific category of the emerging new data sets, the one that is described as user-generated content (UGC) that usually involves citizens' participation and can occur (i) proactively, when users voluntarily generate data on ideas, solve problems, and report on events, disruptions, or activities that are of social and civic interest, or (ii) retroactively, when analysts process secondary sources of user-submitted data published through the web, social media, and other tools [14]. Depending on the level of decision making needed on the part of users generating information, proactive sensing modes can be disaggregated into active and passive sensing modes. In active sensing [1], users voluntarily opt into systems that are specifically designed to collect information of interest (e.g. through apps that capture information on quality of local social, retail, and commercial services or websites that consolidate information for local ride sharing), and actively report or upload information on objects of interest. In passive sensing, users enable their wearable or in-vehicle location-aware devices to automatically track and passively transmit their physical sensations or activities and movements (e.g. real-time automotive tracking applications that measure vehicle movements yielding data on speeds, congestion, incidents, as well as biometric sensors, life loggers, and a wide variety of other devices for personal informatics relating to health and well-being).

Why this specific category of data sets is very important? Cities are made up of both citizens and infrastructures for food, water, energy, transportation, and other activities. Therefore, they are considered as complex social-technological systems, where citizens as human agents operate various functions. As it has been discussed, there are three dimensions that should be comprehensively understood in order to realize a smart city concept: technologies, institutions, and, above all, people [15]. Thus, this citizens' (users') generated content allows the emergence of new patterns of urban governance that integrates citizens' input for decision making. Or, more precisely, it has the potential to embed citizens' input in all major decisions and so to stimulate a more democratic and effective governance, at least theoretically.

To understand better how this potential can (or can't) be realized, let's take an example. One of the most usual applications concerns crowdsourcing: citizens are provided with a free-to-use mobile application that allows them to report several city problems, from litter and traffic jams to accidents, problems in traffic lights and city lighting, etc. Such applications require basic conditions to be met in order to be successful. First, there must be an administrative unit (and a relevant design of the app) that links the reported problems with the responsible municipal departments. Second, there must be a screening and a follow-up in each reported problem, and citizens should be immediately informed regarding the time required to manage the problems (if they can be managed, in many cases reported problems are not under the jurisdiction of the municipalities, or they can't be fixed immediately). Third, the administrative system of the municipality should be rearranged in order to recognize, prioritize, and respond to the problems based on specific criteria. Only if all these conditions are met, such a crowdsourcing application can be used as a tool to improve the municipal services and adapted them to citizens' needs and remarks. However, even if these conditions are met, still the mobile app will be used only by those who have a smartphone and access to Internet, which means finally the poorest part of the population will be actually excluded from the crowdsourcing. Thus, in many cases, this kind of mobile apps is used as an excuse to cut in person municipal services and audits in the most marginalized areas, which results instead of improving the services and the inclusivity model of governance to marginalize further the most vulnerable populations. In addition, there is another important problem. The use of these apps is in many cases supported by a kind of naïve techno-optimism that, with the proper information, city management will be actually improved. However, without underestimating how crucial is the proper information flow for city governance, there are many problems that are reported by citizens that simply can't be resolved because they are not information problems. In many cases, as it happened in Greece during the economic crisis 2010–2018, municipal services collapse by the luck of funds, so the continuous citizens reporting makes the incapability to respond even worst. In other cases, as an example when citizens are reporting smells and pollution incidents by industries, municipalities are simply irrelevant to intervene as this is usually the jurisdiction of regional or national authorities. And in other cases, when there are continuous reports and complains about as an example an old school or stadium that creates risks, things can only be fixed by a new investment.

7.1.3 More Challenges

Table 7.1 presents the main challenges and findings regarding big data sets and their availability according a well-recognized and cited OECD report [16].

Before we go further, we hope that it is already clear that the concept of smart cities in the era of the "digitalization of everything" goes far beyond technological approaches, but still, it requires substantial investments and a crucial role of technologies. Actually, integrating citizens, technologies, and institutions into a new organic network results in a new level of complexity [17], on top of the already complex urban system. Consequently, the new more complex system can present new emerging

Table 7.1 Major challenges and findings regarding big data set.

Challenge	Findings
Massive amounts of digital data are being generated at unprecedented scales and velocity, much of it from new sources such as the Internet. The reliability, statistical validity, and generalizability of new forms of data are not well understood. This means that the validity of research based on such data may be open to question	The expertise and knowledge required to exploit the scientific value of these data and to make them available for reuse is in many cases dispersed across countries and scientific disciplines. Opportunities to gain leverage and build on common international expertise are missed, and costs consequently incurred as a result of duplication
While many countries and cities have vast amounts of more traditional forms of administrative, survey, and census data collected by and held by national statistical agencies and government departments, knowledge about the existence of such data as micro-data records is a precondition for the efficient and effective planning of international research	Many significant activities are underway across the world to make research data easier to find, but not all are documented according to the international standards that now exist for data documentation and interchange. As a result, information about the existence of micro-data and their availability for reuse is often difficult to find
New forms of personal data, such as social networking data, can provide insights into the human condition. However, the use of those data as research resources may pose risks to individuals' privacy, particularly in case of inadvertent disclosure of the identities of the individuals concerned. There is a need for greater transparency in the research use of new forms of data, maximizing the gains in knowledge derived from such data while minimizing the risks to individuals' privacy, seeking to retain public confidence in scientific research that makes use of new forms of data	Many national privacy laws authorize the reuse of personal data for historical, statistical, or scientific purposes, provided appropriate safeguards are put in place. However, there is no internationally recognized framework code of conduct that deals specifically with the use of new forms of personal data for research
Barriers to access to social science data hinder national and cross-national collaboration that could exploit their research value. These barriers relate to a variety of obstacles (legal, cultural, language, proprietary rights of access), all of which have to be identified and removed if cross-national research is to be promoted	A number of activities are underway across the world to develop and provide access to social science micro-data for comparative research purposes. These activities tend to be "domain specific" (i.e. international studies of political behavior, social attitudes and lifestyles, fertility and family formation). They are driven primarily by the interests of leading social scientists in these fields, less so by national statistical agencies

Table 7.1 (Continued)

Challenge	Findings
The drive to address what is increasingly an interdisciplinary and international research agenda requires the use of existing capacity to its full potential	Data resources and capacity to analyze data exist separately in national official statistical agencies and the research community, both within and across countries
Data sharing, including the creation of appropriate metadata to international standards, is fundamental to the process of scientific enquiry. Researchers need incentives to ensure effective data sharing	There are few incentives to encourage researchers to manage, maintain, archive, and share data resulting from their research. Without clear incentives it is unlikely that the benefits of international collaboration will be fully realized

Source: From Ref. [16], with kind permission from OECD.

properties that are not expected, and this is one of the most important challenges of the smart cities concept.

There are many real applications that demonstrate the potential of urban informatics. Let's see some of them.

7.2 ISLANDS OF INFORMATION IN OCEANS OF BIG DATA

In this section we will present some representative cases of big data systems and their applications to urban governance. On purpose, we will not refer too much on waste management, since there are already a lot of examples in previous chapters. Our aim is to show that big data sets, if properly combined and managed, can reveal new dimensions and patterns of the urban complexity and stimulate a deeper understanding of urban sustainability and change.

7.2.1 Understanding Cities

Carlo Ratti, a professor of urban technologies and planning, director of the SENSEable City Lab at MIT, has contributed substantially to our understanding on smart cities, big data, and how they can be used to understand urban change and citizens' behavior. I will refer to some of his works that really demonstrate the relevant potential.

The first one regards a case study in Rome [18]. In this work, a new real-time urban monitoring platform was developed for the real-time evaluation of urban dynamics based on the anonymous monitoring of mobile cellular networks. The platform sets a new unprecedented benchmark for the monitoring of a large urban area, which covered most of the city of Rome, in real time, using a variety of sensing systems and opened the way to a new paradigm of understanding and optimizing urban dynamics. The platform was used to obtain a set of maps describing the vehicular traffic status and the movements of pedestrians and foreigners. Moreover, the system developed for the project acquired the instantaneous location of buses and taxis and the voice and data traffic served by all the base transceiver stations of the mobile cellular network. All these data were combined and updated in real time to realize several information layers on top of a map of Rome. This case study provided some impressive, although experimental, visualizations that answer questions like what is the people's distribution during special events, which landmarks of Rome attract more people and when is public transportation in accordance with the people's geographical distribution, and what are the main traffic patterns in a detailed spatial–temporal scale.

In another study that used digital footprints from UGC to reveal the patterns of city visitors' presence and movements [19], the central idea is explained in a great way: "Today, it's possible to gather every click of every move of every user who interacts with any software in a database and submit it to a second-degree data-mining operation. Along with the growing ubiquity of mobile technologies, the logs produced have helped researchers create and define new methods of observing, recording, and analyzing a city and its human dynamics. In effect, these personal devices create a vast, geographically aware sensor web that accumulates tracks to reveal both individual and social behaviors with unprecedented detail. The low cost and high availability of these digital footprints will challenge the social sciences, which have never before had access to the volumes of data used in the natural sciences, but the benefits to fields that require an in-depth understanding of large group behavior could be equally great."

While traditional studies on urban structure were focused on parameters like population, employment, and commuting or shopping trips, the recent availability of several large-scale data sets on human activities (mobile phones, smart cards, detailed travel surveys) provides several alternative ways to gain insights of the urban metabolism and flow. A relevant application with powerful visualizations can be found in a study about centrality in Singapore [20] that developed a simple quantitative index calculated by combining density and diversity of human activities. Subsequently, the centers of urban activities were identified as spatial clusters of locations with high centrality, allowing to assess the spatial distribution of centers over different years and to trace urban transformation processes.

Another study [21] identified the preferred urban spaces in Alicante province based on social media showcases users' preferences, a trend that can provide an indication as to why some urban public spaces are preferred over others. Pittsburgh in the United States has been studied on a large scale based on 18 million check-ins collected from users of a location-based online social network – the study identified dynamic city clusters much different than the traditional neighborhoods.

7.2.2 Understanding Supply Chains

Since we speak about social media, let's have a look on how Twitter can reveal problems in the beef supply chain and stimulate food waste minimization strategies. The fact is that, on an average, 45 000 tweets are tweeted daily related to beef products to express their likes and dislikes. These tweets are large in volume, scattered and unstructured in nature. Using these tweets as a big data set and working in depth with the relevant hashtags, researchers [22] found out that five major issues were concerning the consumers: loss of color, hard texture, excess of fat and gristle, bad flavor, smell and rotten, and the existence of foreign bodies like plastics, metals, insects, etc. Moreover, the researchers identified the root causes and proposed a waste minimization strategy by linking each of the five issues mentioned above with a specific part of the beef supply chain. It was finally concluded that although major retailers have made a provision for the customers to make a complaint in the store, still, customers are not doing so in most of the cases, but they are using social media like Twitter to express their disappointment, tagging the name of the retailer, which is affecting the reputation of the retailers. There is plenty of useful information available on Twitter, which can be used by food retailers for developing their waste minimization strategy. This information is big in size considering its volume, variety, and velocity but provides very important insights for the identification of problems in the supply chain and the development of waste minimization approaches. Figure 7.1 presents the linkages between the complains in tweets and the supply chain.

Food waste was also the focus of another approach that was aiming to support policy makers in evaluating their policy options to enhance food security through reducing waste within the food chain. To explore and identify the interrelationships between organizational factors contributing to the management of food security, the researchers [23] developed a research design to capture aspects of food distribution and consumption that may impact upon the generation of food waste. It was concluded that to address the wider aspects of the food security challenge and to produce more generalizable findings, the authors need to consider the non-contextual implications of food loss and food waste as part of food supply chain systems; thus the only way was to move from the (relatively) "small" amount of data and information assembled to "big data" needs and implications. In this way, the problem of food losses and food waste becomes a big data challenge that requires the proper framework in order to become effective and efficient, as shown in Figure 7.2.

7.2.3 Understanding Waste Management

An important analysis that demonstrates the big data potential has been implemented on construction waste management in Hong Kong [24]. Investing in big data provides two basic advantages. First, you can avoid the different biases that are usually met in small data sets and thus portray a more realistic picture. Second, by analyzing big

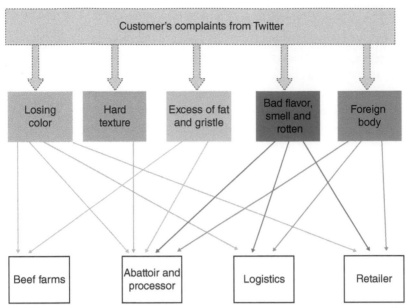

Figure 7.1 Linkages between complains in Twitter and supply chain.
Source: **From Ref. [22], © 2016 The Authors, published with open access at Springerlink.com, adapted by Nick Rigas, D-Waste.**

data, it is possible to discover hidden patterns, unknown correlations, and other useful actionable information. It is clear that big data can lead to actionable information that would not be possible to discover with small data. Let's see how this is outlined in the case of construction waste in Hong Kong. The waste generation rate (WGR) is one of the most usual performance indicators, and it is calculated by dividing waste generation in volume (m³) or quantity (tons) by square meters of gross floor area or by million US$ worth of construction work. The lower the WGR, the better the performance. Analyzing six years' data sets and almost two million records, the researchers found [25] that there was a remarkable performance disparity between the public and private sectors, with contractors performing better in managing both inert and non-inert waste in public projects than they do in private projects. In another effort, by analyzing one day's waste disposal records randomly selected, it was discovered that a considerable number (734 out of 4780) of waste haulers tend to overload than their permitted load weight. Meanwhile, often lorries are underloaded, which is more likely due to poor fleet management. By further analyzing the WGRs of the individual projects, it was found that a handful of companies achieved consistently low WGRs. Perhaps these companies are truly good at managing C&D waste, in which case their experiences should be disseminated to the whole industry. On the other hand, the WGRs of other companies were consistently high, suggesting that a review of the company's poor performance might be advisable. This kind of useful actionable information can only be revealed with big data.

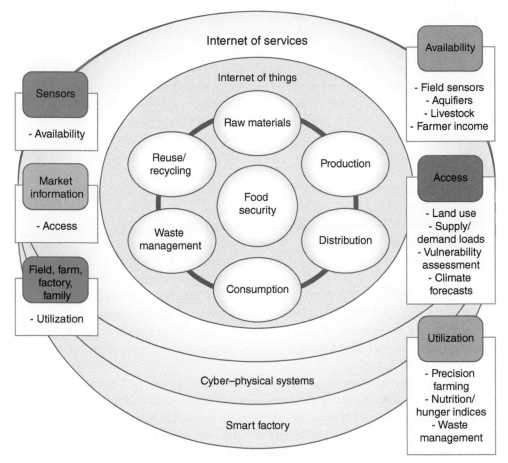

Figure 7.2 Big data framework for food security.
Source: From Ref. [23], © 2017 The Authors, published by Elsevier Ltd. under the CC BY license, adapted by Nick Rigas, D-Waste.

Another team created a big data-based architecture for construction waste analytics [26]. The architecture was validated for exploratory analytics of 200 000 waste disposal records from 900 completed projects. It was revealed that existing waste management software classify the bulk of construction waste as mixed waste, which exposes poor waste data management.

7.2.4 Mobile Phones Are the Key to Create New Big Data Sets

A special role is attributed in cellular phones and mobile apps in the creation of urban big data sets. Cellular phones are gradually becoming a mean of technological improvement that helps people to project their life in their surrounding environment.

This kind of technology is supplementing and even in some extreme cases "replacing" social lives of people [27]. It mirrors activities, desires, anxieties, trends, habits, beliefs, moods, etc. Moreover, it streams experiences outside each person's small world and projects them in the screen of a smart appliance, thus facilitating interventions in what is considered true, false and allowing the shaping of human desires and beliefs [28]. Cellular smartphones empowered by the relevant apps have the unique characteristic of connecting spatial data with time and information on thoughts, moods, activities, and status. This makes mobile phones get correlated with human behavior, becoming a mirror [29] of people's activities and attitudes that can be studied and predicted in a large scale, providing data that can be grouped and used for planning reasons that suit different human groups. Figure 7.3 shows the rapid expansion and the different dimensions of mobile phones and their usage.

More specifically, mobile apps can be very useful in identifying, registering, and addressing recycling problems in urban and rural areas, either in terms of problems in quality and quantity or in terms of recycling attitudes in every neighborhood. They can also be very useful to stimulate or track personal behavior and recycling performance, calculate the benefits involved, and create a better atmosphere that will support behavioral change [30]. Mobile apps can be used to create recycling networks and local context competitions in each and every neighborhood, promote waste exchange initiatives, and involve citizens in decision making regarding major waste management and recycling issues. This will also result in behavioral change since it will allow a more participatory approach to take place. Mobile apps can substantially improve both the individuals' behavior and the systemic performance of a waste management system, but the system must be in place in order to utilize the apps. Last but not least a successful mobile recycling application that delivered results in a certain community can't be replicated in another "audience" without considering the cultural, social, and historical evolution of the new audience. As it is widely recognized and understood, in the waste management community, the only successful way is to combine global principles and lessons learned with local conditions and expertise. The same is true for mobile applications dedicated to recycling and waste management [31].

A characteristic example of the broad use of mobile apps is a study on Brazilian informal recyclers, in which there was an effort to map the internal of special organization of waste collection of informal pickers using tracking devices [32]. Using mobile phones there has been an effort to map the special organization of waste collection and to develop software tools for coordination between waste pickers, clients, and planning operations. Through GPS tracking, web-based mapping, and mobile applications, informal pickers were supported with the aim to collect, manage, and interpret spatial data by themselves and to redesign their own system collaboratively with others. Despite of the difficulties and obstacles that the program faced, due to the particular characteristics of how informal workers work and think, results showed that the experiment was successful. Table 7.2 presents some of the

The world is becoming mobile

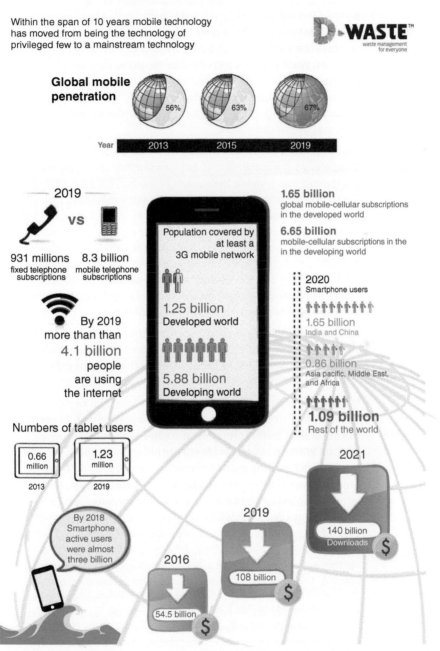

Within the span of 10 years mobile technology has moved from being the technology of privileged few to a mainstream technology

D-WASTE™
waste management for everyone

Global mobile penetration

Year 2013 — 56% | 2015 — 63% | 2019 — 67%

2019

VS

931 millions fixed telephone subscriptions | 8.3 billion mobile telephone subscriptions

By 2019 more than than **4.1 billion** people are using the internet

Numbers of tablet users

0.66 million — 2013 | 1.23 million — 2019

By 2018 Smartphone active users were almost three billion

Population covered by at least a 3G mobile network

1.25 billion Developed world

5.88 billion Developing world

1.65 billion global mobile-cellular subscriptions in the developed world

6.65 billion mobile-cellular subscriptions in the in the developing world

2020 Smartphone users

1.65 billion India and China

0.86 billion Asia pacific, Middle East, and Africa

1.09 billion Rest of the world

2021 — 140 billion Downloads $

2019 — 108 billion $

2016 — 54.5 billion $

Figure 7.3 Mobile phones and usage statistics – data sets by multiple sources compiled by D-Waste.

Source: Based on data sets by multiple sources compiled by D-Waste.

Table 7.2 Mobile phone technology solutions for informal recyclers.

Problems	Solutions
Lack of internal understanding on how they operate	By tracking automatically valuable spatial data about operations of informal pickers, those data can be utilized by the informal workers themselves or can be forwarded to formal recycling systems. Waste pickers can visualize mistakes they do when practicing waste collection, can improve their logistics, and can learn from their mistakes. Special platforms can be made particularly designed for waste pickers in order to get required information for their work
Lack of infrastructure and means for working	Mobile phones are simple in use, have a range of free, or are inexpensive applications that can be used instead of expensive equipment and provide the sense of freedom for actions, outside fixed formal boundaries that informal sector members trying to avoid. They offer applications like camera, maps, location tracker, information about places to drop of their materials, calculator of distance, temperature, access to digital information, scanning of barcodes, meteorological information, etc. They can exchange texts, pictures, and videos and have real-time communication, by phone, video, or instant messaging; it can offer coordination between partners, workers, or possible clients
Lack of public visibility of the service they offer to the community	Informal sector lacks respect regarding the services it provides to society, as its actions mostly remain hidden. Once its activities are mapped, it can help to the recognition of the sector's offering
Lack of trust between informal sector and private entities	Mobile tools can provide valuable data of informal sector activities to formal sector (public and private), as well as register the otherwise lost information and experience of the waste pickers. Waste pickers can identify problems and come up with solutions differently than the formal sector, based on their practical experiences. With a platform of coordination, and information exchange by using mobile technology, business and residents can work with the informal sector while making it easier for the waste pickers to organize themselves. Gathering and utilization of data can help different partners expand and optimize their work. Waste pickers can strengthen their position towards the municipality and support internal training of the new collectors. Also, hybrid systems with the formal systems can be effective, where informal and formal sector can interact, supplementing each other's gaps where necessary

Source: From Ref. [33], with kind permission from D-Waste.

benefits that informal recyclers can have from using proper mobile apps, as presented by D-Waste [33].

7.2.5 Predicting Pandemics

We will finish this list of different case studies with an example of the health sector. A team of researchers [34] combined information from Zika-related Google searches, Twitter microblogs, and the HealthMap digital surveillance system with historical Zika-suspected case counts to track and predict estimates of suspected weekly Zika cases during the 2015–2016 Latin American outbreak, up to three weeks ahead of the publication of official case data. Then, they evaluated the predictive power of these data and used a dynamic multivariable approach to retrospectively produce predictions of weekly suspected cases for five countries: Colombia, El Salvador, Honduras, Venezuela, and Martinique. Models that combined Google and Twitter data with autoregressive information showed the best out-of-sample predictive accuracy for one-week ahead predictions, whereas models that used only Google and Twitter typically performed best for two- and three-week ahead predictions. The authors concluded that "In the absence of access to real-time government-reported Zika case counts, we demonstrate the ability of Internet-based data sources to track the outbreak. Our model predictions fill a critical time-gap in existing Zika surveillance, given that early interventions and real-time surveillance are necessary to curb mosquito transmission. Official Zika case reports will likely continue to be delayed in their release; thus, it is important that health and government officials have access to real-time and future estimates of Zika activity in order to allocate resources according to potential changes in outbreak dynamics."

Now, we are ready to move to the dark side of the moon.

7.3 THE DARK SIDE OF THE MOON

We already mentioned social media as a resource for big data sets. We also referred to examples, from beef supply chain to Zika virus outbreak prediction and deeper understanding of urban life, flows, and metabolism. Now, it's time to face the hard truth that there is no free lunch out there. The same tools that provide the big data sets that can be utilized to understand better and deeper urbanization and urban metabolism can be and are actively used for a massive surveillance of the social media users. As Forbes [35] reported in November 2019, 9 in every 10 Internet users are being actively monitored online. And where this might have been done by armies of analysts in the past, it is now automated. Advances in artificial intelligence and pattern analytics have enabled billions of accounts to be watched in real time. A relevant report [36] that provided the input for the Forbes article stated that "While authoritarian powers like China and Russia have played an enormous role in dimming the prospects for technology to deliver greater human rights, the world's leading social media platforms are based in the United States, and their exploitation by antidemocratic forces is in large part a product of American neglect."

China's WeChat is a site for integrated social interaction. It is simultaneously a form of currency, a dating app, a tool for sporting teams and news provider: Twitter, Facebook, Google Maps, Tinder, and Apple Pay all combined into one. But according to a BBC blog [37], it is also an ever more powerful weapon of surveillance and social control for the Chinese government. Unfortunately, the problem is not restricted to China, but it concerns each and every social media user, each and every Internet user, and each and every mobile app user in the world, which means almost four billion people worldwide. The Cambridge Analytica scandal (in which Ted Cruz used psychological data based on research spanning tens of millions of Facebook users, harvested largely without their permission, for his electoral campaign) is a good example of what's happening. Facebook had allowed someone to extract vast amounts of private information about vast numbers of people from its system, and that entity had passed the data along to someone else, who had used it for political purposes [38].

Daniel Trottier who has studied in depth the social media practices and especially surveillance wrote [39] that "Facebook and other social media increasingly regulate social life. The way they collect, archive, and disseminate personal information is noteworthy for surveillance scholars. The Facebook profile has arguably overtaken the CCTV camera as the primary imagery for surveillance studies. Different surveillance models are manifest through Facebook. This suggests a complexity of social media surveillance. Understanding social media surveillance requires an understanding of the features that add to social media's volatility. Even when talking about one kind of surveillance, or one context, other contexts and practices are not far off. For instance, interpersonal visibility greatly augments state and institution-led surveillance."

Many public and private sector organizations use profiling to determine service levels for different customers [40]. Employers and universities use information from internal information systems and external social networking sites for selection and disciplinary purposes. Insurers and private healthcare providers use biographical and transactional data for checking claims and setting premiums. Law enforcement and intelligence agencies gather a wide range of data for prosecutions and counterterrorism investigations. In the United States, the United Kingdom, and Italy, even well-educated "digital native" university students have been found to have little idea about the potential consequences of sharing personal information on social networking sites. Another research [41], focusing on the oil industry, has demonstrated that while activists contest various sponsorship initiatives in the oil industry through the online mediation of protest activities, corporations monitor and in some cases silence specific individual activists in online media. This significantly impedes the potential of social media for granting visibility to counter-hegemonic discourses in the public sphere. Interestingly, Julian Assange has explained from 2011 [42] that Facebook is "the world's most comprehensive database about people, their relationships, their names, their addresses, their locations, their communications with each other, and their relatives, all sitting within the United States, all accessible to U.S. Intelligence." Assange added that it's not just Facebook, but Google and Yahoo as well as all other major US organizations that have developed built-in interfaces for US intelligence because it helps get around the costly and time-consuming serving of subpoenas.

7.3.1 Surveillance Capitalism?

The critiques regarding the use of the web as a tool for surveillance have started in early 1990s. Already in 1988, Clarke has proposed the term "dataveillance" [43] as the systematic monitoring of people's actions or communications through the application of information technology. In 1990, Poster wrote [44] that "Today's 'circuits of communication' and the databases they generate constitute a Super-panopticon, a system of surveillance without walls, windows, towers or guards." Other authors have argued [45] that there are two modes in the new digital surveillance: the unlimited monitoring independently of geographic borders and the active sorting of people on a real-time basis.

In the very interesting book *Internet and Surveillance: The Challenges of Web 2.0 and Social Media* [46], Fuchs makes things crystal clear: "Social media platforms are web-based platforms that predominantly support online social networking, online community-building, and maintenance, collaborative information production and sharing, and user-generated content production, diffusion, and consumption. ... it is clear that a principle that underlies such platforms is the massive provision and storage of personal data that are systematically evaluated, marketed, and used for targeting users with advertising." In addition, he mentioned that online advertising is a mechanism by which corporations exploit web users who form an Internet prosumer/producer commodity and are part of a surplus value generating class that produces the commons of society that are exploited by capital. Other authors suggest [47] that the new ubiquitous information and communication technologies, in particular recording-enabled smart devices and social media programs, are giving rise to a profound new power for ordinary people to monitor and track each other on a global scale. Along with this growing capacity to monitor one another is a new capacity to explicitly and publicly judge one another – to rate, rank, comment on, shame, and humiliate each other through the net. The result is that a new apparatus of surveillance and control is being generated that threatens individual freedom through a coercion of the will by an anonymous and interconnected crowd.

Probably the most comprehensive work on the aspects of web, social media, and surveillance is Shoshana Zuboff's emblematic book *The Age of Surveillance Capitalism: The Fight for a Human Future at the New Frontier of Power"* [48], a book that was selected by Barack Obama as one of the best books of 2019. Zuboff, a Charles Edward Wilson Professor Emerita in Harvard Business School, analyzes what she calls "information civilization." Building on the example of a smart thermostat, she explains how sensitive household and personal information are shared with other smart devices, unnamed personnel, and third parties for the purposes of predictive analysis, advertisement, and sales. Generalizing this approach, she goes one step further defining surveillance capitalism as a system that unilaterally claims human experience as a free raw material for translation into behavioral data. Even more, although some of these data are applied to product or service improvements, the rest are declared as proprietary behavioral surplus, fed into advanced manufacturing processes known as "machine intelligence," and fabricated into prediction products that anticipate what you will do now, soon, and later. Then, these prediction products are traded in a new kind of marketplace that Zuboff calls "behavioral futures markets." She claims that "automated machine processes not only know our behavior but also shape our behavior at scale. With this reorientation from knowledge to power, it is no longer enough to automate information flows about us; the goal now is to automate us." She also structures another important argument that surveillance capitalism's products and services are not the objects of a value exchange. They are just the "hooks" that lure users into their extractive operations in which our personal experiences are scraped and packaged as the means to others' ends. She concludes that "We are not surveillance capitalism 'customers' …We are the sources of surveillance capitalism's crucial surplus: the objects of a technologically advanced and increasingly inescapable raw-material-extraction operation."

It is obvious now that the huge opportunities to advance urban sustainability using big data sets are the other side of the continuous profiling, population sorting, real-time monitoring, and finally massive surveillance that is in the core of the business models of the web business. There is nothing surprising in that it has become a common sense, although we are just starting to understand that probably it creates a deeper transformation of power and governance patterns for which we are not prepared at all. But why it took us almost 20 years to understand the threat involved? Zuboff provides a good answer. The main problem is that surveillance capitalism is unprecedented, and it was impossible to understand it using the lenses of our experiences, as it was exactly the case of the people who first show a car and named it a "horseless carriage." She writes: "This is how the unprecedented reliably confounds understanding: existing lenses illuminate the familiar, thus obscuring the original by turning the unprecedented into an extension of the past. This contributes to the normalization of the abnormal."

A deeper discussion on the opportunities opened by big data systems can't bypass the serious questions on power, inequality, marginalization, and market abuse of big data sets. Still, the raise of the new science of urban sustainability is still possible, and every day that passes it becomes more necessary.

7.4 FROM CITIES AS MACHINES TO CITIES AS ORGANISMS

As our world becomes rapidly urbanized, new problems and challenges appear that stimulate a new wave of questions that can't be answered using the traditional understanding of cities. Do fundamental urban processes exist that generate a basic kind of city, in whatever context or culture or time period? Can we develop quantitative models that simulate urbanization and urban metabolism? Can we simulate a city and predict its environmental and health performance and impacts? Or maybe the expressions of urbanism are too diverse to be included in a single model. Is the urbanization of our era simply an elaboration upon past urbanization processes, or is it a fundamentally different kind of process [49]? It is this type of questions that stimulate new approaches and concepts in our understanding of cities. In addition, the direct relationship between urbanization, health, and environmental protection creates an urgency to develop predictive models that will allow us qualitative and quantitative forecasts regarding sustainability, pollution, and biodiversity.

Michael Batty has described the current shift of our understanding of cities as a transition from thinking of "cities as machines" to "cities as organisms" [50]. Batty has pioneered to our understanding of cities as complex adaptive systems explaining that cities do not exist in perfect conditions and benign environments, they are always in a continuous interaction with their surroundings and the world, they are always far from equilibrium, and they can't be centrally planned but evolved. Thus, cities are forever changing entities, being closer to biological rather than mechanical systems, and thus, they should be studied and understood using the advances of complexity and systems theory. A major concept that goes through all the relevant work is that location is no longer the key to explaining how cities function – it is interactions. Cities are better understood by the nonlinear dynamic interactions of their many constituent parts. Batty has advocated [51] that we need a new kind of science to understand cities as complex systems of networks and flows between people, goods, resources, and institutions. Networks and flows are the key to understand and design cities, not locations.

In an emblematic 2007 paper [52], Luis Bettencourt and his team presented empirical evidence indicating that the processes relating urbanization to economic development and knowledge creation are very general, being shared by all cities belonging to the same urban system and sustained across different nations and times. Many diverse properties of cities from patent production and personal income to electrical cable length are shown to be power law functions of population size with scaling exponents that fall into distinct universality classes. In addition, Bettencourt and his team predicted that the pace of social life in the city increases with population size, in quantitative agreement with data. As it was written at the discussion part of this paper, "Despite the enormous complexity and diversity of human behavior and extraordinary geographic variability, we have shown that cities belonging to the same urban system obey pervasive scaling relations with population size, characterizing rates of innovation, wealth creation, patterns of consumption and human behavior as well as properties of urban infrastructure. Most of these indicators deal with temporal processes associated with the social dimension of cities as spaces for intense interaction across the spectrum of human activities. It is remarkable that it is principally in terms of these rhythms that cities are self-similar organizations, indicating a universality of human social dynamics, despite enormous variability in urban form." The analysis of the results demonstrated that the unique social interaction in cities can transcend biology and thus redefine "urban metabolism." It was found that most urban indicators (on a per capita basis) scale with the population (city size), so an increase in innovation patents per capita goes hand in hand with the increase of crime or pollution per capita and the relevant decrease of the infrastructure required. Few years later, in another paper [53], Bettencourt explained that "population size plays a fundamental role… the systematic variation of the properties of cities with population size reveals the ways in which cities result in more than the simple agglomerations of people. This is the phenomenon that anthropologist Carneiro described as quality from quantity in his studies of the emergence of organizational forms in small human societies. Scaling laws for cities show systematic effects of spatial densification, temporal acceleration and socioeconomic diversification, that have long been discussed in the social sciences, but that can only now start be appreciated for their quantitative generality. In particular two general aspects of the scaling properties of urban indicators appear systematically across time, and in different urban systems: i) economies of scale in urban material infrastructure and ii) increasing returns in socioeconomic productivity." Now, it's time to focus more on scaling.

7.4.1 Urban Theories and the Role of Scaling

Urban scaling theory draws on different insights, from urban geography, urban economics, and spatial and regional sciences. According to the urban scaling theory, the population size determines the socioeconomic development of a city, thus the pace of change and life. The words of the famous architectural historian Kostof [54] "Cities are places where a certain energized crowding of people takes place" describe exactly the meaning of urban scaling, as the more the population, the more intense, diversified, and dense the social interaction. Scaling relationships in cities are important because they inform us about the way city size and location are constrained by geometry. Geometrical constraints also determine shape, and these in turn relate to issues such as energy use, density of occupation, and circulation in buildings and streets [55].

Urban scaling views cities as integrated socioeconomic networks of interactions embedded in physical space and then derives specific quantitative predictions about the relationship between population size, material output, and areal extent of settlements. The social interactions facilitated by physical proximity – and the lower costs associated with such interactions – drive productivity. As it has been noticed

[49], these relationships (between population size and socioeconomic parameters) have been statistically investigated not only for modern metropolitan systems (in the United States, Western Europe, Brazil, Japan, India, China, and South Africa) but also for Native American farming villages in North America (before the arrival of the Europeans), pre-Hispanic Andean and Central Mexican settlements, ancient Greek and Roman cities, medieval European cities and towns, and cities in Tudor England. The results show a striking similarity in scaling relationships.

In a very recent report [56] titled "Urban Science: Integrated Theory from the First Cities to Sustainable Metropolises," nine important lessons from the evolution of urban science are highlighted. As they are reflecting the state of the art of the advances in urban science, it is useful to recap them:

- The use of urban scaling in the analysis of cities quantifies many of their fundamental general characteristics [57], especially their capacity to create interrelated economies of scale in infrastructure and increasing returns to scale in socioeconomic activities [58]. There are empirical patterns – regarding the relationships among population size, areal extent, productivity, and spatial form – that are common to cities and urban systems across time.

- Despite the long-standing metaphor of the "city as an organism," in many important aspects, cities are very different from biological entities, especially in their capacity to exhibit increasing returns to scale when output increases by more than the proportional change in inputs.

- Urban systems are not like engineered systems, but they should be considered as "complex adaptive systems" where distributed growth and interaction between cities generate emergent properties affecting the development of individual cities in a feedback loop.

- Despite the continuous diffusion of innovation waves from one city to others, cities unevenly adapt their structure to innovation waves [59]; some cities lose their leadership in sectors, and others take advantage of the emergence of new innovative activities.

- Many urban phenomena are network phenomena. For many urban scientists the essential feature of urban areas is interactions – among individuals, households, businesses, economic sectors, and cities. The mathematics of network models have emerged as a ready-made language with which to describe and examine urban interactions across scale. To understand cities, we must view them not simply as places in space but as interactions and flows (of people, goods, and information); to understand flows, we must understand networks – the relations between objects that compose the system of the city.

- Human settlements constitute one of the few ecosystems that are significantly increasing in their extent. Urbanist have come to realize that the functioning of cities has many effects on larger regions, global resources, and human well-being. Urban ecosystems are fundamentally different [60] from their natural counterparts in the dominant influence of human actions, both intentional and unintentional, on ecosystem functions. Urban ecology seeks to understand how urban settings shape the socioecological interactions within ecological systems and their role as both drivers and responders to environmental change. Cities are not only driving changes in fundamental ecological processes; emerging evidence shows that cities are driving rapid evolutionary change with significant implications for human well-being on a contemporary scale [61].

- The impacts of cities on remote areas ("hinterlands") have seldom been considered in urban sustainability assessments, because urban areas are bounded and predefined for the purpose of spatial planning. But urbanization and land change are two global processes with far-reaching spatial consequences – urban land teleconnections [62] is a conceptual framework that explicitly links land use and environmental changes to underlying urbanization dynamics. It links places and regions that although far removed geographically are connected through the various environmental consequences of urban expansion. Urbanization is driven by local processes with global consequences.

7.4.2 From Urban Scaling to Urban Sustainability

Geoffrey West, a theoretical physicist and past president of the Santa Fe Institute, studied the origin of universal scaling laws that pervade biology from the molecular genomic scale-up through mitochondria and cells to whole organisms and ecosystems. This led to the development of realistic quantitative models for the structural and functional design of organisms based on underlying universal principles. Interestingly, in 2010, he published an article [63] with Luis Bettencourt, in which they identified three aspects that are straightforward dependent (scale) with the size of the city in population. First, the space required per capita shrinks, thanks to denser settlement and a more intense use of infrastructure. Second, the pace of all socioeconomic activity accelerates, leading to higher productivity. Third, economic and social activities diversify and become more interdependent, resulting in new forms of economic specialization and cultural expression. What it means in practice? For example, doubling the population of any city requires only about an 85% increase in infrastructure, whether that be total road surface, length of electrical cables, water pipes, or number of petrol stations. This systematic 15% savings happens because, in general, creating and operating the same infrastructure at higher densities is more efficient, more economically viable, and often leads to higher-quality services and solutions that are impossible in smaller places. Interestingly, they found out that similar savings exist in carbon footprints too, which means most large, developed cities are "greener" than their national average in terms of per capita carbon emissions, although this was not checked for cities undergoing extremely rapid development, as in China or India, where data are poor or lacking. As they wrote:

> The bigger the city, the more the average citizen owns, produces and consumes, whether goods, resources or ideas. On average, as city size increases, per capita socio-economic quantities such as wages, GDP, number of patents produced, and number of educational and research institutions all increase by approximately 15% more than the expected linear growth. There is, however, a dark side: negative metrics including crime, traffic congestion and incidence of certain diseases all increase following the same 15% rule. The good, the bad and the ugly come as an integrated, predictable, package. Our work shows that, despite appearances, cities are approximately scaled versions of one another: New York and Tokyo are, to a surprising and predictable degree, nonlinearly scaled-up versions of San Francisco in California or Nagoya in Japan. These extraordinary regularities open a window on underlying mechanism, dynamics and structure common to all cities. [63]

But West went even further, and in 2017 he published his monumental book *Scale: The Universal Laws of Life, Growth, and Death in Organisms, Cities, and Companies* [64]. In this book, West explains that scaling is related to the structural units of cities (and not only). These structural units are the evolving networks of people, material, and energy flows that constitute a city. He identifies three core common characteristics of the hierarchical networks that deliver energy to organisms – whether the diverse circulatory systems that power all forms of animal life or the water and electrical networks that power cities. First, the networks are "space filling," that is, they service the entire organism. Second, the terminal units are largely identical, whether they are the capillaries in our bodies or the faucets and electrical outlets in our homes. Third, a kind of natural selection process operates within these networks so that they are optimized. In chapter 7, paragraph 2 of his book, West puts a question: "So what is the common unifying factor that transcends these differences and underlies this surprising structural and dynamical similarity (of cities)? I have already strongly hinted at the answer: the great commonality is the universality of social network structures across the globe. Cities are people, and to a large extent people are pretty much the same all over the world in how they interact with one another and how they cluster to form groups and communities. We may look different, dress differently, speak different languages, and have different belief systems, but to a large extent our biological and social organization and dynamics are remarkably similar. After all, we are all human beings sharing pretty much the same genes and the same generic social history. And no matter where we live on the planet, all of us emerged only relatively recently from being mobile hunter-gatherers to becoming predominantly sedentary communal creatures. The underlying commonality that is being expressed by the surprising universality of urban scaling laws is that the structure and dynamics of human social networks are very much the same everywhere."

Few paragraphs later he provides a very comprehensive definition about cities: "A city is not just the aggregate sum of its roads, buildings, pipes, and wires that comprise its physical infrastructure, nor is it just the cumulative sum of the lives and interactions of all of its citizens, but rather it's the amalgamation of all of these into a vibrant, multidimensional living entity. A city is an emergent complex adaptive system resulting from the integration of the flows of energy and resources that sustain and grow both its physical infrastructure and its inhabitants with the flows and exchange of information in the social networks that interconnect all of its citizenry. The integration and interplay of these two very different networks magically gives rise to increasing economies of scale in its physical infrastructure and simultaneously to a disproportionate increase in social activity, innovation, and economic output."

There are many cases where a scientific discovery or revolution was fueled by new types of data and new empirical phenomena that provided new insights. "Measurement has played a central role in the development of our understanding of the entire universe around us. Data provide the basis for constructing, testing, and refining our theories and models whether they seek to explain the origins of the universe, the nature of evolutionary processes, or the growth of the economy. Data are the very lifeblood of science, technology, and engineering, and in more recent years have begun to play an increasingly central role in economics, finance, politics, and business" [64]. During the last 20 years, the efforts we described towards a new science on cities were taking place together with the biggest urbanization wave in human history and the big data revolution. As it becomes obvious, the huge new data sets that we described earlier in this chapter are the fuel required to stimulate the rise of the new science on cities. And as this new science of cities is based on complexity and systems theory, there is not only a need for more data but for better in-depth analysis. This is where artificial intelligence advances can help.

7.4.3 Big Data Meets the New Urban Science

We think that now is more than clear that if we combine the emerging new science of cities with the big data revolution, we certainly hope that a new science of urban sustainability can rise. Big data can help us in revisiting classical problems and introduce complexity theory and urban scaling in predictive models. But what we really need is much more than that. As Geoffrey West put it in his book, in chapter 10 [64], we need a grand unified theory of (urban) sustainability:

> Almost all existing approaches to the challenge of global sustainability focus on relatively specific issues, such as the environmental consequences of future energy sources, the economic consequences of climate change, and the social impact of future energy and environmental choices. While such focused studies are of obvious importance and where most of our research efforts should be directed, they are not sufficient. They focus primarily on the trees and risk missing the forest. It's time to recognize that a broad, multidisciplinary, multi-institutional, multinational initiative, guided by a broader, more integrated and unified perspective, should be playing a central role in guiding our scientific agenda in addressing

this issue and informing policy. We need a broad and more integrated scientific framework that encompasses a quantitative, predictive, mechanistic theory for understanding the relationship between human-engineered systems, both social and physical, and the "natural" environment – a framework I call a grand unified theory of sustainability.

Towards this direction, researchers [65] used detailed census data to construct sustainable development indices in hundreds of thousands of neighborhoods and show that their statistics are scale dependent and point to the critical role of large cities in creating higher average incomes and greater access to services within their national context. The results shown that several central features of sustainable development are intimately connected to agglomeration and heterogeneity effects in cities and how the scale of spatial units of analyses matters to measure inequality and inform more effective assessments and policy. Based on empirical analysis using data from Brazil and South Africa, the researchers demonstrated that despite environmental challenges, larger cities tend to exhibit greater economic productivity and expanded access to housing and urban services, which means that they host nucleate solutions for sustainable development within their nations. Second, they found that large inequality typically results as improvements are initially expanded. Such patterns of inequality are scale dependent, being larger between different neighborhoods within metropolitan areas. However, high levels of inequality are not inevitable, and several places offer counterexamples of development with low inequality. They also observed that stronger heterogeneities are associated with more spatially concentrated outcomes at characteristic scales of about a kilometer.

While this type of research shows the way for the further evolution of the new urban science, it is important to notice that all the challenges we already mentioned in this chapter about big data (availability of data sets, ownership and access, uniformity, political biases, capacity to collect and analyze big data, etc.) pose a huge barrier. The needs for IT infrastructure to manage big data sets of different formats, in high volume and real time, should not be underestimated as they require substantial investments and know-how. The preparation of data sets in order to be integrated and combined requires significant scientific resources and concrete methodologies that are not yet available and standardized.

All types of observational big data pose limitations in deriving theoretical insights and in hypothesis generation without adequate cross-fertilization of knowledge between the data sciences and the urban disciplines, but the challenges are greater with certain forms of UGC and sensor data that yield high-value descriptions but are less amenable to explanations and explorations of causality [1]. In addition, there is another, and distinct, source of new urban data that shares some of the techniques reviewed above but remains largely unavailable for scientists and citizens to analyze. This is the proprietary commercial data gathered by large tech giants such as Google or Tencent, suppliers of geodata such as ESRI, and lesser known dedicated surveillance data companies. As it was highlighted [56], "These data are driving the 'big urban data' revolution – which is not only about the sheer quantity of data but also about the individual-level behaviors captured by such data – and supporting scientific publications. But the myriad practical and ethical issues of turning urban data into a commodity, the impediments that proprietary data present to reproducibility, the potential selection bias introduced when research is channeled by the availability of data collected for profit and the use of geospatial data for control and even repression should give urban scientists pause and must be seriously considered."

We will close this chapter presenting a great example for the limitations of big data and the need to combine it with the lens of theory. West points out in chapter 10 of his book [64] that the discovery of the Higgs particle was a result of an experiment that was generating 150 EB of data a day, an equivalent to one billion TV movies of 1 GB each. Another measure is more revealing of the huge size of the data set: the data generated in one day by the experiment in the Large Hadron Collider (LHC) that was built at CERN in Geneva was 50 times bigger that the size of the data by all of the world's computers and other IT devices taken together! West writes:

> Obviously this means that the strategy of naively letting the data speak for themselves by devising machine-learning algorithms to search for correlations that would eventually lead to the discovery of the Higgs mechanism is futile. Even if the machine produced a million times less data it is extremely unlikely this strategy could succeed. How, then, did physicists discover the proverbial needle in this mammoth haystack? The point is that we have a well-developed, well-understood, well-tested conceptual framework and mathematical theory that guides us in where to look. It tells us that almost all of the debris resulting from almost all of the collisions are actually uninteresting or irrelevant as far as searching for the Higgs particle is concerned. In fact, it tells us that out of the approximately 600 million collisions occurring every second, only about 100 are of interest, representing only about 0.00001 percent of the entire data stream. It was by devising sophisticated algorithms for focusing on only this very special tiny subset of the data that the Higgs was eventually discovered.

> The lesson is clear: neither science nor data are democratic. Science is meritocratic and not all data are equal. Depending on what you are looking for or investigating, theory resulting from the traditional methodology of scientific investigation, whether highly developed and quantitative as in the case of fundamental physics, or relatively undeveloped and qualitative as in the case of much of social science, is an essential guide. It is a hugely powerful constraint in limiting search space, sharpening questions, and understanding answers. The more one can bring big data into the enterprise the better, provided it is constrained by a bigger-picture conceptual framework that, in particular, can be used to judge the relevance of correlations and their relationship to mechanistic causation.

An interesting review of West's book in *New York Times* [66] was closing as follows: "In the 16th century, Francois Rabelais, a French scholar, admonished that 'science without conscience is the ruin of the soul'. Mr. West's warning that big data without a theoretical framework is the ruin of science is an important contemporary corollary caution that *Scale* (West's book) will hopefully establish for the next generation of scholars."

Let's see now how the discussion on the rise and the need for a new urban science and the big data revolution is straightforward connected with the implementation of circular economy.

7.5 CIRCULAR ECONOMY: DIGITIZED OR DEAD

In this section, we will argue that the shift to circular economy, on a large scale capable to create a worldwide positive footprint, will not be possible without two main conditions:

- The development of the new science of cities that was already described in Section 7.4,
- The digitalization of the circular loops – in other terms, circular economy either will be fully digitized or will never prevail.

The first condition, the development of a new science of cities, seems rather obvious. If we want to stimulate the shift to circular economy, the level of supply chains, even global ones, is not enough. Cities are the real metabolists of energy, food, raw materials, and any other resource, including the human ones. Without a new theory that will stimulate the unified understanding of sustainability in cities instead of the fragmented thematic approaches, we will not be able to develop the business and governance models required to deliver closed loops, at least for the most important materials. Instead, there is a high risk that while we will do efforts and spend resources to optimize one or another thematic area (e.g. improve water access or reduce air pollution), uncontrolled economic growth or the problems in waste management or food chain will undermine the overall sustainability. This is not theoretical; it is already happening in many developing countries of the world where the economic growth of the last 20 years has really undermined the environmental quality and creates very serious health impacts. It will be easy to refer to one of the thousands examples of the pollution incidents and horrific problems that have been created as developing countries are growing economically, as the infamous air pollution in New Delhi [67]. However, even more important is to understand that chemical pollution threatens the rich world as well, in ways that usually are not expected. Take the example of drinking water in the United States. A recent report by the Environmental Working Group [68], based on unpublished data of the US Environmental Protection Agency, found that at least 110 million Americans may already be contaminated by chemical like per- and polyfluoroalkyl substances (PFAs), which are linked with cancer, liver damages, and low birth weight! PFAs are found in packaged food and several household products (like Teflon, paints, polishes, and cleaning products), carpet, leather and apparel, textiles, paper and packaging, coatings, rubber, and plastics.

So, the need for a new unified theory on urban sustainability is a global one. Without it, circular economy restricted to supply chains (and even worst ignoring the problems of chemical pollution as we explained in Chapters 3 and 4) will only be a partial effort, if not an alibi, that will not create the systemic change required. In addition, as we have already discussed, circular loops, in many cases, will cross countries' and municipal borders and each circular loop will create its own geographical footprint, connecting different locations on a regional, national, or even global scale. Without a unified theory of sustainability, these closed loops will remain fragmented pieces of a puzzle that no one knows its global picture.

The second condition requires some more details. The fact is that circular economy, besides extra energy to close the loops, requires a lot of extra information to connect the dots of the different life cycle phases of a material or a product. As it was mentioned, the lack of feedback information and performance indexes does not allow the measurement of how a product, a process, or a company is really transiting from linear to circular models, and this critical issue limits the further development of other circular activities [69]. Data discrepancies, gaps, and confidentiality issues hamper the measurement of circularity and the application of the relevant models to a cross-industry network. A proper information flow throughout the supply chain enables companies and all the stakeholders to share and develop sustainability capabilities and build competitive advantages.

The new dimensions of product development and reuse concept in circular economy would also significantly increase the operational facility and network complexity. In this line, the massive data generated during the manufacturing, use, and disposal fits the characteristics of big data and requires artificial intelligence to spot the trends and optimize the logistics and asset management [70]. Utilizing big data analytics is a fundamental basis for informed and data-driven decision making in supply chain networks in the framework of circular economy, while stakeholders' collaborative approaches among all supply chain members are a major precondition. However, this requires holistic information processing and sharing along the entire supply chain network that can effectively create a basis for achieving the triple bottom line of economic, ecological, and social benefits [71].

In a very interesting paper [72], researchers mentioned that although quantitative tools, metrics, and indices have been developed to support the decision-making process, these approaches are most easily applied to either single corporates or single supply chain systems. In such cases, decisions are made by a single corporate entity or a coalition of corporates with a clear history of cooperation and mutual trust because of direct supply chain relationships. Furthermore, although these principles can be applied to different corporates in the supply chain or cross-industry networks, such efforts are often hampered by data discrepancies, gaps, and confidentiality issues. Consequently, it becomes obvious that these issues prevent the optimization of opportunities for sustainability enhancement, which naturally leads to the question of whether emerging technologies can be used to provide the necessary solutions.

As an illustrative example, consider the smart waste management system model that was proposed for a city by the authors of the paper "The future of waste management in smart and sustainable cities: A review and concept paper" [73]. This model consists of three elements:

Element 1: Infrastructure for the collection of product life cycle data.

Element 2: Connected and involved citizens for sharing products and services to avoid waste generation and facilitate the adoption of novel business models with the aim of waste prevention, and value creation.

Element 3: Intelligent and sensor-based infrastructure for proper upstream separation and on-time collection of waste when a product reaches its end of life.

It is obvious that such a system requires huge amounts of data from many different sources and in many different formats; thus either it will be a fully digitized big data system, or it will not work. As the authors say, "Overall, implementing the proposed framework requires a high degree of collaboration among different stakeholders involved in the entire product value chain. While developing an integrated database might not be an issue by itself, convincing different players to invest efforts in on-time data collection is a challenge. Therefore, not only different players need to realize the business and sustainability outcomes of such platform but extensive efforts are needed to alleviate political, legal, and commercial barriers towards this integrated process. However, as we move forward to an Industry 4.0 era, it is expected that cloud-based business models are well regulated and better equipped with strategies for handling legal and commercial barriers such as intellectual property and data security."

We believe that it now becomes obvious that digitalization is a basic condition that can boost the transformation towards a more sustainable circular economy. It can help in closing the material loops by providing accurate information on the availability, location, and condition of products. Digitalization also enables more efficient processes in companies, helps minimize waste, promotes longer life for products, and minimizes the transaction costs. Thus, digitalization boosts the circular economy business models by helping to close the loop, slow the material loop, and narrow the loop with increased resource efficiency [74].

Key Take-Outs of Chapter 7

Key take-outs	Why is it important?
A new plethora of big data sets are gradually becoming available from many different sources (apps, sensors, IoT, user-generated content, social media) and in many different formats (photos, videos, texts, unstructured data points, etc.)	Combining the new big data sets with traditional ones, and supported by artificial intelligence analytics, we are capable to monitor, understand, model, and forecast the urban metabolism in a way that was never possible before. This perspective is highlighted by the term urban informatics

Key take-outs	Why is it important?
There are several barriers for utilizing the big data sets, and they include their ownership, their availability, the required high capacities in computing power, and the fractured data taxonomies and preparation. There is a need for a global cooperation and standardization to stimulate better and broader use of big data sets	Overcoming these barriers requires substantial political changes and will, but this will drive the rise of new patterns of urban governance and citizen involvement. However, the role of political theories and ideologies, potential biases based on access to the web and institutional development will always be crucial in the way big data are managed
The billions of smartphones are a new, very important generator of new big data sets that allow a deeper, real-time, and spatially granulated understanding of the daily patterns of city life	The access to these data sets is seriously restricted by the big software and telecommunication companies, and this requires legal and institutional measures to allow their use for scientific research purposes
The same tools that provide the big data sets that can be utilized to understand better and deeper urbanization and urban metabolism can be and are actively used for a commercially driven massive – and individual – monitoring of 90% of the users	The deeper discussion on the opportunities opened by big data systems can't bypass the continuous profiling, population sorting, real-time monitoring, and finally massive surveillance that is in the core of the business models of the web business
There is an ongoing discussion that drives the rise of a new urban science that will face cities as an emergent complex adaptive system resulting from the integration of the flows of energy and resources that sustain and grow both its physical infrastructure and its inhabitants with the flows and exchange of information in the social networks that interconnect all of its citizenry	The rise of this new urban science is actually fueled by the big data revolution, opening the way for the development of a new grand unified theory of urban sustainability that will integrate economic, environmental, and social aspects. Urban scaling is a very important tool that can utilize big data and stimulate predictive models for better governance and planning
The big data revolution creates an ocean of data that must be transformed to meaningful islands of information	No matter how important the advances of artificial intelligence are, such a transformation requires the guidance from theoretical models and concepts to avoid mechanistic causation
The implementation of a circular economy on a massive scale that will create a positive environmental and health impact requires (i) the development of the new unified theory on urban sustainability and (ii) full digitalization and big data systems	Without these conditions, circular economy will only be a partial effort, if not an alibi, that will not create the systemic change required

REFERENCES

1. Thakuriah, P., Tilahun, N., and Zellner, M. (eds.) (2017). Big data and urban informatics: innovations and challenges to urban planning and knowledge discovery. In: *Seeing Cities Through Big Data*, 11–45. Cham: Springer International Publishing.
2. Gantz, J. and Reinsel, D. (2012). *The Digital Universe in 2020: Big Data, Bigger Digital Shadows, and Biggest Growth in the Far East*. IDC.
3. MacGillivray, C. and Reinsel, D. (2019). Worldwide global datasphere IoT device and data forecast, 2019–2023. https://www.idc.com/getdoc.jsp?containerId=US45066919 (accessed 3 May 2020).
4. Bunn, J. (2011). How big is a petabyte, exabyte, zettabyte, or a yottabyte? High scalability. http://highscalability.com/blog/2012/9/11/how-big-is-a-petabyte-exabyte-zettabyte-or-a-yottabyte.html (accessed 30 January 2020).
5. Chojecki, P. (2019). How big is big data? https://towardsdatascience.com/how-big-is-big-data-3fb14d5351ba (accessed 30 January 2020).
6. Tozzi, C. (2017). How big is big data? A look at the definition of big data and examples of it in action. https://blog.syncsort.com/2017/07/big-data/how-big-is-big-data-definition-examples (accessed 30 January 2020).
7. Hashem, I.A.T., Chang, V., Anuar, N.B. et al. (2016). The role of big data in smart city. *International Journal of Information Management 36* (5): 748–758.
8. Gibbs, D. and Krueger, R. (2007). Containing the contradictions of rapid development? New economy spaces and sustainable urban development. In: *The Sustainable Development Paradox: Urban Political Economy in the United States and Europe*, 95–122. New York: The Guilford Press.
9. Ahvenniemi, H., Huovila, A., Pinto-Seppä, I., and Airaksinen, M. (2017). What are the differences between sustainable and smart cities? *Cities 60*: 234–245.
10. Evans, J., Karvonen, A., Luque-Ayala, A. et al. (2019). Smart and sustainable cities? Pipedreams, practicalities and possibilities. *Local Environment 24* (7): 557–564.
11. Computer Hope (2019). How many sheets, rows, and columns can a spreadsheet have? https://www.computerhope.com/issues/ch000357.htm (accessed 30 January 2020).
12. Batty, M. (2013). Big data, smart cities and city planning. *Dialogues in Human Geography 3* (3): 274–279.
13. Mavropoulos, A., Purchase, D., Nitzsche, G., and Ramola, A. (2019). *How IND4.0 Transforms the Waste Sector*, 90. Vienna, Austria: International Solid Waste Association ISWA.
14. Thakuriah, P. and Geers, G. (2013). *Transportation and Information: Trends in Technology and Policy*. New York: Springer.
15. Nam, T. and Pardo, T.A. (2011). Conceptualizing smart city with dimensions of technology, people, and institutions. *Proceedings of the 12th Annual International Digital Government Research Conference on Digital Government Innovation in Challenging Times - dg.o '11*. College Park, MA: ACM Press, p. 282. https://doi.org/10.1145/2037556.2037602.

16. OECD Global Science Forum (2013). *New Data for Understanding the Human Condition: International Perspectives. Report on Data and Research Infrastructure for the Social Sciences*. Paris, France: OECD.
17. Jennings, P. (2010). Managing the risks of smarter planet solutions. *IBM Journal of Research and Development 54* (4): 1–9.
18. Calabrese, F., Colonna, M., Lovisolo, P. et al. (2011). Real-time urban monitoring using cell phones: a case study in Rome. *IEEE Transaction on Intelligent Transportation System 12* (1): 141–151.
19. Girardin, F., Calabrese, F., Fiore, F.D. et al. (2008). Digital footprinting: uncovering tourists with user-generated content. *IEEE Pervasive Computing 7* (4): 36–43.
20. Zhong, C., Schläpfer, M., Müller Arisona, S. et al. (2017). Revealing centrality in the spatial structure of cities from human activity patterns. *Urban Studies 54* (2): 437–455.
21. Marti, P., Serrano-Estrada, L., and Nolasco-Cirugeda, A. (2017). Morphological and functional attributes of preferred urban public spaces in Alicante province. *City and Territory in the Globalization Age Conference Proceedings*, Valencia, Universitat Politècnica de València (27–29 September 2017), pp. 1147–1155.
22. Mishra, N. and Singh, A. (2018). Use of Twitter data for waste minimisation in beef supply chain. *Annals of Operations Research 270* (1–2): 337–359.
23. Irani, Z., Sharif, A.M., Lee, H. et al. (2018). Managing food security through food waste and loss: small data to big data. *Computers and Operations Research 98*: 367–383.
24. Lu, W., Webster, C., Peng, Y. et al. (2018). Big data in construction waste management: prospects and challenges. *Detritus 1*: 129–139.
25. Lu, W., Chen, X., Ho, D.C.W., and Wang, H. (2016). Analysis of the construction waste management performance in Hong Kong: the public and private sectors compared using big data. *Journal of Cleaner Production 112*: 521–531.
26. Bilal, M., Oyedele, L.O., Akinade, O.O. et al. (2016). Big data architecture for construction waste analytics (CWA): a conceptual framework. *Journal of Building Engineering 6*: 144–156.
27. Zhang, L., Pentina, I., and Kirk, W.F. (2017). Who uses mobile apps to meet strangers: the roles of core traits and surface characteristics. *Journal of Information Privacy and Security 13* (4): 207–225.
28. Hsiao, C.-H., Chang, J.-J., and Tang, K.-Y. (2016). Exploring the influential factors in continuance usage of mobile social apps: satisfaction, habit, and customer value perspectives. *Telematics and Informatics 33* (2): 342–355.
29. Giordano, C., Salerno, L., Pavia, L. et al. (2020). Magic mirror on the wall: selfie-related behavior as mediator of the relationship between narcissism and problematic smartphone use. *Clinical Neuropsychiatry 16* (5–6): 197–205.
30. Jog, Y., Venkatesh, R., Pandit, A., and Bhadauria, R.S. (2017). Understanding role of mobile apps in smart city services. *IJUNESST 10* (4): 27–38.

31. Mavropoulos, A., Tsakona, M., and Anthouli, A. (2015). Urban waste management and the mobile challenge. *Waste Management and Research 33* (4): 381–387.

32. Offenhuber, D. and Lee, D. (2012). Putting the informal on the map: tools for participatory waste management. *PDC '12: Proceedings of the 12th Participatory Design Conference: Exploratory Papers, Workshop Descriptions, Industry Cases – Volume 2* (August 2012), p. 4. https://doi.org/10.1145/2348144.2348150.

33. D-Waste (2014). *White Paper: Mobile Applications & Waste Management: Recycling, Personal Behavor, Logistics*, 44. Athens, Greece: D-Waste.

34. McGough, S.F., Brownstein, J.S., Hawkins, J.B., and Santillana, M. (2017). Forecasting Zika incidence in the 2016 Latin America outbreak combining traditional disease surveillance with search, social media, and news report data. *PLOS Neglected Tropical Diseases 11* (1): e0005295.

35. Doffman, Z. (2019). Your social media is (probably) being watched right now, says new surveillance report. https://www.forbes.com/sites/zakdoffman/2019/11/06/new-government-spy-report-your-social-media-is-probably-being-watched-right-now (accessed 3 February 2020).

36. Shahbaz, A. and Funk, A. (2019). *Freedom on the Net 2019 The Crisis of Social Media*, vol. *32*. Washington, DC: Freedom House.

37. McDonell, S. (2019). WeChat and the surveillance state. *BBC News* (7 June 2019).

38. Wong, J.C. (2019). The Cambridge analytica scandal changed the world: but it didn't change Facebook. *The Guardian* (18 March 2019).

39. Trottier, D. (2011). A research agenda for social media surveillance. *Fast Capitalism 8* (1): 59–68.

40. Brown, I. (2015). Social media surveillance. In: *The International Encyclopedia of Digital Communication and Society* (eds. R. Mansell and P.H. Ang), 1–7. Wiley.

41. Uldam, J. (2016). Corporate management of visibility and the fantasy of the post-political: social media and surveillance. *New Media and Society 18* (2): 201–219.

42. Assange, J. (2011). WikiLeaks revelations only tip of iceberg. https://www.rt.com/news/wikileaks-revelations-assange-interview (accessed 3 February 2020).

43. Clarke, R. (1988). Information technology and dataveillance. *Communications of the ACM 31* (5): 498–512.

44. Poster, M. (1990). *The Mode of Information: Poststructuralism and Social Context*. University of Chicago Press.

45. Graham, S. and Wood, D. (2007). Digitizing surveillance: categorization, space, inequality. In: *The Surveillance Studies Reader* (eds. P.H. Sean and J. Greenberg). McGraw-Hill Education.

46. Fuchs, C., Boersma, K., Albrechtslund, A., and Sandoval, M. (2013). *Internet and Surveillance: The Challenges of Web 2.0 and Social Media*. Routledge.

47. Weissman, J. (2019). P2P surveillance in the global village. *Ethics and Information Technology 21* (1): 29–47.

48. Zuboff, S. (2019). *The Age of Surveillance Capitalism: The Fight for a Human Future at the New Frontier of Power: Barack Obama's Books of 2019*. Profile Books.
49. Smith, M.E. and Lobo, J. (2019). Cities through the ages: one thing or many? *Frontiers in Digital Humanities 6*: 12.
50. Batty, M. (2011). Building a Science of Cities. UCL Working Papers Series, Paper 170, ISSN 1467-1298, 15 p.
51. Batty, M. (2013). *The New Science of Cities*. MIT Press.
52. Bettencourt, L.M.A., Lobo, J., Helbing, D. et al. (2007). Growth, innovation, scaling, and the pace of life in cities. *Proceedings of the National Academy of Sciences 104* (17): 7301–7306.
53. Bettencourt, L.M.A., Lobo, J., Strumsky, D., and West, G.B. (2010). Urban scaling and its deviations: revealing the structure of wealth, innovation and crime across cities. *PLoS One 5* (11): e13541.
54. Kostof, S. (1999). *The City Shaped: Urban Patterns and Meanings Through History*. Thames & Hudson.
55. Hudson-Smith, A., Milton, R., Smith, D., and Steadman, P. (2007). Geometric scaling and allometry in large cities. *Proceedings, 6th International Space Syntax Symposium, İstanbul, 2007,* Istanbul, Turkey (12–15 June 2007), p. 17.
56. Lobo, J., Alberti, M., Allen-Dumas, M. et al. (2020). *Urban Science: Integrated Theory from the First Cities to Sustainable Metropolises*, SSRN Scholarly Paper ID 3526940. Rochester, NY: Social Science Research Network.
57. Batty, M., Carvalho, R., Hudson-Smith, A. et al. (2008). Scaling and allometry in the building geometries of greater London. *European Physical Journal B 63* (3): 303–314.
58. Bettencourt, L.M.A. and Lobo, J. (2016). Urban scaling in europe. *Journal of the Royal Society Interface 13* (116): 20160005.
59. Davis, D.R. and Dingel, J.I. (2020). The comparative advantage of cities. *Journal of International Economics 123*: 103291.
60. Pataki, D.E. (2015). Grand challenges in urban ecology. *Frontiers in Ecology and Evolution 3*: 1–6.
61. Alberti, M. (2017). Grand challenges in urban science. *Frontiers in Built Environment 3*: 1–5.
62. Seto, K.C., Reenberg, A., Boone, C.G. et al. (2012). Urban land teleconnections and sustainability. *Proceedings of the National Academy of Sciences 109* (20): 7687–7692.
63. Bettencourt, L. and West, G. (2010). A unified theory of urban living. *Nature 467* (7318): 912–913.
64. West, G. (2018). *Scale the Universal Laws of Life, Growth, and Death in Organisms, Cities, and Companies*, 1e. Science/Business; Penguin Random House.
65. Brelsford, C., Lobo, J., Hand, J., and Bettencourt, L.M.A. (2017). Heterogeneity and scale of sustainable development in cities. *Proceedings of the National Academy of Sciences of the United States of America 114* (34): 8963–8968.

66. Knee, J.A. (2017). Review: how laws of physics govern growth in business and in cities. *The New York Times* (26 May 2017).

67. Page, L. As Delhi elections approach, what will it take to clean the capital's air? https://caravanmagazine.in/environment/what-will-clean-delhi-air (accessed 7 February 2020).

68. Reuters (2020). US drinking water contamination with 'forever chemicals' far worse than scientists thought. *The Guardian* (22 January 2020).

69. Bianchini, A., Pellegrini, M., Rossi, J., and Saccani, C. (2018). A new productive model of circular economy enhanced by digital transformation in the fourth industrial revolution: an integrated framework and real case studies. *Proceedings of the XXII Summer School "Francesco Turco" – Industrial Systems Engineering*, Palermo, Italy (12–14 September 2018), p. 7.

70. Ramadoss, T.S., Alam, H., and Seeram, R. (2018). Artificial intelligence and internet of things enabled circular economy. *The International Journal of Engineering and Sciences (IJES) 7* (9 Ver.III): 55–63.

71. Gupta, S., Chen, H., Hazen, B.T. et al. (2019). Circular economy and big data analytics: a stakeholder perspective. *Technological Forecasting and Social Change 144*: 466–474.

72. Tseng, M.-L., Tan, R.R., Chiu, A.S.F. et al. (2018). Circular economy meets Industry 4.0: can big data drive industrial symbiosis? *Resources, Conservation and Recycling 131*: 146–147.

73. Esmaeilian, B., Wang, B., Lewis, K. et al. (2018). The future of waste management in smart and sustainable cities: a review and concept paper. *Waste Management 81*: 177–195.

74. Antikainen, M., Uusitalo, T., and Kivikytö-Reponen, P. (2018). Digitalisation as an enabler of circular economy. *Procedia CIRP 73*: 45–49.

Chapter 8

Stairway to Heaven or Highway to Hell?

Life can only be understood backwards; but it must be lived forwards

— Søren Kierkegaard

Recommended Listening

Bohemian Rhapsody, **Queen**

Because this song includes all the colors of the past and the uncertainties of the future. Because the future will not arrive without a guiding narrative for and from our past.

Recommended Viewing

The World in 2050 (The Real Future of Earth), **https://youtu.be/g_1oiJqE3OI**

Because it provides a good preview of the main trends and problems that humanity will face in the next 30 years.

Industry 4.0 and Circular Economy: Towards a Wasteless Future or a Wasteful Planet?, First Edition.
Antonis Mavropoulos and Anders Waage Nilsen.
© 2020 John Wiley & Sons Ltd. Published 2020 by John Wiley & Sons Ltd.

As the book goes to its end, it's time to recap the main issues that were discussed, identify issues that were not addressed and questions that were not answered, and try to highlight some useful insights. But before that, it is obvious that there are several important issues that were not covered by the book in details; as an example we did not discuss in depth about obsolescence, eco-design, business models, supply chains, products as services, circular economy indicators, and many other aspects. This is not because we underestimate the importance of these issues, it's because we do not believe that a good book, especially for contemporary and continuously flowing essentially contested concepts (ECCs) like the circular economy and IND4.0, should or can cover everything – this is possible only for mature science textbooks. We decided to emphasize on the most important aspects from our own experiences and pay attention more to the unfolding linkages and relationships and less to a full coverage.

The starting point of this book was the main question we put in Chapter 1, the so-called trillion-dollar question. Will IND4.0 meet with circular economy delivering not only better resource efficiency but a more sustainable future for everyone in the planet? Or IND4.0 will evolve according to the business as usual linear model, leaving circular economy a wishful thinking, resulting in faster resource depletion, acceleration of environmental degradation, and deeper inequality?

To answer this question, we started, in Chapter 2, with a review of the history of industrial revolutions trying to understand their dynamics and their relationship with the rise of pollution in our planet. Based on this review we described the basic concept and content of IND4.0, as well as the main technologies and their functions. The role of information and innovation was highlighted.

In Chapter 3, we tried to explain the dynamics and the dialectics of circular economy as an unfolding policy narrative that creates already a global trend. We discussed technical, economic, and social aspects, and we identified systemic restrictions and scientific challenges. A major problem that was discussed was the tension between unlimited economic growth and the biophysical limits of our planet.

Continuing our trip to the interlinkages between circular economy and IND4.0, in Chapter 4, we presented how IND4.0 redefines both "resources" and "waste." We opened the discussion regarding waste hierarchy and its suitability, and we discussed the need to reconsider recycling in the framework of circular economy. We also explained why waste management goes beyond waste and the important role of final sinks.

In Chapter 5, we presented the current state of the art of what we called Waste Management 4.0, which means the ongoing transformation of the waste industry due to the applied IND4.0 technologies. We discussed inspiring examples and new dilemmas, and we also demonstrated the ongoing changes on the "hardware" and "software" of waste management.

The lessons learned, the experiences gained, and the need for a new leadership to stimulate the digitalization of the waste industry are discussed in Chapter 6. The importance of strategic awareness, the emerging risks and business models, and the alternative innovation methodologies were addressed as basic conditions for riding the wave of IND4.0. We also stressed the importance of political leadership and policy making for shaping a sustainable future.

Chapter 7 describes the rise of the new science of urban sustainability and how it is related and fueled by the big data revolution. Challenges and opportunities were presented, and the emergence of "surveillance capitalism" was spotted as a major threat. The importance of urban scaling was highlighted, and the dependence of circular economy on big data was discussed.

In all the chapters mentioned above, there is a permanent theme: the challenges and the opportunities we face concern the core of our economic and social structures, we are not talking about partial improvements or easy fixes, circular economy and IND4.0 have a revolutionary potential, and their realization is already creating radical transformational waves of change.

Both circular economy and IND4.0 signal an end. Circular economy signals the end of the current dominant economic system that is based on the continuous, endless, and infinite growth that stimulates more and faster resource consumption and lethally increasing pollution. IND4.0 signals the end of the simple integration of information technologies in the current industrial landscape – the ongoing advances of information technologies open new unimaginable opportunities, but these opportunities can't be realized without a serious transformation of our economies and the industrial ecosystem.

But circular economy and IND4.0 are not only signaling the end, but they also allow us to make a new start and to preview the future. Circular economy has the potential to drive a more sustainable and wasteless future if it goes beyond the current dominant technocratic narrative. Two conditions should be met: (i) circular economy has to move from good intentions to science, and (ii) it has to integrate the missing social aspects and put labor at its core. IND4.0 provides the technological alternatives required to realize circular economy, deliver new working and ownership functions, reduce inequality, improve human societies, and manage or resolve global problems in a new context. But this requires (i) political leadership that will drive the world towards sustainability in economic, social, and environmental terms and (ii) new forms of governance that will reallocate the power distributions between governments and multinational giants and between rich and poor countries.

Unfortunately, in human history, every end does not necessarily mean an immediate new start. As Antonio Gramsci has written [1], "The crisis consists precisely in the fact that the old is dying and the new cannot be born; in this interregnum a great variety of morbid symptoms appear." We are exactly in this period. Even worst, there is no guarantee that the new start will create a better future; in contrast the possibility for a future with much more pollution, inequality, poverty, and massive refugee streams is very high. A more systemic approach is required to understand exactly where the world stands this period.

8.1 CIRCULAR ECONOMY OR SPACE RACE?

On Monday, 16 November 2015, the US Congress voted (and later President Obama signed) the "Space Resource Exploration and Utilization Act of 2015" [2]. According to this bill, the President, acting through appropriate federal agencies, shall "(1) Facilitate commercial exploration for and commercial recovery of space resources by United States citizens; (2) Discourage government barriers to the development in the United States of economically viable, safe, and stable industries for commercial exploration for and commercial recovery of space resources in manners consistent with the international obligations of the United States; and (3) Promote the right of United States citizens to engage in commercial exploration for and commercial recovery of space resources free from harmful interference, in accordance with the international obligations of the United States and subject to authorization and continuing supervision by the Federal Government." Representative Lamar Smith, Texas Republican and chairman of the Science Committee, commented that "This bill encourages the private sector to launch rockets, take risks and shoot for the stars." Peter Diamandis, co-founder and co-chairman of the company Planetary Resources, Inc., said, "A hundred years from now, humanity will look at this period in time as the point in which we were able to establish a permanent foothold in space. In history, there has never been a more rapid rate progress than right now." According to Peter Diamandis, this is effectively the largest piece of resource legislature ever signed by a US president.

It's really interesting that while EU and China are dealing with different versions of circular economy models, trying to recapture as an example metals from waste streams and drive them back to industrial activities, the United States is preparing a new space race to catch valuable materials from the resource – rich asteroids (and not only) that are expected to come very close to Earth in the next 50 years. Those asteroids include fuels (hydrogen and oxygen), construction materials (nickel, iron, and cobalt), and platinum group metals (platinum, palladium, osmium, iridium) for strategic uses like electronics. If humans were ever able to get their hands on just one asteroid, it would be a game changer because the value of many asteroids is measured in the quintillions of dollars, which makes the market for Earth's annual production of raw metals – about $660 billion/year – look paltry in comparison [3].

Besides the impressions and the huge, but not unmanageable, challenges involved in the effort to utilize metals from asteroids, it is important to draw three main conclusions. First, the world's most powerful governments are getting prepared for a future where crucial resources might be in relevant or absolute scarcity. Second, the world can respond to scarcity either by expanding the linear economy to non-Earth resources or by advancing the circular economy. Third, the choice of each government is also relevant to the technological means available – the United States is rushing towards asteroid mining because, simply, they can do it before the rest.

We think that the example above illustrates also another important aspect: the biggest governments and of course the biggest companies in the world recognize that the world and the world economy get close to its limits, or it might already be beyond them. This comes as a critique not only from anti-systemic thinkers but also from people that are in the core of capitalism, such as John Mackey, the CEO of Whole Foods company. Mackey is asking for a serious reform towards "conscious capitalism," a mode of doing business that attempts to create value for all stakeholders – employees, customers, community, and shareholders – rather than sublimating the needs of the first three to those of the last [4]. Robert Reich, a university professor and former Secretary of Labor in the Clinton administration, in his book *Saving Capitalism* provides a roadmap of what must be done: rebuild the malfunctioning economic system and restore democracy [5]. As he explains, capitalism is not a suprahuman force whose "invisible hand," left to its own devices, will work for the good of the world; rather, it is a human construct requiring human direction. There has been a concerted "consolidation of market and political power to overwhelm regulatory governance," he asserts, to the extent that no matter how hard 99% of Americans work, their earnings are no longer commensurate with their merit [6].

Other thinkers believe that we are getting closer to the rise of new social structures that go beyond capitalism. Paul Mason, author of the book *PostCapitalism: A Guide to Our Future* [7], believes that we are entering a postcapitalism era because of three major changes that information technologies (or IND4.0 if you prefer) are bringing. First, they have reduced the need for work, blurred the edges between work and free time, and loosened the relationship between work and wages. Second, information is corroding the market's ability to form prices correctly because markets are based on scarcity while information is abundant. Third, because we can already watch the spontaneous rise of collaborative forms of production, there is a trend to produce goods, services, and organizations that no longer respond to the dictates of the market and the managerial hierarchy. As he writes, "the biggest information product in the world – Wikipedia – is made by volunteers for free, abolishing the encyclopedia business and depriving the advertising industry of an estimated $3bn a year in revenue" [8]. Naomi Klein, the influential author of the book *This Changes Everything: Capitalism vs. the Climate* [9], believes that our way of looking at the world has been influenced by the enlightenment paradigm that the Earth is not a living system, a mother to be feared and revered, but rather this inert ball from which wealth could be extracted infinitely. The impact of climate change forces us to discard this perspective, along with our "market-logic" belief "that our greatest power is as consumers." The more we consume, the more fossil fuel we burn, and the more waste we dump, the greater the impact on our environment of carbon pouring into the atmosphere. The unusually high floods, intense and frequent storms, droughts, wildfires, and extinctions we are experiencing are "telling us that we need an entirely new economic model and a new way of sharing this planet."

There is a plethora of similar signaling alarms that demonstrate a kind of growing systemic inconvenience [10]. In any case, from the point of view of this book, the fact is that both circular economy and IND4.0 can unfold in two basic ways. Either they can reform radically and thus revitalize capitalism, or they can corrode its foundations and drive the emergence of new social structures. What really matters is that there is a growing consensus that the global economic system can't continue its current trajectory and serious shifts are required to rebalance the human footprint on the planet.

Both IND4.0 and circular economy show the ways that such a rebalance can be achieved. But also, both IND4.0 and circular economy face barriers relevant to the dominance of the markets in decision making. Markets and big companies will be seriously disrupted if, as an example, a serious change in the ownership patterns (e.g. abandoning car ownership and having car as a service) happens. The big software companies that control artificial intelligence (and all our computers, tablets, and mobile phones) are seriously threatened to lose their monopoly if a more open and democratic approach to intellectual properties (IP) is adopted allowing artificial intelligence to be built through free software licensing and used by local communities to deal with their problems. The big profits are still in the expansion of the linear model for as long as possible, although a long-term approach shows the serious advantages of circular business models. But what is the meaning of long term in a world that can't make, for five years in a row after the Paris Agreement, a single step towards the reduction of carbon dioxide emissions because the fossil fuel giants still dominate the global politics? What is the meaning of long term when CEOs are taking high bonuses for firing thousands of people? What is the long-term impact when private banks that drove hundreds of thousands of people to poverty are pushing their own private debts to citizens as it happened to 10 million Greeks? Even worst, what is the meaning of adopting a circular economy for materials and building walls to avoid the circulation of climate, war, and poverty refugees?

What is clear is that both IND4.0 and circular economy can be interpreted in many different ways, but finally all the ways can be categorized in two mainstreams. The first stream includes the definitions and interpretations in which circular economy and IND4.0 are the perfect vehicle to change everything (in the ways the economic system is functioning) under the condition that nothing will change (in terms of the market dominance and the profitability of the multinational giants and the banking sector). The second stream concerns the understanding that the roots of the current economic, social, and environmental problems are deeply planted in capitalism's economic and political structures and a radical change can't happen without reinventing capitalism or going beyond it. Both circular economy and IND4.0 are ECCs.

8.2 CIRCULAR ECONOMY AND IND4.0 AS ESSENTIALLY CONTESTED CONCEPTS

The term essentially contested concepts gives a name to "a problematic situation that many people recognize that in certain kinds of talk there is a variety of meanings employed for key terms in an argument, and there is a feeling that dogmatism ('My answer is right and all others are wrong'), skepticism ('All answers are equally true (or false); everyone has a right to his own truth'), and eclecticism ('Each meaning gives a partial view so the more meanings the better') are none of them the appropriate attitude towards that variety of meanings" [11]. The term was introduced by Walter Bryce Gallie [12] to facilitate an understanding of the different applications or interpretations of the sorts of abstract, qualitative, and evaluative notions such as "art," "philanthropy," and "social justice." ECCs involve widespread agreement on a concept but not on the way that it should and would be realized. They are concepts the proper use of which inevitably involves endless disputes about their proper uses on the part of their users, and these disputes cannot be settled by appeal to empirical evidence, linguistic usage, or the canons of logic alone [13].

In a beautiful article [14] dealing with the concept of sustainability, John R. Ehrenfeld, executive director of the International Society for Industrial Ecology, remarks that all ECCs are emergent properties of complex systems and are subjective in the sense that they arise through an assessment by some observer looking on the whole system. ECCs are unquantifiable but can be described via qualities coming from the observer's assessment. In addition, he mentions that "ECCs ... cannot be managed in the deterministic sense that 'management' implies: that a manager operates according to some set of rules describing the behavior of the system being managed, and further that the outcome can be measured according to some quantifiable metric." In his relevant book on sustainability, Ehrenfeld defined sustainability as "the possibility that human and other life will flourish on the Earth forever" [15]. Here "flourishing" is the emergent property, and the system producing flourishing is the

Earth. And he explains [14]: "I chose flourishing as the quality that encompasses all three legs of sustainable development because it conjures up a vision of a desirable future state and, thus, can be assessed as being present or not. It is certainly not going to be easy to get there, but it is not something 'that never can be achieved.' Flourishing is a metaphor for many things, but always connotes aliveness, joy, health, and many other qualities related to being. The challenges we face today…are different from those related to managing sustainable development. Our goal should be to attain sustainability because it exists now only in tiny bits and patches, if at all. Even if we continue to disagree on the meaning of sustainability, we are largely in agreement that the present state of the Earth is unsustainable. We can come to terms here because we do define unsustainability in quantitative measures and rules."

We do believe that this is exactly the case for circular economy too. We can continue to disagree about what exactly is the definition of circular economy, but we can agree that the linear model stimulates a serious decline of the ecosystems that support humanity. Although we are still at the beginning to develop indicators to measure circularities, we have a lot of scientific tools to measure environmental degradation, loss of biodiversity, lethal pollution, inequality, and global warming. And it will be very difficult to find a more successful term than "flourishing" for the future of our societies. So, our ambition should be to advance circular economy and IND4.0 to increase the possibility that human and other life will flourish on Earth forever.

In Chapter 1, we wrote that none of the concepts of circular economy and IND4.0 are well defined, so we prefer to introduce them as flowing ideas that will create tangible results, depending on their final content and social interpretation. We think that now it is clear: both circular economy and IND4.0 can easily be characterized as ECCs. This explains the plethora and the conflicts between the definitions. This explains why they can be viewed and used either as concepts that are advancing the business as usual approaches or as serious disruptions of the dominant systemic approaches. This also explains why both concepts are, and in the future will be much more, a subject of political struggle and debate. Consequently, when we talk about IND4.0 and circular economy, we can't avoid the role and the biases of political beliefs and ideology; both concepts are not technocratic or neutral, but they are shaped by not only the personal but also the social dominant views.

That is a clear call to arms. A positive social footprint of the circular economy will not happen automatically; we have to contribute and shape it. Greenwashing practices in the name of sustainability or circular economy will continue to dominate the practices of all the entities that have a primary interest in preserving the business as usual as long as possible. Putting the role of working people and human rights in the center of the discussions about circular economy and IND4.0 is necessary, but it will happen only as an outcome of political struggle. In a way, circular economy and IND4.0 oblige all of us to become less technocrats and more citizens and to think more about the society (as a whole) and its future and less about our narrow expertise. Both circular economy and

IND4.0 can't be understood and discussed without substantial support from economics, sociology, philosophy, history, biology, industrial ecology, and complexity science. Holistic and systemic approaches are becoming a condition to understand, plan, and realize the transition towards a circular economy within the IND4.0 landscape. And this is really missing, as we explained in the previous chapters, in the current mainstream narratives. Let's see a real and representative example of the dynamics between the existing business models, circular economy, and IND4.0.

8.3 SQUEEZING IND4.0 – UNDERMINING CIRCULAR ECONOMY

First of all, we will make the assumption that whatever is advertised or said by multinational companies (that are stronger than most of the national governments) is true, so there are no fake news or purposeful mistakes. This assumption is certainly questionable. It means that we assume that Volkswagen was only an exception in bombarding consumers with greenwashing propaganda while intentionally cheating about its cars' emissions. Second, we will speak by example; theories in this case have the risk to be accused for ideological or political biases. A good example is the one of e-waste, as this is an emblematic waste stream where circular economy meets IND4.0.

We have already discussed the rising tide of e-waste that is creating huge environmental and health impacts. The fact is that IND4.0 provides powerful technologies that allow us to resolve the problem of e-waste. Take as an example Daisy, Apple's 11-meter-long robotic station, which was announced in 2018 that is capable to disassemble 1.2 million iPhones (15 different models) per year or almost 200/hour [16]. Daisy can recover 1.9 tons of aluminum, 770 kg of cobalt, 710 kg of copper, 42 kg of tin, just under 1 kg of gold, and 100 g of palladium for every 100 000 end-of-life iPhones [17]. Apple celebrated the Earth's day announcing a major expansion of its recycling program quadrupling the number of locations US customers can send their iPhone to be disassembled by Daisy, its recycling robot [18].

In the traditional e-waste recycling (which can be done either manually or mechanically), electronics are smashed (and in the case of informal recycling plastics are burned) in an effort to uncover valuable materials and collect them. Shredders are also used to tear up the devices in order to make manual sorting possible. However, in such a process, full recovery of valuable materials is impossible as, for example, rare earth minerals from the iPhone speaker are usually attached to other metals during the process, and so they are lost. Daisy disassembles iPhones in many pieces, and recovering much more of an individual material – like cobalt, a key battery material – makes it possible for Apple to have enough volume to convince a recycler to take its materials so it can be recycled back into new batteries. As the company's website explains [18] for cobalt, Apple sends iPhone batteries recovered by Daisy upstream in its supply chain. They are then combined with scrap from selected manufacturing sites, and the cobalt recovered through this process is now being used to make brand-new Apple batteries – a true closed loop for this precious material. Apple also uses 100% recycled tin in a key component of the main logic boards of 11 different products. The company's engineering of an aluminum alloy made from 100% recycled aluminum allows the new MacBook Air and Mac mini to have nearly half the carbon footprint of earlier models.

So, it seems reasonable that the company claims that its devices are built to last as humanly possible and explains [19]: "You count on your device day after day. So it's designed to withstand years of use and be ready for more. It's made of durable materials, supported by ongoing OS updates, and backed by a network of more than 5000 Apple-certified repair locations you can count on if something unexpected happens. Because the longer you use your device, the better it is for the planet." Let's keep this in mind, we will call it the CLAIM, and we will come back at the end of this section.

We can rightly argue that, based on innovations like Daisy, the end of e-waste is possible. Since Apple did it, it's a matter of time that all the similar companies will follow, and technologically speaking, we are capable to divert all the e-waste, more or less, from dumpsites and landfills, saving almost $60 billion/year in resources. The argument is, in principle, correct. However, the reality is completely different; once again the devil is hidden into the details. Let's see them.

How many iPhones has Daisy recycled? We really do not know, but we know that Apple announced that it has received one million devices back, through its recycling program, in 2018. In 2018, Apple announced that there were 1.4 billion active Apple devices – of those devices, 900 million were active iPhones in use. So, receiving back 1 million devices per year, it will take 900 years to take back and recycle all the iPhones that were in use in 2018 only. With two Daisy robotic stations, in their full capacity of 2.4 million devices per year, we will reduce this time to 375 years, and with 4, we will only need 187.5 years. But, if we make and run in full capacity 100 robotic stations, then it will only take 7.5 years to recycle all those 900 million iPhones. However, while Apple's robot can only work with newer iPhones (and could potentially be used by other companies, since Apple has said that it is willing to share the IP with anyone who asks), it doesn't actually work with the iPhone 4, which is the phone that is most likely to be old enough to actually be at its end of life now [20]. Besides the iPhones, some products, like AirPods, can't be recycled now because the battery can't be taken out, so it is not possible to run them through a traditional recycling stream because they create serious risks of fire. In addition, the company's laptops are also much more difficult to repair and recycle than its phones.

However, the real problem is that the company does not take back as much iPhones as it can. There are two reasons for that: first the value of secondhand phones and second the company's limited recycling network. Users aren't giving Apple back their phones, because old iPhones are worth dramatically more on the used market than the material cost or Apple's trade-in value program that gives customers credit towards a new phone when they send in an old one. As an example, the 4-plus-year-old iPhone 6 can be sold for $100, while Apple will give you a $90 gift card. Even the iPhone 4 can get $40, while Apple offers nothing [21]. So, economically speaking, it will be crazy to destroy a perfectly functional $100 product to recover $0.23 worth of cobalt. As Kevin Purdy explains in iFixit website, "The recyclers we regularly speak with rarely see newer-model iPhones, and when they do, they repair and resell them. They certainly don't grind them up for raw materials. Even non-functional phones are parted out and sold to people like us, who are in desperate need of original iPhone service parts."

Still, the idea that Apple makes a specialized robot that can dismantle its products is fascinating, since Apple and its partners are the only ones that have the full know-how to deliver such a product. This brilliant idea could work under one condition: that Apple will be obliged to take back its devices in all its shops around the world through a global extended producer responsibility (EPR) scheme. But in any other case, Apple's products will more possibly end up to dumpsites and sanitary landfills or to anyone of the e-waste recycling facilities around the world than at one of Apple's 500 stores that take back products. However, it seems that Apple, together with nine other relevant companies including Motorola, Samsung, HP, and Vivo, does not like at all the application of EPR schemes as they are legislated in India and the EPR supervising authority in India suspended its import license on 4 April 2019 [22].

In addition, the company has repeatedly opposed laws for the right to repair. Following the car companies that fought hard to defend their monopoly in car repairing, computer and cellular phone companies are doing the same by not providing original parts to third partners or by making their products very difficult to be repaired. As it was reported at Vox website [23], in California, Apple successfully urged lawmakers in California to abandon right-to-repair legislation, arguing that customers could hurt themselves by doing their own iPhone repairs. The company also hired lobbyists in New York and testified in Nebraska, among other efforts, to resist similar legislation. Microsoft is also following the same patterns and opposes any laws that promote the right to repair.

On Wednesday, 24 October 2018, *The Guardian* reported [24] that Apple and Samsung were being fined €10 and 5 million, respectively, in Italy for the "planned obsolescence" of their smartphones. An investigation launched by the Italian competition authority found that certain smartphone software updates had a negative effect on the performance of the devices. The investigation followed accusations that operating system (OS) updates for older phones slowed them down, thereby encouraging the purchase of new phones. The Italian authorities stated that "Apple and Samsung implemented dishonest commercial practices" and that OS updates "caused serious malfunctions and significantly reduced performance, thus accelerating phones' substitution." Interestingly, almost one year ago, Apple has admitted [25] that it actually slows down older iPhones, a feature introduced to protect against problems caused by aging batteries. In a public letter [26], it was explained that the company decided to manage how much power the iPhone draws from older batteries in order to prevent unexpected shutdowns. But the public didn't find out about it until some iPhone owners used an app called Geekbench to test the speed of their iPhones. Geekbench wrote a blog post about the issue, and Apple admitted it was intentionally slowing down some iPhones with older batteries a few days later.

Obsolescence seems to be a serious problem for several Apple products. From July to October 2019 the company has been sending out full-screen alerts warning iPhone 5 users to update their OS to the latest iOS 10.3.4 before 12:00 a.m. UTC on 3 November 2019, in accordance with the global positioning system (GPS) time rollover that happened on 6 April 2019 and affected devices relying on date and time to tell your location [27]. The ones that would not update would not be possible to go online. With the iPhone 5 having a nonremovable battery and practically everlasting, this new update could be considered another way for Apple to make older models obsolete. A similar move was recently done by Sonos, a company famous for its speakers. The company developed a proprietary wireless communication system that enabled multiple speakers to sync up for easy whole-home audio. But in January 2020, Sonos announced that it will end support for many of its oldest wireless speakers, although they were perfectly functional [28].

Kamila Pope, author of the book [29] *Understanding Planned Obsolescence: Unsustainability Through Production, Consumption and Waste Generation*, in a blog post at ISWA's President's blog, commented [30] that systemic planned obsolescence is typical of the IT sector and happens with the replacement or updating of a system by modifying and introducing new functions to make its continued use difficult, if not impossible. Software updates are a prime example of this practice. In many cases, such changes render both the previous software and the hardware obsolete. The product becomes obsolete due to, among others, incompatibilities, file/data reading disability, and exhaustion of memory. All these malfunctions make the product's use redundant, forcing the replacement of the hardware or, worse, the device as a whole. A relevant study in Austria [31] found three forms of perceived obsolescence, being either related to a phone's (i) basic functionality, (ii) up-to-dateness, or (iii) ability to keep up with social practices. Furthermore, it was found that the perceived speed of obsolescence is key to considerations of phone repair and reuse.

AppleInsider tested a pair of AirPods purchased in 2016 against a pair from 2018 and found that the older pair died after 2 hours and 16 minutes, less than half the stated battery life for a new pair [32]. The batteries used in AirPods charge faster, last longer, and pack more power into a small space than other types of batteries do. But they die faster, too, often after just a few years, because every time you charge them, they degrade a little. They can also catch fire or explode if they become damaged, so technology companies make them difficult, if not impossible, for consumers to replace themselves. The result is: a lot of barely chargeable AirPods and wireless mice and Bluetooth speakers are ending up in the trash as consumers go through products – even expensive ones – faster than ever. The keyboard of the 2015 MacBook Pro was characterized as a repairability flop [33] by iFixit engineers, based on questions like "How time consuming (and therefore expensive) is it to open?" and "Can broken components be replaced individually, or will you have to swap out more expensive larger modules?"

We think that now few things became clear. The invention of Daisy, the robotic disassembly station, is a fantastic technological innovation that can stimulate the end of e-waste. The first problem is that Daisy remains in a pilot–demo phase and without a massive investment, it will take hundreds of years to recycle Apple's 900 million iPhones if they will come back to Apple. The second problem is that the vast majority of these iPhones will never go back to Apple for the reasons we explained above; thus Daisy remains a tangible but still decorative innovation. At the same time, Apple's business models advanced planned obsolescence, as the CEO of the company officially accepted, and the company fights back EPR schemes and laws for the right to repair. So, although in principle the innovation of Daisy creates a realistic potential for the end of e-waste, it's the applied business model that seeks for profitability in the classic linear economy that blocks this fascinating potential. In a more general way, IND4.0 can drive substantial advances and stimulate circular economy, but the existing business models undermine this potential and stimulate the extension of the linear business models.

How about the CLAIM? Well, a thorough examination of the CLAIM provides that there is nothing wrong in it; it is true that Apple's products are designed to last and perform well for a long time. The problem is in the business model that is pushing shorter lifespans of the products. And, second, the most important problem is at the design of the products that should incorporate recyclability, modular design, and easier repair in order to make the CLAIM a tangible contribution to a circular economy.

Speaking about the role of design, there is a lot of potential for improvements, especially in the case of mobile phones. An approach closer to circular economy could combine product modularity, product–service systems (PSS), and design for attachment [34]. With a modular design, it becomes easier to repair the product or to replace parts, allowing for an extended lifespan. In combination with PSS, the modules, when no longer used by one customer, can return to the market to be reused by another one. Such a concept would probably directly reduce the environmental impact of the smartphone on the production phase, which accounts for most of the emissions throughout its life cycle.

The idea of modular mobile phones has already been explored as a way to increase the eco-efficiency of mobile phones, despite the fact that most companies seem to move in the opposite direction to enclose all the mobile phone life cycle in a single package that excludes the possibility of repair or upgrade by the users. In order to quantify the actual value of eco-efficiency, a specific component upgrade strategy should be determined. A study that analyzed different strategies suggested that component exchange pattern including main board is always more eco-efficient than the replacement of the whole product [35]. Throughout the study, it has been suggested that there is a need for an eco-efficiency index that will allow to evaluate alternative strategies of component modularization aiming good combinations of component upgrade and reuse. However, the business models regarding modular phones and the customers' convenience remain serious barriers for the further development of the idea, considering also the high competitive landscape of the industry [36].

To complete the picture, a study that analyzed the mobile phone repair markets in the Netherlands, Poland, and China [37] found that to maintain a strong position in the local market and to sustain the trust of customers, independent mobile phone repair shops offer a range of customized services based on direct contact with customers. In China, the increasing prices of spare parts and falling prices of mobile phones constitute the most important challenges, whereas in the Netherlands and Poland, the most important challenges are the competitive pressures from informal repair activities and new repair shops. The research revealed that repairability strongly depends on the global manufacturers' circularity choices. The study concluded that the future of local independent repair shops dynamically depends on the circular choices (durability, design, and reparability) of global actors (e.g. mobile phone manufacturers) as key players and outer-circle remanufacturing and recycling industries of a circular economy. While there are local economy and (digital) platform elements (of online repair manuals), mobile phone repair is not purely local because of the nonlocal availability of product components and materials.

We already discussed that the only way to advance innovations like Daisy is through EPR schemes, although the big tech companies rather dislike the idea and do everything possible to avoid it. We also discussed that the repair and reuse market is also based mostly on the repairability and lifespan policies of the global big tech companies. Obviously, this means that the role of governance and institutional development becomes crucial for the transition to a circular economy.

8.4 GOVERNANCE IS THE KEY – CITIES ARE THE LOCK

According to ISWA's report "How IND4.0 transforms the waste sector" [38], the rise of IND4.0 creates new opportunities for waste management and circular economy governance. Location information plays an essential role in the far majority of digital platforms, and the importance of location intelligence is reaching new heights thanks to more than 25 billion devices that will be connected to the "Internet of Things" (IoT) by 2021. Digital platforms enable multidirectional network effects and value creation and allow platform owners – managers – to address the markets. The near-zero marginal cost curve of digital platforms creates the possibility of market domination by one platform, and the high barriers to entry or exit lead to monopolistic and oligopolistic market structures.

On the other side developing digital platform businesses require a shift in design thinking from resource ownership to resource orchestration. Several recommendations have been developed by EU for how digital platforms could more easily apply and expand in the context of government environments. The main ones concern that governments should focus on orchestrating and reusing existing government services as a starting point for developing their platforms; they should start building IoT capabilities and optimize the use of their open data services by defining a service delivery approach that matches the needs of their target consumers. OECD has recommended "that governments develop and implement digital government strategies" to assist and guide them to achieve that digital transformation. The recommendation emphasizes the crucial contribution of digital technologies as a strategic driver to create open, participatory, and trustworthy public sectors to improve social inclusiveness and government accountability and to bring together government and nongovernment actors and develop innovative approaches to contribute to national development and long-term sustainable growth.

The greatest challenge of collective intelligence lies in engaging citizen. It is extremely difficult to mobilize a large range of the population that has a diversified base of knowledge on specific public policy issue. This is compounded by the challenge of finding the right balance of key stakeholders and citizens to provide political legitimacy to the consultation. ISWA suggested [38] several innovative powerful tools for civic innovators allowing greater collaboration, transparency, and participation for civic projects that could enhance the work of civic innovators, particularly government innovation teams and labs, looking for more decentralized open ways of working in waste management. The use of mobile applications for better governance, and especially for improving waste management and recycling [39], is a key tool that is already applied in hundreds of cases worldwide.

The application of EPR and other recycling schemes and their supervision and monitoring are becoming easier through the digitization of the waste sector. Tracing hazardous materials and substances, restricting illegal treatment and disposal, and preventing waste trafficking are also advanced due to digital footprints, IoT, and modern surveillance techniques.

There is an ongoing debate about the best options of recycling. It seems that the dilemma is presented like this: EPR or deposit return (DR) systems. The discussion is fueled by the successes [40] and failures [41] of different schemes in different countries [42]. In addition, the rise of IND4.0 brings new opportunities and challenges for both options [43].

This debate should be properly framed. To start with, waste management situation, industries, and stakeholders are different in each country, so there is nothing like a "silver bullet." As an example, in EU, different EPR systems present different rates of success or failure for the same products in different countries. Different EPR schemes demonstrate huge differences in recovery rates in the same country [44]. Consequently, the feasibility of introducing a DR system should be based on the specific local, industrial, and waste management landscape. A second important point is that both EPR and DR systems should be faced as dynamic solutions and not static ones. This means that they have to fit in a continuously changing consumer market, so they have to be both flexible enough to adapt to market changes and robust to deliver the required recycling rates.

A third point is that either EPR or DR will not be successful in isolation; they have to be combined with broader policy and economic initiatives. Especially as we move towards a circular economy, one thing to be considered is that circularities will not be restricted within countries' level; regional integration and global trade will still be part of the solution. This is usually a problem for DR systems because part of the collection and recycling potential is lost because consumers buy the products in one country and consume it in another [42].

In principle, it has to be clear that EPR and DR should work together because they serve different purposes. In an EPR system producers and the value chain of retailers and others work together in a system that supports the placement and the recovery of market products. The primary driver of EPR is to deliver a collective aligned system that supports the value chain and consumer. A DR scheme, on the other side, is a system designed to influence the behavior of consumers by requiring them to pay a deposit on items they purchase, with that deposit returned to them when they take the product to a redemption point. A DR scheme has to be supported by a system of administration and redemption points, material collection, and other services (like EPR), but its primary purpose is influencing consumer behavior.

Looking at reuse and prevention, much of this happens outside the waste sector and thereby cannot be properly influenced with waste-related legislation alone. Policy makers and regulators must therefore be willing to address these topics by setting appropriate objectives and creating incentives or passing related regulations for the production and service sectors and by making sure that those are being enforced. Only if waste that in principle is reusable is handed over to waste management companies, they can ensure to enable reuse by having the right facilities and processes in place. Proper end-of-waste criteria need to be defined, put in place, pursued, and enforced in order to achieve waste management objectives by enabling reuse whenever environmentally sound and preventing the reuse of goods that contain contaminants that should be removed from the anthroposphere (for example, asbestos tiles).

An important element is to develop policies that will allow a second life of goods – recently evolving systems that allow for secondhand use of products as well as buyback and return systems outside of the waste sector deserve mentioning. These systems allow for preventing waste through extending the product lifetime or incentivize the return of waste products and feeding them into a closed loop for component reuse or material recycling, overcoming the planned or perceived obsolescence. Internet-based marketplaces allow for easy selling of used products to other consumers who might perfectly be satisfied with the still functioning product. In these cases, products that would have become waste are reused, and waste generation is thereby prevented.

Remanufacturing models pose a high potential to reduce waste. In remanufacturing approaches the collection plays a major role, and in general terms it is preferable to establish direct connections between the owner of the used product and the remanufacturing/recycling company. This reduces the consumer's effort to manage the end-of-life material or product while offering potential revenue as well and secures that the remanufacturer/recycler obtains the best quality of incoming product. While this collection system allows for a maximum of reuse, remanufacturing, and recycling, it heavily relies on information sharing and spare part availability; therefore and more general, in order to allow for fully exploiting reuse/recycling potential equipment, manufacturers should have a legal obligation to provide this type of information. This is where policy makers and regulators are asked to provide appropriate boundary conditions and enforcement.

A continuously increasing complexity is the basic characteristic of the advances in waste management policies and legislation. More regulations, separate management of different waste streams, emphasis to resource recovery and end-of-waste criteria, linkages with global markets, and supply chains are some of the factors that drive the complexity of legal instruments higher. At the same time, the introduction of circular economy and the Sustainable Development Goals (SDGs) require much better statistics and reporting than today. The advances of circular economy are expected to further increase the complexity of the waste management systems as they will become more integrated to supply chains and they will embed the chains' uncertainties too. As a result, the development of proper information systems to monitor the evolution of waste management in different geographical levels becomes a key condition for the development of sustainable waste management policies. These information systems can serve different objectives, and their development should capture historical data too in order to develop the baseline conditions.

A comprehensive report on the governance of circular economy [45] suggested ten principles that should shape the new governance patterns, as shown in Table 8.1 – the table has been created by the authors based on the report's content.

There are three more crucial remarks for unlocking the potential of circular economy and IND4.0.

The first regards the higher importance of regional and municipal authorities and the need to develop national, regional, and municipal approaches well interconnected. For each specific product or material, circularities will be developed in different geographical scales, and in many cases transnational flows would be required. Finding

Table 8.1 Governance principles for circular economy from the report "Governance for the Circular Economy – Leadership Observations" [45].

Principles	Content
1. Put purpose in the center	Governance of organizations, business, and government need to identify the purpose that drives decision making, business strategy, and communication in all forms (with a CE agenda). The purpose identifies a "bigger picture" and defines strategic objectives precisely. A positive contribution purpose is often found in family-owned enterprises, as continuity is often seen as the most important organizational value. Governance with a purpose as a central focus will support collaboration across the network
2. Find the generating motor	Create governance that is self-seeding and self-generating to develop your mission. Develop an ecosystem, a generating motor that will support new ideas to grow and to succeed
3. Develop regenerative wealth definitions	Redefine value to focus on environmental integrity, across finance, stakeholder, and shareholder systems. See time as an asset, continuity as a value. Explore how "temporal space" can be communicated as a global value
4. Identify, use, and create conditions	Be local. Relate your operations and decisions to the quality of places. Apply direct feedback and physical consequence. Avoid trying to "win" solutions but engage solutions that work. By sharing and adapting, the learning will accelerate and expand
5. Apply hard and soft boundaries – identify boundaries for collaboration and control	Be bold and tough on what you do not want, such as toxicity. To achieve what you do want, facilitate a soft environment for collaboration. Include participation of all actors in services and products including labor and customers. Develop an agenda that will facilitate what you want to achieve
6. Accelerate collective performance with mechanism design	Make systems and market adjustments when necessary to achieve the purpose through mechanism design. Create conditions to help people and organizations to behave as if they were sustainable. Control and adjust based on new learning and new possibilities
7. Activate the value chain	Identify and connect the value chain actors. Pay attention to a common language, enhance reciprocity, create meaning, and share profits across the value chain. Employ new technology to bring parties together. Make community approach count. Define "buy-in" (product, shareholders, leadership), a value-based result, not a short-term return

Table 8.1 (Continued)

Principles	Content
8. Engage the customer	Reframe consumption as a service and production as a value chain of good practice. Regarding product and pricing, act multilayered, be transparent, and make reporting available. Educate customers to increase understanding of value and performance. Promote dialogue with consumers, labor, and communities to build understanding in multiple directions
9. Harvest generic approaches from regional success	Develop regional sets of incentives, collaboration, and legislation. Embrace local conditions and reinforce natural capital as well as human capital. Recognize identity and the regional biography. Make use of proximity to develop smart solutions. Observe working solutions to develop generic lessons and make them available to other regions
10. Redesign and monitor performance	Find new value in environmental integrity through redesign and reconsidering the possibilities. New technologies and science can open new approaches and new business models (for example, service systems versus products to control material use). Communicate performance

the right balance between local initiatives that create an impact to citizens and scaling up them to become meaningful is one of the key problems that should be faced.

The second remark is about leadership. We have already discussed about it in Section 6.2.4. As we already wrote, "policymaking for the circular and digitized world requires a tailored approach to problem-solving and setting goals specific to a region. A general challenge is creating an appropriate regulatory environment around a technological innovation that ensures the trust of stakeholder while encouraging innovation." In addition, looking at recent advances, it is important to say that governments require to be bolder. The recent global wave of regulations regarding single-use plastics should teach us that people are ready for more global coordination, and certainly they will encourage it as long as it is meaningful and proves the expected benefits. This wave was both a response to local pressures and policies and the result of a global discussion and coordination. For advancing circular economy, even bolder decisions should be made. Here we want to mention only two of the many potential needs: (i) the need to develop worldwide EPR schemes for selected products that are designed, developed, retailed, and managed through worldwide supply chains and (ii) the need to develop technical specifications for material flows, instead of waste flows.

The third element regards the socialization of circular economy. There is a need to shift towards a social circular economy [46]. Social enterprises use business principles to achieve societal good and seek to make a positive change in the world. The social circular economy unites the circular economy and social enterprise concepts to deliver benefits for people, planet, and profit. It allows a fully systemic view by drawing on the environmental principles of the circular economy and the societal vision of social enterprise, both of which are underpinned by a pursuit for economic prosperity. It thus aligns well with enhancing well-being for people and planet and the UN's SDGs.

8.5 BEYOND BUSINESS AS USUAL OPTIMIZATION

We wrote in the previous section that governments need to be bolder. The truth is that in the transition period we are living, as the fourth industrial revolution is unfolding in geographies, industries, working and human relationships, governance, and politics, we all need to think and act bolder. A phrase that symbolizes what we must do is to think, plan, and act beyond business as usual optimization.

Optimization, by definition, is an act, process, or methodology of making something (such as a design, system, or decision) as fully perfect, functional, or effective as possible. Spiced with some mathematics, optimization is the procedure that aims to find the optimal solutions to the objective function or functions under constraints. As an example, investors are trying to find the optimal solution for their money: the investment that will provide the maximum benefits with the minimum cost – in this case the objective function is the ratio benefits/costs, the higher the ratio the better. Add some restrictions, let's say within a 3-year period, or 2 restrictions like within a 3-year period in automobile industry, or three restrictions like within a 3-year period, in automobile industry and a minimum profit of 50%, etc. The point here is that when we deal with optimization processes, we need to have (i) a well-defined system, (ii) an objective function, and (iii) specific restrictions.

When the business as usual determines our lives, optimization offers a lot of solutions and opens the way for performance enhancement. As an example, cars, as we know them today, are the result of a 100-year period of continuous optimization, and this is why if you categorize them per different type, they look quite similar. There are three reasons that explain why within every generation of vehicles, there are striking similarities between the designs of different cars, and all three reasons are related with different paths of optimization. First, it's the rule of business (optimization of revenues): when a particular design is a commercial success, it's just a matter of time that it will be copied, in one or another way. Second, cars involve a lot of technological advances, and their shape has to follow their functionalities based on the best available technologies of their era – the more they use the same technologies, the more the shape in particular parts becomes similar (technological optimization). Third, until few decades ago, improvements and innovations required 5–15 years to become available and applied in mainstream design. Today, in many cases, from the innovative concept till its commercial application, only some months are required (optimization of time from concept to commercialization). All these three reasons drive the rapid expansion of good ideas across the entire automobile industry. But they also lead to a certain homogeneity in terms of the design. Thus, they put the systemic design of cars in a continuous optimization, but they actually restrict new, radically innovative designs to emerge.

The example of cars will help us to understand what we mean by going beyond business as usual optimization. When I was a student, one of our professors was teaching us aerodynamic optimization for cars, one of the most elementary requirements for the design of a vehicle. Aerodynamic drag is one of the main obstacles to accelerate a solid body when it moves in the air. About 50–60% of total fuel energy is lost only to overcome this adverse aerodynamic force [47]. So, our work was to take an initial shape and try to reduce the drag by adding components (like vortex generators in the roof or a pair of flaps in the spoiler) or by extending the rear body of the car or by reshaping the curves of the car. We were quite happy if we achieved a 10–20% improvement (based on the initial shape), but in difficult cases an improvement of 1–2% was also substantial because it might be enough to win a race with half a second difference. The system was well defined: a car running on a specified road with certain power and friction between the tires and the road. The objective function was to reduce the aerodynamic drag. The restrictions were usually financial (you can spend up to a certain amount of money for improvements) or related to the car's use or to the time available for improvements. This is an exercise that has been done millions of times from thousands of engineers and contributed substantially to the evolution of cars as we know them today, from the first steam-powered automobile of Nicolas-Joseph Cugnot in 1796 to the Formula 1 technological masterpieces.

This aerodynamic optimization will continue to happen; for each new model of car that appears, this is exactly the business as usual optimization. However, today, in the transition to IND4.0, this is not the optimization we need; we have to go beyond that. Traditional optimization techniques are losing their meaning when you see the bigger picture [48]:

- The average private car in the United States carries only 1.6 people.
- 28% of car trips are a mile or less.
- Cars are parked 95% of the time.

Now, you realize that with the advances of software, sensors, big data analytics, and artificial intelligence, you can utilize car as a service, reducing the time it is parked by 60–70%, serving 6–8 people per car, instead of 1.6, and finally using less cars you can serve much more people, reducing traffic jams, accidents, and carbon dioxide emissions.

So, here is the fact. In a transition period in which new technologies redefine the industrial landscape, focusing on the business as usual optimization distracts you from the new, possibly unimaginable opportunities that are becoming possible. The business as usual optimization becomes misleading because (i) the system is not well defined, but we are in a huge technological transition that changes what we call "system"; (ii) the objective function that must or should be optimized might be completely different than the usual one; and (iii) restrictions are uncertain because they are related both with the (changing) definition of the system and the objective function.

This is exactly our era: an era in which we have to think bold and full of ambition, an era that makes all scenarios possible, although with different possibilities, and an era that will redefine the systems, the objective functions, and the restrictions. When people showed a car for the first time, they gave it the name "horseless carriage" [49] because this was what their lens from the past allowed them to do. In our era, the future can't be forecasted with the lens of the past because what is coming is unprecedented in many ways. The invention of the light bulb did not result from optimizing the candles. In this transition period, strategic thinking and complexity management are becoming the keys to the future – business as usual optimization restricts the options because it's about optimizing the past when the future runs faster than its shadow.

8.6 ENVIRONMENTAL INCREMENTALISM? NO, THANKS

Incrementalism is a method of working by adding to a project using many small incremental changes instead of a few (extensively planned) large jumps. In politics, incrementalism is the theory, or implementation thereof, that change should be introduced gradually or by increments.

Incrementalism is a serious risk for the environmental movement today – systemic changes like the circular economy can't be introduced and become viable incrementally because the dominant linear ecosystem will swallow them. This discussion is very relevant to the discussion on the business as usual optimization of the previous section. Incrementalism can be considered as an effort to improve – optimize the business as usual without disturbing the business and without diverting too much from the usual. In this view, an incremental view for the transition to circular economy is equivalent to a sterilization of a huge systemic change in order to make sure that the dominant linear ecosystem will not be seriously affected. In such a view, circular economy is incrementally identified with a continuous improvement of recycling, which as we discussed is one of the least preferable stages in a circular economy context.

The case of global warming has a lot to teach us about incrementalism. As we already mentioned in Chapter 1 in October 2018, the Intergovernmental Panel on Climate Change (IPCC) warned that only a dozen years are available to keep global warming to a maximum of 1.5 °C, beyond which even half a degree will significantly worsen the risks of drought, floods, extreme heat, and poverty for hundreds of millions of people. In November 2019, more than 11 000 scientists from all over the world declared that "planet Earth is facing a climate emergency." The know-how and the technologies to take meaningful actions are all here, the vast majority of governments agree that this is an existential threat for humanity, the Paris Agreement was celebrated as a monumental achievement, and so, what? Carbon dioxide levels are rising fast, and they will continue to rise in the atmosphere [50]; there is not even a single sign of hope that the required actions will be implemented in time. Even worst, after the COP 2019 in Madrid, it was written [51] that "Two weeks of talks ended on Sunday afternoon with a formal recognition of the need to bridge the gap between greenhouse gas targets set in 2015 in Paris and scientific advice that says much deeper cuts are needed. Current targets would put the world on track for 3C of warming, which scientists say would ravage coastal cities and destroy agriculture over swathes of the globe." For sure, incrementalism in global warming seems to ignore the urgency of both mitigation and adaptation; it seems to be a response that does not consider the dimension of time.

So, what's going wrong? Dr. Jennifer Allan, lecturer in International Relations in the School of Law and Politics of the Cardiff University, argues reasonably [52] that the Paris Agreement itself is a dangerous form of incrementalism in two ways. First, it repackages existing rules that have already proven inadequate to reduce emissions and improve resilience. Taken as a repackaged set of institutions, the Paris Agreement most probably will be unable to reverse the trend of rising emissions. Emission reductions stemming from the Paris Agreement will slow the rate of emissions growth, although aggregate emissions will rise, suggesting a global temperature increase between 2.6 and 3.1 °C by 2100. Second, state and nonstate actors celebrate the Agreement as a solution, conferring legitimacy on its rules; however, developing countries and nongovernmental organizations accepted the Paris Agreement to secure the participation of the United States and to uphold previous agreements. Given the reification of existing rules, the ratchet-up mechanism and nonstate actors offer the last remaining hopes in global efforts to catalyze climate action on a scale necessary to safeguard the climate. She concluded that "The Paris Agreement is designed to endure, offering no opportunities to revisit its fundamental design. Parties are locked in the cycles of submission, reporting, global stocktake, and resubmission of NDCs. Even if those reviews and the many other reports on the state of the global climate show that the global response is inadequate, parties remain in the same circle."

When the vulnerability for populations, cities, and resources becomes so high that even advanced human settlements are suffering by floods and fires, then you can't spend five years doing nothing. With the ongoing climate emergency, what is required is a rather transformational response instead of an incremental one. Even regarding adaptation, a transformational adaptation is urgently required because the incremental adaptations are not working [53]. In the article "Policymaking Under Pressure: The Perils of Incremental Responses to Climate Change" [54], it is explained that many proponents of climate change action favor incremental steps in the hope that they will improve the environment or at least serve as a basis for more comprehensive policies. The authors address why ad hoc responses to climate change may well be no better than, and possibly will be worse than, no action at all. Incremental climate change policies can give rise to predictable and nontrivial problems, such as non-effect, leakage, climate side effects, other side effects, lock-in, and lulling. Such problems not only can undermine the interim policies themselves but also may delay the adoption of a more comprehensive climate change policy.

Brendan Edgerton, the director of Circular Economy at the World Business Council for Sustainable Development, says that complacent incrementalism means that we are moving forward while looking only back [55]. He describes how companies are used to measure their relevant progress comparing today with yesterday, and he mentions that there is an inherent danger in a company measuring its success against yesterday. Backward-looking metrics without ambitious forward-looking targets commensurate with the scale of change required may give leadership a false sense of doing enough. Describing our response to global warming, he writes: "We're currently chasing a Tesla going 150 mph in a golf cart cruising at 15 mph, convincing ourselves we have a chance with persistence. Every day we maintain the status quo we fall further behind, further magnifying the insurmountable tasks of mitigating climate change."

Brad Zarnett, one of the top Medium writers on climate change that explores why corporate sustainability and our attempts to address climate change are a massive systemic failure, believes that business is at the epicenter of the harm. In his words [56], "An objective look at the evidence strongly suggests that the greatest efforts of corporations is to resist change that may impact their short term earnings while simultaneously finding new and inventive ways to 'go green' that have little or no impact on the climate plundering system, that is to say, that maintain the status quo... So business follows it's north star goal to maximize profits, and much like government, they make promises, pledges, declarations and commitments but at the end of the day little changes and incrementalism reigns."

There are two fundamental flaws here. First, the status quo will always choose the way to change in order to ensure that its status will not change. This is what we are living today regarding global warming. Multinational fossil fuel giants spent millions of dollars in advertisement to persuade us that a small demonstration project they created in the middle of nowhere is a serious contribution to carbon dioxide reduction, as a nice alibi that will allow them to continue the business as usual. Volkswagen and Bosch [57] selected to pretend they are "green" and cheat the emissions using a proper software. Powerful governments undermine even the incremental commitments of the Paris Agreement to provide more time and better control to the fossil fuel industry. And while global warming accelerates extreme weather phenomena, the fossil fuel industries and their partners slow down any meaningful response and prolong their global dominance.

Still, there is a plethora of calls to individuals to do their own share by recycling a little bit more, buying green products, and making sustainable lifestyle changes. Although everyone who consciously is trying to contribute to less carbon emissions deserves our respect, the shift of the responsibility to individuals is a serious and misleading distraction, and this is the second fundamental flaw.

First, because individuals buy, recycle, wear, and eat what the markets make available for them, they can choose between products, but they do not decide which products are available; this is decided and imposed by the multinational giants and their advertisement channels. As an example, individuals did not suddenly decide that they prefer coffee in capsules; there was a well-orchestrated campaign, for years, to make coffee capsules sexy and trendy and develop a new market segment.

Second, because individuals do not control how products are created and supply chains work. What is really the importance of a personal choice when the biggest company in the world of aircrafts ignores the most elementary safety rules, bypasses all the official mechanisms, does not inform the pilots for the existence of a very problematic software, and admits its mistake only after 2 planes crashed and 346 victims? In a report on nonfinancial risk management, it was mentioned that [58] "... the companies had failed to report material information related to the scandals in the preceding years. Boeing almost never refers to product and service safety when compared to its peers, Volkswagen lacked transparency around air emissions, and Nissan had the lowest, or near lowest, transparency of all the major car manufacturers around governance issues."

Third, and maybe most important, because in the case of global warming (but not only) small actions contribute to a big change only in a meaningful time frame and as a part of a holistic coordinated approach – but as the time goes by and the big emitters choose the path of least change, to put the burden on individuals and small actions is another way to avoid the responsibility and bare the relevant costs. Michael Mann, professor of atmospheric science of the Earth System Science Center at Penn State University, wrote in the magazine *Time* that [59] "There is a long history of industry-funded 'deflection campaigns' aimed to divert attention from big polluters and place the burden on individuals. Individual action is important and something we should all champion. But appearing to force Americans to give up meat, or travel, or other things central to the lifestyle they've chosen to live is politically dangerous: it plays right into the hands of climate-change deniers whose strategy tends to be to portray climate champions as freedom-hating totalitarians. The bigger issue is that focusing on individual choices around air travel and beef consumption heightens the risk of losing sight of the gorilla in the room: civilization's reliance on fossil fuels for energy and transport overall, which accounts for roughly two-thirds of global carbon emissions."

In her very interesting book *Recycling Reconsidered* [60], Samantha MacBride, director of Research at New York City Department of Sanitation, criticizes incrementalism and writes that in many cases incrementalism might actually intensify the problems that are supposed to be solved, as an example by boosting bogus "green" consumption. She also comments that although, in some cases, small changes might empower people to scale up their interventions on an institutional or economic level, it is not always sure that the environmentalists are ready to depart from the "small and beautiful" demonstration projects, as a result of the deep depoliticization of the environmental movement in the United States. MacBride highlights that environmentalism should not be a conflict-free, win-win incrementalism and movements like Zero Waste, that embed deep incrementalism in them, need to be critically reexamined.

In the era of climate emergency, with the massive loss of biodiversity, when marine litter plagues the oceans and pollution kills millions every year, sticking to small incremental changes is a clear signal of ignorance or of political manipulation by corporate interests; in both cases, "small and beautiful" choices will finally result in "big and nasty" damages on human health and the environment. Some problems can't be solved by incremental changes; systemic radical innovation, social or technical, is required. Let's see a great example.

8.7 IND4.0 MEETS THE HORSE MANURE CRISIS

In 1898 the first international urban planning conference convened in New York. One topic dominated the discussion: manure. Cities all over the world were experiencing the same problem. Horses were a staple of nineteenth-century cities: all transportation, whether of goods or people, depended on them. London, for example, had 11 000 horse-drawn cabs; several thousand buses, which required 12 horses/day, added extra 50 000 horses, and a large number of carts, drays, and wains, which were constantly making deliveries, added an immeasurable amount of extra horses to the streets daily. On average, a horse can produce from 7 to 16 kg of manure and around 1 l of urine per day – which leaves New York city alone dealing with 1150 tonnes of horse manure per day. And London wasn't better off, either.

Unable to see any solution to the manure crisis [61], the delegates abandoned the conference after 3 days instead of the scheduled 10 days. It was made clear to all attendees that they were dealing with a persistent waste management nightmare; the bigger and richer the cities became, the more horses they needed to function. Rises in living standards drove demand and consumption of more goods, products, and amenities, which in turn necessitated more horse-drawn transport. In 10 major US cities, the horse population rose by 328%, whereas the population increased by 105%. In turn, the rising number of horses led to more stables, more land, more food, and eventually more manure. Not to mention that when these overworked horses would die in the streets, their carcasses were often left there until they began to putrefy.

The manure and decomposing horse carcasses that lined the city streets attracted flies and helped spread typhoid fever and other diseases. One writer in *The Times* of London predicted: "In 50 years, every street in London will be buried under nine feet of manure."

Participants had hoped to hammer out a solution to the horse problem and its smelly attendant consequences, but instead, seeing no way out of the morass, they disbanded and went home in frustration. Albeit living in a transitional period, the world's top urban planners weren't exactly wrong to abandon their posts; any step against horses would essentially challenge the basic means of transporting people, goods, and any type of cargo, which consequently meant the conference had to challenge the whole business model in its core.

Few years later, not only the horse manure apocalypse was prevented, but horse transport itself became nonessential – and it was thanks to Henry Ford. When Ford created a new way to produce motor cars at affordable prices [62], cars, electric trams, and motor busses began appearing on the streets, replacing horse-drawn buses for good. Horses began to be phased out of daily life. As prices for hay, oats, and land rose, and fears of horse pollution became more urgent, the masses began to adopt the fledgling technology.

By 1912, the number of cars on New York City's roads had surpassed the number of horses. Buyers found cars to be cheaper to own and operate and much more efficient – not to mention more sanitary. The once-essential horse came under fire from magazines like *Harper's Weekly* and *Scientific American*, which praised the automobile for its economic sustainability and ability to reduce traffic. The second industrial revolution opened the way for the third one, the era of mass car production that electrification had permitted. Ford had drastically reshaped urban life by introducing the modern car, along with a new production scheme. This was not brought about by regulation or by government policy. Instead, it was technological innovation that made the difference. With electrification and the development of the internal combustion engine, the horse manure crisis was solved once and for all.

As we are living in the era of the fourth industrial revolution, as we live the transition towards a new economic paradigm that will transform radically all the industrial sectors (and certainly it will wipe out some of them), the horse manure crisis must be discussed in detail. The story demonstrates three major issues [63]:

"First, some of the most important problems we face, especially in the field of circular economy and waste management, it's not possible to be resolved by simply optimizing the current system. They could be resolved only by radical technological innovation, and the solutions that will be offered might be unimaginable even for the best waste management experts.

Second, waste management problems are interwoven within the structure of the economy and the daily habits of people. Therefore, trying to find solutions to serious waste management problems (e.g. hazardous waste in Africa, marine litter) is impossible in the narrow view of waste management – you have to address and influence the economic and social patterns first, and then maybe you can tackle waste management problems.

Third, the solution to the horse manure crisis did not arrived from the horse and carriage experts, not even from the sanitation experts. It came from a completely different sector that typically kept the wheels of the carriages and changed all the rest elements, including the business models. It came from the invention of a new product and a new ecosystem between citizens, industries, transportation companies, etc."

When horses were completely replaced by cars, it sure seemed like pollution was something of the past. But knowing what we know now about greenhouse emissions and combustion engines, we wouldn't be so quick to clap. Almost 100 years later, pollution by cars has become a big health and environmental problem. According to recent reports, in at least 15 cities, air pollution has now become so dangerous that the expected health impacts of half an hour of cycling outweigh the benefits of exercise altogether [64]. Virtually all diesel cars produce toxic nitrogen oxides (NOx), leading to at least 38 000 annual premature deaths due to heart and lung disease and strokes [65]. Most of the deaths are in Europe, where highly polluting cars are the main culprit, and in China and India, where dirty trucks cause most of the damage. In the United States, air pollution causes about 200 000 early deaths each year. In the United States, emissions from road transportation are the most significant contributor, causing 53 000 premature deaths, followed closely by power generation, with 52 000 [66]. Our overreliance on cars is becoming too costly in maintaining a healthy ecosystem; we are thus called to use our genius in order to find alternatives.

Frankly speaking, if you substitute manure with air pollutants and horses with cars, we are getting close (or maybe we already are in) to a modern "horse manure crisis." In practice, as it has happened so many times in human history, we resolved one problem (the horse manure crisis) by delivering an innovative solution that gradually created a new problem that initially was unknown and unexpected, but finally it might be harder to be resolved! "Zero horse manure" turns to be different than "zero pollution." There is no free lunch out there; there are no magical solutions. When we talk about the environment and human health, each and every choice creates impacts, independent of our ability to forecast or understand them.

This will probably be the case with the fourth industrial revolution too. It will deliver great innovations that will resolve hard problems, but it will finally create new more complex problems. When it comes to waste management, the dialectics of technology get tricky for two main reasons. First, there are still too much things to do with the already existing stocked waste pilling up in the world's dumpsites or the incredible amounts of plastic in the oceans. Second, we keep on producing more and more materials, more complicated, sophisticated, and personalized materials that are harder to be reused and recycled. We are often inundated with problems we cannot identify, and we come up with solutions impregnated with different sets of unidentified problems. Sometimes our solutions become problems in themselves, but science and knowledge can be used to rethink and reinvent our social and economic systems on the base of sustainability, even though there is no historical guarantee for succeeding a wasteless future. Sustainability, after all, is not simply an objective fact articulated from a planet reaching its "natural limits," but first and foremost a human and humanitarian request for justice, health, and access to basic living standards around the globe.

No matter how transformative new technologies will be, there are steps and phases that can't be avoided if you want to build concrete and lasting solutions regarding health and environmental protection. Let's see an example.

8.8 NO SANITATION = NO SUSTAINABILITY

One crucial issue that has been repeated several times in this book, in Chapters 1, 3, and 4, is the importance of sound waste management for the protection of human health and the environment, as well as the need to have final sinks for the different types of pollution that will be generated during the transition to circular economy. In this section, we would like to emphasize that no matter how much we talk or act on circular economy, no matter how fast the advances of IND4.0 are progressing, our priority should always be to ensure sanitation, especially for the poor countries and cities all around the world. On purpose, we refer not only to countries but also to cities, because there are cities even in rich countries that suffer from the lack of sanitation. The examples of Flint [67] and the findings in Lowndes County [68] in the United States are just the top of the iceberg. According to a recent post in Global Citizen's website [69], in the United States at least two million people lack running water and basic indoor plumbing. Millions of the country's most vulnerable communities – including low-income people in rural areas, people of color, Native communities, and immigrants – lack access to sanitation and water, in part due to discriminatory practices embedded in past water infrastructure development initiatives, according to the 2019 report "Closing the Water Access Gap in the United States: A National Action Plan" [70].

The global picture, according to the World Health Organization [71], is as follows:

- In 2017, only 45% of the global population (3.4 billion people) used a safely managed sanitation service.

- Only 31% of the global population (2.4 billion people) used private sanitation facilities connected to sewers from which wastewater was treated.

- 2.0 billion people still do not have basic sanitation facilities such as toilets or latrines.

- Of these, 673 million still defecate in the open, for example, in street gutters, behind bushes, or into open bodies of water.

- At least 10% of the world's population is thought to consume food irrigated by wastewater.

- Some 827 000 people in low- and middle-income countries die as a result of inadequate water, sanitation, and hygiene each year, representing 60% of total diarrheal deaths. Poor sanitation is believed to be the main cause in some 432 000 of these deaths.

Why is this important and do we have to refer on it in this book that deals with circular economy and IND4.0? Because, first, sanitation is directly linked with the development of proper waste management practices and still at least one-third of the world lacks proper waste collection services and 35–40% of the waste is managed through uncontrolled disposal practices (dumpsites, open burning, leakages to the environment). Second, because there is an ongoing discussion, usually in the name of recycling, circular economy, and "zero waste" practices, that ignores, undermines, or even worst opposes any new infrastructure that will provide safe waste disposal, a core element of any sanitation system. The following story aims to stress the importance of sanitation and its role regarding sustainable development.

On 7 November 2018, a very interesting article was published in *Financial Times*, on the successes of Africa's health revolution [72]. In 2011, infectious diseases stopped being the leading cause of death in Africa. By 2015, only 44% of African deaths were the result of diseases such as dysentery, pneumonia, malaria, tuberculosis, and HIV. That is still high; however, for the first time in recorded history, noncommunicable diseases, such as cancer, heart disease, and stroke, are the leading cause of death in Africa too, as in all other areas of the world. Importantly, the rate of decline in Africa has been the most rapid in history, with deaths in recent decades falling at three or four times the pace they did in advanced nations. It seems that Africa is undergoing a rapid health revolution.

Although this is certainly a success story, there are some crucial details. As Thomas Bollyky explains in his great book *Plagues and the Paradox of Progress* [73], recent reductions in infectious disease have not been accompanied by the same improvements in income, job opportunities, and governance that occurred with these changes in wealthier countries decades ago, so "the world is getting healthier in ways that should make us worry." There is also another problem: many people in poor countries are contracting noncommunicable diseases at younger ages than in rich countries. The risks of early death from heart disease, diabetes, and other afflictions commonly referred to as "diseases of affluence" are, in reality, becoming diseases of poverty, in countries without proper health systems in place.

Bollyky says that we are watching something unprecedented. Poor countries are achieving remarkable results, in terms of health protection, much faster than rich nations did the last 100 years. But they follow a different route. The rich Western countries combated infectious diseases by developing housing laws, municipal water and waste services, sewage systems, and proper institutional and administrative frameworks. As an example, the introduction of water filtration and chlorination in major US cities between 1900 and 1940 accounted for about one-half of the 30% decline in urban death rates during those years [74]. However, today, poor countries are eliminating infectious diseases mainly through medicine supported by international assistance funds. The result is a paradox: less children killed by infectious diseases, but more sick adults without access to health, sanitation, and employment opportunities. Still, in 2018, 96 new disease outbreaks were reported across 36 of the 47 member states of the World Health Organization in Africa. The most commonly reported disease outbreak was cholera, which accounted for 20.8% of all events, followed by measles and yellow fever [75].

We think that the lesson is more than clear. Scientific advances in medicine accelerated progress and deliver early successes in the case of infectious diseases in Africa. However, without proper sanitation systems in place, there is no guarantee that these successes would result in permanent improvements regarding health protection; in contrast it seems that we have healthier children with sicker adults. The road to sustainability, whatever is the meaning the reader gives to this ECC, goes through proper sanitation; you can't bypass it.

This is a friendly request to all the readers of this book. Next time you will talk about circular economy and recycling practices in any poor country or city, please make sure that before going into the details, you put the right question: what are you doing with sanitation and how you will close the dumpsites? And help them to find the proper answer and develop their own roadmap [76] about it or at least tell them to contact ISWA's #CLOSEDUMPSITES Task Force [77].

8.9 ASK SISYPHUS THE TRILLION-DOLLAR QUESTION

In all the industrial revolutions till today, humans were capable of increasing their productivity and efficiency to produce more with less resources, more energy efficiency, and cost effectiveness. In each industrial revolution, humans had the opportunity to rebalance their relationship with nature and ecosystems to sustain better lives with less effort due to the productivity gains of the industrial revolutions. However, the serious productivity gains were counterbalanced by the continuous increase of the population, the prolonged life expectancy, and the serious increase of the average consumption rates. The result was that the industrial revolutions, the ones that were responsible for all the positive transformation of human lives from the medieval misery to the modern world, were also stimulating and accelerating resource depletion and pollution.

For the last 500–600 years, continuous economic and population growth seemed to be negligible on a planetary level; the Earth and its ecosystems seemed to be too big to be affected. Economic growth fueled by the industrial revolutions seemed to be the only way towards prosperity. Today, we know well that the dominant paradigm of economic growth violates the Earth's biophysical limits and brings planetary level pollution and environmental problems that threat the very existence of the modern way of life. It has therefore become clear that the ongoing industrial revolution will not be like the previous ones. If IND4.0 will, as the previous ones, accelerate resource depletion and pollution, the consequences will be astonishing for the human civilization, and the only way to counterbalance them will be either a serious decrease of the population or of the living standards worldwide.

IND4.0 brings unimaginable opportunities for energy and resource efficiency. It brings technological advances that can deliver prosperity, better living conditions, and a more sustainable world for everyone on this planet.

For the waste management sector, IND4.0 makes the potential of a wasteless future more realistic than ever. As it was explained already, IND4.0 opens new ways to prevent, reduce, and even eliminate waste from specific sectors and streams to advance resource recovery, to achieve high standards of treatment and disposal, and to substantially reduce pollution and environmental impacts. At the same time, it provides new tools to stimulate stakeholders' interaction, awareness, and citizens' participation to apply the "polluter pays" and the "EPR" principle on a global scale, to make the governance of cities (including waste management) more inclusive and participatory, and to reduce or eliminate "dirty–dangerous–difficult" jobs and improve occupational safety and working conditions. However, all those new opportunities should be faced more as a potential and less a realistic expectation, and this is because IND4.0 also creates substantial risks and challenges. The same advances that open the way towards a more sustainable and circular economic system can be easily turned to drivers for further resource depletion, pollution, and a more wasteful future, as it has already happened in previous industrial revolutions.

This book started by putting the trillion-dollar question. Will IND4.0 and the circular economy converge, hence delivering not only better resource efficiency but a more sustainable future for everyone on the planet? Or will IND4.0 evolve according to the business as usual linear model, leaving the circular economy a mere flight of fancy, resulting in faster resource depletion, acceleration of environmental degradation, and deeper inequality?

The reader who managed to read all the chapters will have to formulate his/her own opinion about the answer. The authors' opinion is that the answer is not yet clear. Although the necessity of the convergence between circular economy and IND4.0 is of planetary importance, the course of IND4.0 till now and the weaknesses of the mainstream narrative of circular economy show that divergence is more possible than convergence. In this case, as we have already discussed, the health and environmental consequences will be unprecedented. However, there are decades of millions of people all around the world that are inspired by circular economy and they want to contribute to it, there are positive developments in policies and supply chains and a global awareness of the environmental challenges, there are plenty of new unimaginable opportunities that should be utilized by social and technological innovation, and the rise of the urban sustainability science is visible. To put it in a different way, the opportunity of a huge systemic change by the combination of circular economy and IND4.0 is too big to be lost – history is written by human hands moved by necessity. If we want our societies to catch this opportunity, three conditions should be met:

- First, there is an urgent need to integrate the narrative of circular economy with its social and environmental components to move it further than the busy corporate corridors of the executives and ground it to reality, human rights, and science.

- Second, the development of suitable governance and institutional models, in accordance with IND4.0 and the material flows (which means going beyond the politically dominant view of recycling as circular economy), is the only way to stimulate the convergence of circular economy and IND4.0. These models should impose new rules and regulations that would stimulate market adaptation to a systemic circular economy concept and not the adaptation of circular economy to what markets are willing to do.

- Third, we need to think bold, go beyond the business as usual optimization, and abandon incrementalism, because the ongoing environmental crisis requires urgently transformational policies and responses.

There is no guarantee that these conditions will be met, but if we do not try hard and struggle for them, more environmental degradation and extra million deaths by pollution are guaranteed. We live in a great transitional crossroad. On the one side, technological advances and the rise of new business models based on IND4.0 are making a wasteless future more realistic than ever. On the other side, the ongoing waste crisis in the developing world seems to be increasing, creating global health and environmental impacts, and resource extraction and depletion seem to be accelerated. As Carl Sagan has said, "Never before has there been a moment so simultaneously perilous and promising. We are the first species to have taken evolution into our own hands." The dilemma we face is like this: are we going to stimulate IND4.0 to converge with circular economy setting the scene for a more sustainable future (Stairway to heaven) or we will leave IND4.0 to unfold as all the previous industrial revolutions creating lethal impacts for people and ecosystems (Highway to hell)?

In this crossroad, we all have to contribute towards a better future, in which IND4.0 and circular economy will be combined to deliver more prosperity, less inequality, and a more sustainable planet. In the meantime, we have to be sure that not only we will advocate for the need to converge circular economy with IND4.0, but we will also be able to create examples of this convergence, examples that will demonstrate that this is realistic and beneficiary in many terms. The waste management sector can be a role model that will exemplify the realistic possibilities and the benefits of the convergence. A role model in which we should test the new business models required to combine IND4.0 with CE in order to deliver better environmental and health protection with a positive social footprint, less inequality, and more prosperity. There are many reasons that make the waste management sector suitable for this role. Waste management is directly linked with the environmental quality and the health of everyone on Earth; any substantial change in waste management creates immediate positive or negative impacts. Waste management links resource management, social and cultural context, and policy making and represents the interface in which IND4.0 and circular economy meet the daily lives of billions of citizens. It's a sector heavily regulated, where public and private partnerships are already tested and a diverse array of stakeholders and technologies are active. Add on the previous that the ongoing digitization of the sector, as presented in Chapter 5, provides great examples of the expected benefits (although there are still huge challenges to manage), and the conclusion becomes obvious: waste management can catalyze the convergence of circular economy with IND4.0 because it grounds the circular economy to reality and uses IND4.0 to advance waste and resource management.

Still, there is one more crucial reason that makes waste management the perfect field for stimulating a positive answer to the trillion-dollar question. The readers of this book might be engineers, researchers or professors, consultants or municipal officers, landfill operators, or recycling organizers, but at the end, they (we) are all unified by what we are doing. We are what we do, so let's see what we do in the waste management sector. We manage waste, which means that we deal with the day-to-day leftovers of the society. We manage waste delivering better health and environmental protection, upgrading the life quality standards. That means that we manage the past protecting the future!

But our work has another important element: we are goalkeepers. Goalkeepers because we finally receive the end-of-pipe outcome of the economic life and the daily consumption and we are obliged to manage it, no matter what it is, in order to maintain health and environmental protection standards. And we are goalkeepers that never stop. Not only because waste management is a 24 hours, 365 days a year service, but mainly because every time we think we got the solution, a new waste stream appears and we have to deal with it, and we have to start from the beginning. By mid-1990s we considered that finally we knew what to do with municipal waste. Then we realize that packaging waste requires special handling, and we started to develop the relevant know-how and technologies. Before we found out the best way to deal with packaging, we have also to deal separately with massive streams like batteries, end-of-life cars and car tires, light bulbs, and construction–demolition waste. Then we realized that e-waste in general requires special treatment and management, and we are still trying to develop the relevant technologies. Waste management is a never-ending story – the same is true for circular economy. The waste we produce changes continuously in accordance with the production means, the technical innovation, the level of income, the new needs we are trying to cover, and the social–political status. If there will be a day in which waste will remain the same for 10 years, then this is going to be an almost dead society.

Waste managers are like the primary sinner, the ancient Greek king Sisyphus that was punished by the Gods to push every day of his life a huge rock towards the top of a hill. Every sunset, when he was almost there and ready to push the rock at the top, his strength and resources were over, the rock was sliding down, and he has to start again, from the bottom of the hill, next day. We are like Sisyphus; we need to overcome the fast consumption gravity every single day and push the society's waste towards a never-reaching final destination. But we have also a difference from Sisyphus. He was punished because he cheated the Gods. We are struggling to make sure that the modern Gods, the markets' invisible hands, and its deeply religious zealots will not ruin the planet to save their temples. We hope that this book was a small contribution towards this global effort.

REFERENCES

1. Gramsci, A. (2011). *Prison Notebooks. Columbia University Press 1*.
2. Mavropoulos, A. (2015). Circular economy or space race? *Wasteless Future* (10 December 2015).
3. Business Insider (2016). The value of asteroid mining. https://www.businessinsider.com/the-value-of-asteroid-mining-2016-11 (accessed 10 February 2020).
4. Fortune (2015). Whole foods' John Mackey: the conscious capitalist. https://fortune.com/2015/08/20/whole-foods-john-mackey (accessed 9 February 2020).
5. Reich, R. Saving capitalism. https://robertreich.org/SavingCapitalism (accessed 10 February 2020).
6. Reich, R. *Saving Capitalism: For the Many, Not the Few*. Vintage.
7. Mason, P. (2017). *Postcapitalism: A Guide to Our Future*, Reprint. Farrar, Straus and Giroux.
8. Mason, P. (2015). The end of capitalism has begun. *The Guardian* (17 July 2015).
9. Klein, N. (2015). *This Changes Everything: Capitalism vs. The Climate*, Reprint edition. New York: Simon & Schuster.
10. Mason, P. (2019). The problem for capitalism is that the future may no longer bail out the present. *New Statesman* (2 October 2019).
11. Garver, E. (1978). Rhetoric and essentially contested arguments. *Philosophy and Rhetoric 11* (3): 156–172.
12. Gallie, W.B. (1955). Essentially contested concepts. *Proceedings of the Aristotelian Society 56*: 167–198.

13. Wikipedia (2019). Essentially contested concept. https://en.wikipedia.org/wiki/ Essentially_contested_concept (accessed 8 May 2020).

14. Ehrenfeld, J.R. (2008). Sustainability needs to be attained, not managed. *Sustainability: Science, Practice and Policy 4* (2): 1–3.

15. Ehrenfeld, J.R. (2009). *Sustainability by Design: A Subversive Strategy for Transforming Our Consumer Culture*. New Haven: Yale University Press.

16. Martin, J. (2019). How Apple's Daisy iPhone recycling robot works. https://www. cnet.com/news/how-apples-daisy-iphone-recycling-robot-works (accessed 10 February 2020).

17. Digital (2018). Apple News Blog: IPhone: What can Apple's recycling robot "Daisy" do? https://newsbeezer.com/germanyeng/apple-news-blog-iphone-what-can-apples-recycling-robot-daisy-do-digital (accessed 8 May 2020).

18. Apple (2019). Apple expands global recycling programs. https://www.apple. com/newsroom/2019/04/apple-expands-global-recycling-programs (accessed 10 February 2020).

19. Apple (2019). Environment. https://www.apple.com/environment (accessed 10 February 2020).

20. Peters, A. (2019). Apple's iPhone recycling robot can take apart 200 iPhones an hour: can it dismantle the company's footprint? https://www.fastcompany. com/90413038/apples-iphone-recycling-robot-can-take-apart-200-iphones-an-hour-can-it-dismantle-the-companys-footprint (accessed 10 February 2020).

21. Purdy, K. (2019). Apple's Daisy robot is still a stunt, but their other recycling ideas are good. https://www.ifixit.com/News/15794/apple-daisy-cobalt-recycling (accessed 10 February 2020).

22. Digit (2019). Apple, Samsung, Vivo and other tech companies found violating EPR rules in India, import permits suspended for flouting e-waste rules. https://www.digit.in/ news/mobile-phones/apple-samsung-vivo-and-other-tech-companies-found-violating-epr-guidelines-in-india-import-permits-s-47446.html (accessed 10 February 2020).

23. Root, T. (2019). Apple effectively has a monopoly on fixing your iPhone. There's now a fight to change that. https://www.vox.com/the-goods/2019/7/3/18761691/right-to-repair-computers-phones-car-mechanics-apple (accessed 10 February 2020).

24. Gibbs, S. (2018). Apple and Samsung fined for deliberately slowing down phones. *The Guardian* (24 October 2018).

25. Gibbs, S. (2017). Apple admits slowing older IPhones because of ageing batteries. *The Guardian* (21 December 2017).

26. Kovach, S. (2017). Apple apologizes for slowing down iPhones with older batteries. https://www.businessinsider.com/apple-apologizes-for-iphone-slowdowns-2017-12 (accessed 10 February 2020).

27. Tech Times (2019). iPhone 5 update: planned obsolescence? https://www. techtimes.com/articles/245832/20191028/gps-time-rollover-cause-planned-obsolescence-iphone-5-update.htm (accessed 10 February 2020).

28. Abuelsamid, S. (2020). Why dead Sonos speakers mean you'll never own a driverless car. https://onezero.medium.com/why-dead-sonos-speakers-mean-youll-never-own-a-driverless-car-9ec2d8fa3763 (accessed 11 February 2020).

29. Kamila, P. (2017). *Understanding Planned Obsolescence: Unsustainability Through Production, Consumption and Waste Generation*. Kogan Page.

30. Kamila, P. (2019). Let's talk about planned obsolescence. *ISWA President's Blog* (8 July 2019). https://www.iswa.org/media/publications/presidents-blog/prezsezmore/article/guest-blog-lets-talk-about-planned-obsolescence/1383 (accessed 8 May 2020).

31. Wieser, H. and Tröger, N. (2018). Exploring the inner loops of the circular economy: replacement, repair, and reuse of mobile phones in Austria. *Journal of Cleaner Production 172*: 3042–3055.

32. Semuels, A. Your AirPods will die soon. https://www.theatlantic.com/technology/archive/2019/03/your-airpods-probably-have-terrible-battery-life/585439 (accessed 10 February 2020).

33. Wiens, K. (2018). Apple engineers its own downfall with the Macbook Pro keyboard. https://www.ifixit.com/News/10229/macbook-pro-keyboard (accessed 10 February 2020).

34. Frantz Schneider, A., Matinfar, S., Martino Grua, E. et al. *Towards a Sustainable Business Model for Smartphones: Combining Product-Service Systems with Modularity*, 82–63. EasyChair.

35. Hirose, K. and Mishima, N. (2019). Eco-efficiency evaluation of modular design smartphones. *Procedia CIRP 84*: 1054–1058.

36. Rayner, T. (2018). Modular phones haven't lived up to the initial hype, here's why. https://www.androidauthority.com/modular-phones-hype-851085 (accessed 11 February 2020).

37. Türkeli, S., Huang, B., Stasik, A., and Kemp, R. (2019). Circular economy as a glocal business activity: mobile phone repair in the Netherlands, Poland and China. *Energies 12* (3): 498.

38. Mavropoulos, A., Purchase, D., Nitzsche, G., and Ramola, A. (2019). *How IND4.0 Transforms the Waste Sector*, 90. Vienna, Austria: International Solid Waste Association ISWA.

39. Mavropoulos, A., Tsakona, M., and Anthouli, A. (2015). Urban waste management and the mobile challenge. *Waste Management and Research 33* (4): 381–387.

40. Richter, J.L. and Koppejan, R. (2016). Extended producer responsibility for lamps in Nordic countries: best practices and challenges in closing material loops. *Journal of Cleaner Production 123* https://doi.org/10.1016/j.jclepro.2015.06.131.

41. Lee, D. (2018). Monitour: tracking global routes of electronic waste. *Waste Management 72*: 362–370.

42. Spasova, B. (2019). *Deposit-Refund Systems in Europe for One-Way Beverage Packaging*. ACR+.

43. Gu, F., Guo, J., Hall, P., and Gu, X. (2019). An integrated architecture for implementing extended producer responsibility in the context of Industry 4.0. *International Journal of Production Research 57* (5): 1458–1477.

44. Monier, V., Hestin, M., Cavé, J. et al. (2014). *Development of Guidance on Extended Producer Responsibility (EPR)*, Final Report, 227. Brussels: European Commission, DG Environment.

45. Stuebing, S. and Vries, C. *Governance for the Circular Economy: Leadership Observations*, 72. The Netherlands: Origame.

46. Robinson, S. (2017). *Social Circular Economy: Opportunities for People, Planet and Profit*, 43. Social Circular Economy.

47. Srinivas, V. (2016). Shape optimization of a car body for drag reduction and to increase downforce. Thesis, J. N. T. University Anantapur, Anantapuramu.

48. Schmitt, A. (2016). It's true: the typical car is parked 95 percent of the time. *Streetsblog USA*. https://usa.streetsblog.org/2016/03/10/its-true-the-typical-car-is-parked-95-percent-of-the-time (accessed 8 May 2020).

49. Wikipedia (2019). Horseless carriage. https://en.wikipedia.org/wiki/Horseless_carriage (accessed 3 May 2020).

50. Harvey, F. (2019). Paris climate deal: world not on track to meet goal amid continuous emissions. *The Guardian* (4 December 2019).

51. Harvey, F. (2019). The UN climate talks are over for another year: was anything achieved? *The Guardian* (5 December 2019).

52. Allan, J.I. (2019). Dangerous incrementalism of the Paris agreement. *Global Environmental Politics 19* (1): 4–11.

53. Kates, R.W., Travis, W.R., and Wilbanks, T.J. (2012). Transformational adaptation when incremental adaptations to climate change are insufficient. *Proceedings of the National Academy of Sciences of the United States of America 109* (19): 7156–7161.

54. Coglianese, C. and D'Ambrosio, J. (2008). Policymaking under pressure: the perils of incremental responses to climate change. *Faculty Scholarship at Penn Law* 2008.

55. Edgerton, B. (2019). Don't ignore the perils of complacent incrementalism. https://www.greenbiz.com/article/dont-ignore-perils-complacent-incrementalism (accessed 11 February 2020).

56. Zarnett, B. Climate change: incrementalism is no longer a viable option. https://medium.com/swlh/climate-change-incrementalism-is-no-longer-a-viable-option-34bb9b0a2d82 (accessed 11 February 2020).

57. Levin, D. German parts maker Bosch gets off with relatively light $100 million fine from VW Dieselgate. https://www.forbes.com/sites/doronlevin/2019/05/23/german-parts-maker-bosch-gets-off-with-relatively-light-100-million-fine-from-vw-dieselgate (accessed 12 February 2020).

58. Datamaran (2019). Corporate scandals: Boeing, Nissan and VW could have seen it coming. *Datamaran* (29 May 2019).

59. Mann, M. (2019). Lifestyle changes aren't enough to save the planet. Here's what could. *Time* (12 September 2019).

60. MacBride, S. (2011). *Recycling Reconsidered the Present Failure and Future Promise of Environmental Action in the United States*. The MIT Press.

61. Davies, S. The great horse-manure crisis of 1894. https://fee.org/articles/the-great-horse-manure-crisis-of-1894 (accessed 13 February 2020).

62. Goss, J. (2020). Henry Ford and the Auto Assembly Line. https://www.thoughtco.com/henry-ford-and-the-assembly-line-1779201 (accessed 3 May 2020).

63. Mavropoulos, A. (2017). The horse manure crisis and the 4th industrial revolution. Part 1: Wasteless Future. *Wasteless Future* (22 September 2017). https://wastelessfuture.com/the-horse-manure-crisis-and-the-4th-industrial-revolution-part-1 (accessed 8 May 2020).

64. Mavropoulos, A. Happy birthday Galileo Galilei: greetings from Mariana Trench. https://medium.com/@antonismavropoulos/happy-birthday-galileo-galilei-greetings-from-mariana-trench-83bf5f034c5 (accessed 13 February 2020).

65. Carrington, D., Topham, G., and Walker, P. (2016). Revealed: nearly all new diesel cars exceed official pollution limits. *The Guardian* (23 April 2016).

66. MIT News. Study: air pollution causes 200,000 early deaths each year in the U.S. http://news.mit.edu/2013/study-air-pollution-causes-200000-early-deaths-each-year-in-the-us-0829 (accessed 13 February 2020).

67. Flint water crisis isn't the first – and won't be the last: activists. https://www.nbcnews.com/storyline/flint-water-crisis/flint-water-crisis-isn-t-first-won-t-be-last-n500626 (accessed 13 February 2020).

68. Mavropoulos, A. (2017). Wasted health in Alabama…. *Wasteless Future* (8 October 2017). https://wastelessfuture.com/wasted-health-in-alabama (accessed 8 May 2020).

69. Rodriguez, L. (2019). Millions in the United States' most vulnerable communities lack access to water and sanitation. https://www.globalcitizen.org/en/content/water-sanitation-us-dig-deep-report (accessed 13 February 2020).

70. Roller, Z., Gasteyer, S., Nelson, N. et al. (2019). *Closing the Water Access Gap in the United States: A National Action Plan*. US Water Alliance.

71. WHO (2019). Sanitation. https://www.who.int/news-room/fact-sheets/detail/sanitation (accessed 13 February 2020).

72. Pilling, D. (2018). Africa's warp-speed health revolution has an old threat. *Financial Times* (7 November 2018), p. 3.

73. Bollyky, J. (2018). *Plagues and the Paradox of Progress*; Biomedical Sciences – Global & Public Health. The MIT Press.

74. Population Reference Bureau (2005). Clean water's historic effect on U.S. mortality rates provides hope for developing countries.

75. Mboussou, F., Ndumbi, P., Ngom, R. et al. (2019). Infectious disease outbreaks in the African region: overview of events reported to the World Health Organization in 2018. *Epidemiology and Infection 147*: e299.
76. Mavropoulos et al. (2016). *A Roadmap for Closing Waste Dumpsites The World's Most Polluted Places*. Vienna, Austria: International Solid Waste Association ISWA.
77. ISWA. Let's close the World's biggest dumpsites! closingdumpsites.iswa.org (accessed 24 December 2019).

Epilogue
The Future Starts
with You

Understanding the circular economy is like learning how to bike: you have to keep your balance by cycling and look as far as possible, and certainly not between your legs, to keep your path under control. As soon as you start dealing with the circular economy, you realize the bigger picture that is usually ignored, and then it is impossible to unsee it. The futility dressed with stupidity becomes obvious: an industrial paradigm based on continuous resource depletion and aggregation of pollutants in ecosystems. If this is value creation, we seriously need to redefine value.

Products are temporary, but resources are forever. The next paradigm of the economy will not be created by the innovators polishing the surface of consumer capitalism. It will be created by young entrepreneurs in start-up co-working spaces – experimenting with low footprint business models, wild ideas for product–service systems, and new closed-loop programs. It will be created by brave politicians that challenge the status quo, designing new regulatory constraints and incentives, making room for the connected systems of tomorrow to emerge. It will be created by the people in waste trucks, workers along conveyor belts, creative administrators in the back office of municipal waste management organizations, or street smart waste pickers in the informal economy of many developing countries.

This book has been inspiring to write, because we wrote it for you. You do an incredibly important job. To keep up with an economy dependent on a form of growth tightly coupled to exploitation of fossil fuels and nonrenewable resources is simply impossible. In our current regulatory paradigm, the responsibility almost magically disappears the very moment a product dies and turns into waste. The system is broken. We need to fix it. Together.

Industry 4.0 and Circular Economy: Towards a Wasteless Future or a Wasteful Planet?, First Edition.
Antonis Mavropoulos and Anders Waage Nilsen.
© 2020 John Wiley & Sons Ltd. Published 2020 by John Wiley & Sons Ltd.

Those that are hands on, and with their own eyes see the overwhelming tsunami of mixed-up, low-value waste generated by the contemporary economic system, have a unique insight in both the problem and the possible solutions. I will use these last words to encourage the people working with resource management to take pride in their work and their knowledge. In the coming revolution, we will need to redesign the models and challenge the status quo. Your contributions have never been more important. But you need to consider yourself designers and take part in the creative process.

Waste is valuable resources with lacking metadata. Digital technology will be important in the years to come. If we manage to keep track of how a product is made and from which materials, who made it, and how it should be managed to get back into the loop, a lot of the stuff that today generates costs will turn out to generate income. If we can democratize information management solutions, we will make an important step towards making the circular dream a profitable reality.

But machines do not have their own will. We need visionary humans to lead the way in this process. The people of the evolving resource management industry need to connect, form open networks and alliances, learn from each other, make their voice heard, create new solutions, and make them spread. It will not be easy. But if enough people pull together, we can bend the linear logic all the way around – until it, one beautiful day, turns into a circle.

Anders Waage Nilsen

Epilogue
Towards Irreversible Wastelands

> You cannot say, or guess, for you know only
> A heap of broken images, where the sun beats,
> And the dead tree gives no shelter, the cricket no relief,
> And the dry stone no sound of water. Only
> There is shadow under this red rock,
> (Come in under the shadow of this red rock),
> And I will show you something different from either
> Your shadow at morning striding behind you
> Or your shadow at evening rising to meet you;
> I will show you fear in a handful of dust.
> — T.S. Eliot, Wasteland (with kind permission from Faber & Faber Ltd.)

T.S. Eliot published this famous poem, one of the most important of the twentieth century, in 1922. This was another period of massive arrivals of new technologies like the motored powered airplanes and helicopters, the modern assembly lines for cars, the zeppelins, the electric household refrigerator, the vacuum cleaners and air conditioning, radio broadcasting, the neon lamps, the first talking motion picture, and many others. It was also a period of great scientific discoveries like the relativity theory, the invention of the flip-flop circuit and the first sonar, the inventions of plastics and cellophane, the artificial creation of nitrates that revolutionize agricultural activities, and the discovery of superconductivity.

Many people believe that this particular part of the 434 lines poem refers exactly to this transition period where science made great leaps of technology, however the spiritual and cultural sectors of the world lay forgotten, according to Eliot. "A heap of broken images" shows the fragmented nature of the world, and the snapshots of what

Industry 4.0 and Circular Economy: Towards a Wasteless Future or a Wasteful Planet?, First Edition.
Antonis Mavropoulos and Anders Waage Nilsen.
© 2020 John Wiley & Sons Ltd. Published 2020 by John Wiley & Sons Ltd.

the world has become further serves to pinpoint the emptiness of a world without culture, a world without guidance or spiritual belief.

I do believe that this part of the poem, besides demonstrating the lack of cultural references in periods of massive new technological change, speaks also about the leftovers of human exploitation, and I recall it every time I see a new ecological disaster, either by accident or by intention.

I do believe that this part of Elliot's poem calls for rethinking our views and practices on a global scale. It reminds me that Wasteland is the landscape that is already formulated and will be left behind after the mining operations will be completed in one of the most beautiful places in Skouries, a location in Halkidiki, Greece. A beautiful old-growth natural forest, with crystal-clear waters will be wiped out in order to build a massive open-pit gold and copper mine, along with a processing plant, and a large underground mine. I am not going to discuss technical details and problems that make the specific mining operations a very serious risk for the local ecosystem, nor about the fact that you should not build an underground mine and hazardous waste repositories in an active seismic fault. I just want to highlight that the loss of the beautiful landscape (and its mental impacts on people) is an irreversible linear damage; it also happens in almost all the mining activities around the world.

Or take another example, the ocean Wasteland in the Gulf of Mexico. This was the leftover of the Deep Horizon environmental disaster that threatened more than 400 species that live in the Gulf islands and marshlands, including the endangered Kemp's ridley turtle, the green turtle, the loggerhead turtle, the hawksbill turtle, and the leatherback turtle. In addition, about 34 000 birds were in immediate risk including gulls, pelicans, roseate, egrets, terns, and blue herons. The area of the spill includes 8332 species, including more than 1200 fish, 200 birds, 1400 molluscs, 1500 crustaceans, 4 sea turtles, and 29 marine mammals. Threatened species ranged from whale sharks to seagrass. The loss of biodiversity is also an irreversible linear damage. And it is happening all around the planet due to human activities.

The melting of the glaciers of West Antarctica is accelerated. These glaciers hold 6% of the world's fresh water – their collapse could send sea levels up by at least half to 1 m by 2100. Recently, a team of scientists showed that over the past century, human-driven global warming has changed the character of the winds that blow over the ocean near some of the most fragile glaciers in West Antarctica. Sometimes, those winds have weakened or reversed, which in turn causes changes in the ocean water that laps up against the ice in a way that caused the glaciers to melt. The collapse of the glaciers of West Antarctica is another irreversible linear damage, and in combination with other phenomena related to global warming, it is expected to wipe out cities like Lagos and Manila and islands like Haiti and Kiribati and put in high risk more than 200 million people.

Deforestation in Amazon, which was increased by 85% during the last two years, is not only affecting the global climate. It could affect water supplies in Brazilian cities and in neighboring countries while harming the very farms that farmers are trying to expand by burning the rainforest. More massive deforestation might alter water supplies as far away as Africa or California. The region has been so degraded that even a small uptick in deforestation could send the forest hurtling towards a transition to something resembling a woodland savanna, according to an analysis. A recently proposed bill promotes mining, further expansion of agriculture, and energy production in Amazon areas that have been home to indigenous people for thousands of years. The loss of the Amazon forest and its indigenous civilizations is also an irreversible damage with global impacts.

Mining, a so crucial activity for the modern way of life, is one of the deadliest industries. The death toll is at least 14000 people per year. Worldwide, the International Labour Organization estimates that we have 6000 deaths per day (2.3 million/year) and 340 million accidents per year due to occupational accidents, while the victims of work-related illnesses are around 160 million annually. Hazardous substances alone are estimated to cause 650000 deaths per year.

Next time we will discuss on circular economy, it will be better to remember that we might be able to develop circularities for gold and cooper, we might be able to reduce e-waste, but still, some of the most valuable natural miracles and human achievements are not circulated. There is no circularity for the wastelands we created, and we continue to create, for the landscape in Skouries, the loss of biodiversity in the Gulf of Mexico and the melting of the glaciers in Antarctica. There is no circularity for the loss of the rainforest in Amazon and the indigenous civilizations that will be lost to advance mining and agriculture.

There is no circularity for the millions of workers that are killed every year due to poor working conditions. There is no circulation for the mothers lost in the textile factory that collapsed in Dhaka, killing 1134 people that were preparing textiles for the most famous Western brands. You can't recycle the pain of the relatives of the 270 people that lost their lives from the 2019 collapse of the Brumadinho dam in Brazil, nor the feeling of desperation of the people that lost their villages and their rivers and 19 people from the 2015 collapse of the Mariana dam – you can recycle only the responsibilities because both dams were owned by the same company.

The path we walk wipes out forever people, civilizations, natural miracles, biodiversity, and climate stability in an irreversible, traditional linear way. Either we will be able to stop before it's too late (although it might already be) or a serious collapse of human civilization is waiting in any of the next corners. Last but not least, this course will not change as long as the main decisions are made by the ones who know everything about stock exchange prices but nothing about human and social values. We need to get out of our convenience zones now and act immediately in all the levels.

Antonis Mavropoulos

Index

A

abundance, 69, 102, 107

academia, 75, 269

academics, 79, 89, 99, 116, 174, 271

accelerometers, 221, 223

Accenture, 3

acceptability, 110

access-controlled
  containers, 219, 224
  infrastructure, 280
  inlets, 227

accessibility of data, 41

accumulation
  materials, 90
  wealth, 136
  things, 80
  waste, 169

acid, 25, 96, 241

acidification, stratospheric, 10

acid rain, 9

active involvement, 181

active sensing, 300

active sorting, 313

activism, 198

activists, 312, 371

actuators, 37, 38, 40, 226, 227

adapters, 239

additional
  energy, 87
  growth, 112

jobs, 2
pollution, 10
utilization, 88

additive manufacturing, 39, 41, 45, 218, 220

additives, 14, 88, 162, 163

ad hoc networks, 48

Adler, 242

administrative
  boundaries, 98
  frameworks, 363
  system, 301

advertisements/advertising, 29, 71, 108, 116, 142, 204, 313, 314, 338, 356, 357

advice, 271, 355

aerodynamic, 353

aerosol, 10, 11

Africa, 5, 8, 24, 65, 67, 144, 146, 194, 196, 284, 309, 317, 321, 360, 362, 363, 371, 376

agglomeration, 321

aggregate, 96, 137, 233, 319, 355

agile/agility, 5, 19, 46, 261, 264, 265, 66, 291

agriculture, 4, 25, 26, 73, 74, 95, 112, 131, 180, 183, 355, 377

agro-biofuels, 102

Aich, Samit, 197

aircraft, 30, 144

AirPods, 342, 344, 345, 369

Alipay, 233

Industry 4.0 and Circular Economy: Towards a Wasteless Future or a Wasteful Planet?, First Edition.
Antonis Mavropoulos and Anders Waage Nilsen.
© 2020 John Wiley & Sons Ltd. Published 2020 by John Wiley & Sons Ltd.

global positioning system (GPS), 219, 222, 224, 239, 308, 344
gold, 106, 144, 145, 341, 376, 377
Google, 66, 105, 229, 231, 256, 266, 311–13, 321
governance, 3, 5, 9, 49, 63, 98, 111–14, 117, 153, 168, 176, 189, 218, 253, 269, 282, 283, 298–301, 303, 314, 323, 326, 336, 337, 346, 347, 349, 350, 352, 357, 362, 364, 365
government, 29, 36, 55, 64–67, 165, 168, 169, 240, 241, 271, 287, 288, 302, 311, 312, 336, 337, 347, 350, 356, 359
Graedel, T.E., 164
Gramsci, Antonio, 336, 367
granulate, 50
graphite, 130
gravel, 90, 137, 138
gravity, 225
Greece, 194, 301, 376
green
  bags, 242
  growth, 99, 101, 102, 123, 270, 292
  revolution, 31
  technology, 102
  washing, 108, 115, 340, 341
greenhouse gases (GHG), 7, 8, 18, 58, 87, 90, 99, 121, 128, 131, 139, 150, 162, 163, 165, 166, 175, 355
grid, 66, 129
growth, 2, 5, 7–9, 11, 12, 17, 26–31, 49, 51, 52, 63, 68, 73, 77, 79, 82, 84–86, 89, 90, 93, 99–104, 107, 112, 116, 117, 126, 127, 129, 131, 136, 138, 141, 142, 145, 149, 151, 157, 163, 167, 170, 182, 215, 216, 218, 231, 253, 254, 263, 270, 272, 275, 284, 299, 317, 319, 320, 323, 334, 335, 347, 355, 364, 373
Guangzhou, 67
Guardian, The, 71, 128, 142, 343
guidance, 147, 233, 326, 375
Gulf of Mexico, 82, 132, 376, 377
Gutenberg's, 38
gyroscope, 223

H

Haas, W., 92, 93
habitat, 33, 102
habits, 257, 261, 271, 308, 359
Hafnium, 130
hair, 105, 106
Haiti, 376
handheld material scanners, 225
hardware, 36, 37, 41, 57, 174–76, 178, 221, 222, 226, 229, 231, 234, 236, 238, 243, 244, 246, 257, 286–88, 335, 344
harmful materials, 12
harmful waste, 189
Harvard Business Review, 230, 273
hatch(es), 221, 224, 227, 239
hay, 359
hazardous
  materials, 14, 26, 186, 189, 190, 347
  residues, 186
  substances, 186, 200, 377
  waste, 5, 14, 33, 135, 147, 189, 190, 192, 193, 360, 376
helium, 130
hierarchical, 268, 319
hierarchies, 106
high-resolution, 128
history, 4, 9, 24, 25, 29, 34, 42, 44, 52, 56, 57, 69, 70, 86, 106, 123, 133, 136, 231, 319, 320, 324, 334, 336, 341, 357, 360, 362, 365
Holland, 134
homogeneity, 188, 353
homogenization, 193
Honduras, 311
Hong Kong, 305, 306, 328
hosting, 237
households, 7, 27, 95, 126, 127, 134, 138, 144, 181, 185, 210, 233, 238, 242, 248, 279, 280, 288, 317
human, 9, 15, 24, 28, 37, 38, 40, 41, 74, 107, 116, 236, 253, 256, 257, 260, 337, 363, 374

humanity, 10, 13, 25, 34, 101, 104, 336, 340, 355
human–machine interfaces, 37, 41
hydrocarbons, 186
hydrogen, 102, 337
hydrosphere, 185
hygiene, 26, 56, 362

I

ideal, 56, 87, 158, 225, 243, 283
idealism, 258
ideate, 264, 266
ideation, 266
identification, 180, 223, 224, 233, 238, 305
ideological, 341
ideology(ies), 106, 326, 340
IDs, 221, 238, 239
IEA, 20, 200
iFixit, 66, 343, 345
illegal, 169, 347
immigrants, 194, 195, 361
immigration, 193
implementations, 2, 14, 49, 62, 79, 111, 112, 115–17, 156, 180, 188, 217, 285, 323, 326, 354
import(s), 68, 109, 111, 130, 161, 174, 180, 191, 253, 343
import bans, 253
imported, 30, 68, 96, 111
improvements, 5, 14, 26, 34, 36, 42, 45, 50, 52, 57, 65, 74, 82, 95, 102–5, 113, 117, 167, 178, 189, 252, 253, 260, 272, 273, 307, 314, 321, 335, 345, 352, 353, 354, 362, 363
impurity(ies), 13, 171, 180, 187
incentives, 52, 112, 140, 163, 217, 220, 227, 233, 238, 254, 269, 271, 278, 303, 348, 351, 373
incentivize, 271, 349
incinerated, 29, 140, 144, 150, 189, 191–93, 242
incineration, 6, 27–29, 64, 110, 151–54, 156, 158, 159, 169, 180, 188, 190, 191, 223

incinerators, 29, 71, 158, 169, 170, 189–91, 193
inclusion, 154, 172
inclusive, 124, 195, 253, 269, 290, 364
inclusiveness, 347
inclusivity, 109, 176, 197, 301
incomes, 3, 70, 101, 103, 111, 124, 139, 168, 297, 307, 315, 321, 367, 374
incremental, 155, 253, 258, 260, 272, 273, 354–56, 358
incrementalism, 334, 354–58, 365
incubators, 267
India, 5, 65, 67, 131, 195–97, 309, 317, 319, 343, 360
indicators, 47, 64, 74, 107, 157, 162, 298, 306, 316, 334, 340
Indonesia, 5, 195
industrial
  agriculture, 74
  ecology, 34, 68, 76, 78, 339, 341
  ecosystems, 267
  parks, 46, 64, 240, 241
  plant, 46
  recycling, 71, 233
  robots, 50, 129, 225, 226
  symbiosis, 46, 50, 55, 71, 77, 220, 240, 241, 271
industrialization, 25, 26, 33, 36, 56, 69, 70, 75, 136
industrial revolutions, 2, 9, 12, 13, 17, 24, 25–28, 30, 34–36, 45, 52, 56, 57, 70–74, 85, 102, 116, 126, 127, 132, 134, 135, 146, 147, 161, 174, 182, 186, 188, 214, 255, 285, 334, 352, 359, 360, 363–65
infectious diseases, 362, 363
informality, 195–98
informatics, 295–98, 300, 303, 325, 328
information technology(ies), 38, 197, 251, 313, 335, 337
infrared, 223

Jevons paradox, 9, 82, 89
jobs, 2, 9, 66, 99, 104, 108, 110, 112,
    195, 253, 256, 258, 271, 362,
    364, 373
justice, 339, 361

K

Kabadiwalla Connect, 196
Kaeser, 54
Kallis, G., 103, 123
Kalundborg, 55, 240, 248
Kasa, Sjur, 9, 20
Kennedy, Christopher, 182
Kenya, 111, 196
Kenyan, 196
kerosene, 126, 132
Kirchherr, Julian, 78, 79, 120
Kiribati, 376
Kivu, 107
Klein, Naomi, 295, 338, 367
knowledge, 12, 16, 24, 26, 38, 49, 103,
    198, 214, 240, 245, 255, 261, 266, 267,
    270, 274, 284, 285, 288–90, 296, 298,
    302, 314, 315, 321, 347, 361, 373
knowledge economy, 285
knowledge transfer, 49, 284, 290
Kodak, 273
Korea, 67
Korhonen, J., 78, 79, 87, 98, 120
Kral, U., 183, 186
Kraussman, Fridolin, 90, 91, 92,
    137, 189
Krier, James, 49
Kuznets curve, 104

L

labor, 2, 9, 13, 16, 27, 47, 51, 70, 71, 74, 89,
    105–7, 109, 112, 114, 116, 117, 135,
    139, 194, 195, 241, 242, 252, 260, 271,
    335, 337, 350, 351
labor
    costs, 47
    intensity, 114

markets, 194
    productivity, 2, 27, 71, 74, 135
laboratories, 225
Lagos, 376
land degradation, 3, 4
landfills, 6, 28, 29, 33, 63, 77, 87, 91,
    138–40, 143, 144, 148, 150, 151,
    153, 154, 156, 158, 168–70, 173, 175,
    185–89, 191–93, 196, 200, 253, 275,
    342, 343, 366
land scarcity, 138
Lansink, Ad, 147, 151, 153, 154
Laos, 67
laptops, 342
Large Hadron Collider (LHC), 322
Latin America, 146, 284
Latin American, 311
latrines, 361
lead (material), 4, 5, 138, 148, 149, 162
leadership, 16, 48, 56, 106, 113, 218, 245,
    252, 255, 257, 259, 262, 263, 269,
    270, 283, 284, 289, 290, 317, 335,
    336, 351, 356
leakages, 81, 87, 97, 143, 151, 174, 175,
    177, 356, 362
lean startup, 264–66
learning algorithms, 37, 224
leather, 323
ledger, 41, 42
legacy systems, 229, 230, 239,
    243, 252, 255
legal
    frameworks, 109, 175
    instruments, 173, 199, 349
    standards, 188
    structures, 257
legislations, 31, 63, 64, 68, 133, 134, 142,
    144, 145, 147, 199, 217, 287, 343,
    348, 349, 351
Lemille, Alxandre, 107, 167
Leyen, Ursula von der, 63
Libya, 194
licenses, 244, 285, 288
licensing, 281, 287, 338
Liechtenstein, 185, 210